THE ENVIRONMENTAL PROTECTION AGENCY: ASKING THE WRONG QUESTIONS

THE ENVIRONMENTAL PROTECTION AGENCY: Asking the Wrong Questions

MARC K. LANDY
Boston College

MARC J. ROBERTS
Harvard School of Public Health

STEPHEN R. THOMAS
The Commonwealth Fund

Foreword by Congressman Morris K. Udall

New York Oxford
OXFORD UNIVERSITY PRESS
1990

Oxford University Press

Oxford New York Toronto
Delhi Bombay Calcutta Madras Karachi
Petaling Jaya Singapore Hong Kong Tokyo
Nairobi Dar es Salaam Cape Town
Melbourne Auckland

and associated companies in
Berlin Ibadan

Published by Oxford University Press, Inc.,
200 Madison Avenue, New York, New York 10016

Oxford is a registered trademark of Oxford University Press

Library of Congress Cataloging-in-Publication Data
Landy, Marc Karnis.
The Environmental Protection Agency:
Asking the Wrong Questions
Includes bibliographical references.
1. United States. Environmental Protection Agency.
I. Roberts, Marc J.
II. Thomas, Stephen R. (Stephen Richard), 1943–
III. Title.
TD171.L36 1990 363.7'00973 89-25565
ISBN 0-19-505021-5

9 8 7 6 5 4 3 2 1

Printed in the United States of America
on acid-free paper

For Sam Beer
Our Teacher

Foreword

MORRIS K. UDALL

Twenty years ago, on April 22, 1970, more than 20 million people participated in the celebration of Earth Day. This was the first time that international attention was focused on the environment. In the United States, nearly 10,000 elementary and high schools, 2,000 colleges and universities, and virtually every community staged activities to focus attention on the state of the world's environment. The U.S. Congress stood in recess so that members could devote the day to discussing environmental concerns in their states and districts. All three television networks devoted substantial coverage to events around the country, and the Public Broadcasting System devoted its entire daytime programming to Earth Day coverage.

The wave of support for the environment from Earth Day participants led to the creation of the Environmental Protection Agency, to passage of the Clean Air Act and the Clean Water Act, as well as other major environmental legislation.

As we approach the twentieth anniversary of Earth Day, it is appropriate that we reflect upon the successes and failures of our environmental policies. What follows is an insightful look at the internal workings of the Environmental Protection Agency. The observations of Marc Landy, Marc Roberts, and Stephen Thomas are often tough, sometimes even harsh, but they force policymakers, environmentalists, the media, the courts, and private citizens to rethink the mandate we have given the EPA. The authors suggest, moreover, that ecology and the "quality of life" in urban America should again be part of the mandate, front and center.

More importantly, they use examples of failures within the EPA to teach us about the shortcomings of government as a whole. This book is neither an indictment nor an endorsement of the EPA. Rather it is a frank look at governing from a technical, managerial, and political standpoint. The lessons learned from *The Environmental Protection Agency: Asking the Wrong Questions* can be easily translated to other federal agencies. Are we overburdening our bureaucrats so that they have become largely ineffectual at shaping the policies of the federal government? Is Congress too vague in its legislating or too heavy-handed? Have the courts intervened in the traditional role of policymaking? These and many other questions are addressed in the following pages. One could take the case studies in this book and find similar cases in the State Department, Interior Department, Health and Human Services, NASA—virtually every federal program.

Perhaps it has been too easy for policymakers to blame the failures of our environmental policies on the Reagan years. Certainly, Anne Gorsuch, Rita Lavelle, and Jim Watt did not help things. But, the authors show that the shortcomings in our policies result from a multitude of problems rather than just insensitivity of top administration officials.

I don't always agree with the authors' view of the reasons behind the failings of the EPA; in fact, I would argue that the EPA has been more beholden to industries than to the environmental movement. And I must disagree with their contention that not enough is known about global warming to begin mitigating its effects. Nevertheless, I am intrigued by the management questions they ask about the agency.

A seventeenth-century statesman once wrote to his son, "Do you know with how little wisdom the world is governed?" The authors might argue that this is still true today. The problems faced by government are more complex than ever. Bureaucrats are struggling just to keep their heads above water in many cases. Congress is often obsessed with symbolic issues like flag burning and the pledge of allegiance rather than substantive legislation. The courts are drowning in a sea created by an overly litigious society. And interest groups often choose easily understood, politically sexy issues rather than those that are cumbersome and more difficult to solve. Yet the resources we have available today are more numerous and diverse than at any time in human history.

Information is readily available to guide policymakers. In fact, perhaps too much information is available. Somehow we must bring together the expertise, the political realities, and the managerial experience, to address the many problems we face. We need to free decisionmakers from the "statistics wars" created by studies financed by competing constituencies. We need to relieve top government leaders from the duties of day-to-day management of their agencies so that they may become long-term strategists—anticipating future issues and educating the public about them. Decisionmakers should do just that—make decisions rather than micromanage their agencies.

The Environmental Protection Agency: Asking the Wrong Questions warns us of the folly of inaction. The problems we face today are more serious than ever.

The United States is home to as many as 27,000 toxic waste dumps. We are running out of places to put our garbage. By 1990, half of our nation's cities will be without any landfill capacity for solid waste. Of existing landfills, many violate environmental laws and are leaking pollutants into our water systems. The food we eat and the water we drink is laden with pesticides whose dangers are still unknown. Acid rain has destroyed untold natural resources. And smog in our cities, spewed by smokestacks and automobiles, corrupts the very air we depend on for life.

The authors argue that the public has not been educated about the environmental dangers of an industrial society. I agree. We can never achieve a wholly clean environment. We can not guarantee that Americans will enjoy both one of the highest standards of living in the world and a pristine environment. For one thing, much of the environmental damage we have incurred is irreversible. Nevertheless, we can do better.

As John Maynard Keynes once said, "The important thing for government is not to do things which individuals are doing already, and to do them a little better or a little worse; but to do those things which at present are not done at all." There is no better example of appropriate government involvement than environmental protection. The lessons we learn from *The Environmental Protection Agency: Asking the Wrong Questions* can help us to do that job better.

Washington, D.C.
September 1989

Preface

The collaboration that produced this book started several years ago when the three of us taught together in an executive program for environmental officials at the Harvard School of Public Health. The senior agency officials who participated in that program were an impressive group; intelligent, energetic, and deeply committed to public service. Yet as we worked through a series of public management case studies with them, we realized, to our dismay, that the technical discussion of risks, costs, and benefits was not being effectively connected to either the managerial or political problems these officials faced. Not only was this link difficult for us to establish in the classroom, it seemed to be equally difficult for the program participants to establish it in their work-a-day lives.

We spent several years trying to design the right research project and to acquire the needed foundation support. Then, in 1981, we got our opportunity. Douglas M. Costle, the EPA administrator under President Carter, was preparing to spend a semester at Harvard. We were able to secure his help in developing a detailed research plan that was generously funded by the Alfred P. Sloan Foundation as part of a larger environmental program grant to the Harvard Public Health School. Thus we were able to proceed with a series of case studies that delved into various kinds of EPA activities that occurred during Costle's tenure.

We were left with the realization that our teaching was incomplete. To an extent, we were impeded by our own professional training. Roberts is an economist, and Landy and Thomas are political scientists. Economists are taught to look upon environmental policy as a question of evaluating "outcomes" based on the values of individual citizens (e.g., how much would they be willing to pay for cleaner air, better scenery, etc.). But the policy making process itself shapes what people want, as well as responding to those wants. The conventional wisdom among political scientists is that the politics of public policy is best understood as a clash between different units of government and different organized segments of the public. But the policy making process is more than the competition among interests. We want to understand how arguments are formulated, discussions are conducted, and the specific role of public officials and public institutions in shaping and influencing these activities. We believe that leadership, deliberation, and citizenship are key concepts for achieving this understanding.

This study reflects these broader concerns. We want to know what role technical evidence and public values both do and should play in determining environmental outcomes. We want to understand how organizational structures and political circumstances shape the use of such information and the impact of such values, and how the process of policy development in turn affects citizen attitudes and long-run changes in the character of the polity. Only then can we offer practical advice to public officials about how to do a better job.

Mr. Costle was unstinting in his contribution of time and energy to the project. He provided us with access to agency data and personnel and, more importantly, with many insights into the dynamics of environmental policymaking and implementation. The hours of discussion with him were a vital addition to our education. In fact, our choice of specific cases to focus upon was shaped by his judgment about which ones would prove most illustrative.

For each case, we constructed a list of the key EPA participants who were then invited to Harvard for day long joint interview sessions which were recorded and transcribed. In addition to the three of us, several other faculty and graduate students participated in these sessions—adding their insights and helping us advance the conversation. Based on the leads they provided, we went on to conduct many other interviews as well with people both inside and outside of government. Our gratitude to all those who gave of their time and their insights is great indeed. The resulting transcriptions are one of the most important sources of information for the case studies that form the heart of this book.

These new encounters with federal environmental regulators served only to deepen our puzzlement. Again, we were struck by their talent, perceptiveness, and high mindedness. Yet, as we probed deeper into the details of our case studies, we became increasingly convinced that, in subtle but important ways, EPA was not attaining some of what we considered to be its most important objectives.

Our response to this mismatch between dedication and performance is suggested by our subtitle, *Asking the Wrong Questions*. This book seeks to uncover the mistaken premises that have clouded and distorted debate about environmental policy. Often, neither the EPA nor the public at large has had a clear sense of the choices to be confronted, nor of the strategies that would be most appropriate. By asking the wrong questions, legislators, agency officials, and issue advocates alike missed opportunities that were at the same time intellectual and political.

We believe that fostering public debate and public understanding should be an agency official's highest calling. Hence, he or she must be prepared to incur significant political risks to ensure that the public is presented with a clear vision of what is at stake.

Effective public discussion and the ability of the policital process to hold agencies accountable require that agency officials clearly state the essence of their understanding of a policy problem and of their approach to dealing with it. To frame questions in a way that encourages such strategic thinking, agency officials must free themselves from the day to day crisis management that can be all consuming in the Washington milieu.

We recognize that the clarity we advocate can be both difficult and risky to achieve. The technical merits of a case are often genuinely intellectually obscure.

It may also be politically dangerous to challenge the terms of reference of the ongoing public debate or to call into question the language of a widely accepted statute. There are also bureaucratic costs to clarity. Ambiguity can preserve agency autonomy. Focusing the political debate on minutiae may enable the agency to pursue its own long-run plan.

To require so much would appear to make us naïve. We propose to take away all the political advantages that derive from ambiguity, inexplicitness, and overstatement. We seem to be urging senior agency officials to become the bureaucratic equivalents of Anwar Sadat, risking everything to achieve a fundamental political transformation.

In America, however, a significant constituency does exist for moderation and truth telling. Like any constituency, it remains inchoate until galvanized by leadership. The contemporary mood of budgetary austerity and revulsion against "special interests" provides just such a leadership opportunity. Senior public officials should take advantage of it.

We do not ask bureaucratic leaders to abandon their concern for survival, only that they place such concerns in proper perspective. We believe it is even more naïve to expect that the public interest will automatically be served by the unfettered pursuit of narrow individual and institutional objectives and by unlimited combat among bureaucratic, ideological and economic interests.

Because debates about environmental protection are highly contentious, the reader will want to know which side we are on. Do we want more environmental protection or less? Do we view strict regulation as a necessary corrective for the irresponsibility of industrial polluters, or as an ideologically inspired barrier to the march of economic progress?

At times, our analysis will give comfort to environmentalists; at other times, it will seem to support those who want to ease regulation. But we are not on one "side" or the other. This book does not join the debate as it is currently formulated. Neither "safety," nor "economic efficiency" (nor even the two combined) is a broad enough basis for thinking about environmental problems. Environmental issues cannot be reduced to a formula, an exercise in analytical economics, or to the rigid application of a few dogmatic precepts. We seek to alter the contours of environmental debate so that the choices at stake can be better illuminated.

This book is both politics and economics. It is a collaboration in the deepest meaning of the term. No one of us, no two of us, could have written it. The alphabetical listing of author's names reflects the full and equal participation of all three of us in formulating its fundamental concepts. Its design and its arguments are joint property.

Each of us took the lead regarding specific topics: Landy on Superfund and Chapters 2 and 8; Roberts on Ozone and RCRA; and Thomas on Steel and IRLG.

His coauthors found Marc Roberts to be a gifted editor of everyone's prose. His questions provoked other questions, and the fights that ensued forced clarity on all three of us. This kind of editing eventually became a kind of deliberative discipline, disclosing aspects of the story that none of us had seen in the beginning.

The book divides into three parts. Part One presents an introduction to the ideas that dominate the book and a discussion of the origins of EPA and its development up to the arrival of Costle. Part Two presents the case studies and a summary of their major findings. Part Three first presents a look at EPA during the Reagan years in order to see how the subsequent history of the agency informs those findings. It then provides a general interpretation of what we have found, and, finally, a discussion of what are better questions for EPA's leaders to ask.

We again say thanks to Douglas Costle. Although he may not agree with many of our interpretations and conclusions, his assistance to us is a testament to his generosity and fairmindedness. Academic hindsight is far easier than the actual practice of agency leadership.

We are also grateful to those who participated in the taped interview sessions from which we learned so much: Eckhart Beck, Toby Clark, Swep Davis, Gary Dietrich, William Drayton, John Froines, David Hawkins, Richard Heller, Thomas Jorling, Susan King. Andrew Mank, Joseph Rodricks, James Rogers, and Charles Warren.

We thank Margaret Gerteis and Valle Nazar who coauthored the IRLG and Ozone cases, respectively, and Glenn Roberts and Christopher Dunne for their help on the RCRA and Steel cases. John Bailar gave us a painstaking and insightful critique of an early draft of the manuscript, and Shep Melnick and Terry Davies gave close and thoughtful readings of later drafts. Elaine Rossignol provided great secretarial help and Andrea Methven supplied invaluable word processing and text editing assistance. We also appreciate the aid of the Gordon Center for Public Policy at Brandeis University in producing the final manuscript. We would also like to thank our editor at Oxford, Al Ritchie, for his discerning editorial advice and his many kindnesses.

The Alfred P. Sloan Foundation provided the financial support that made the case study research possible. The Harvard School of Public Health and Dean Howard H. Hiatt provided further financial and institutional support vital to the completion of the project. We greatly appreciate the generosity, patience and forebearance they showed towards us.

We dedicate this book to Sam Beer. During a long and distinguished career as a professor of government at Harvard University, he introduced thousands of students to the study of politics. Our approach to this project has been decisively influenced by him, particularly by his insistence upon confronting theory with experience, to the enrichment of both. In addition, his active and thoughtful engagement with nature as a mountaineer, hiker, and woodsman has shown us what it means to be an environmentalist.

Boston
April 1989

M.K.L.
M.J.R.
S.R.T.

Contents

Part I

1
Introduction

This book employs a small canvas to paint a large picture. It concentrates on a single policy arena, environmental protection, and a single agency, the U.S. Environmental Protection Agency (EPA) to explore critical problems of modern government and democratic politics.

Its central argument is that senior officials of executive branch agencies are responsible for *more* than the programs they administer. It departs from the narrow focus upon specific outcomes that has dominated the academic study of public policy. Instead, it argues that public administrators have a deeper responsibility to preserve and promote the constitutional democracy of which they are the agents.

As the Preamble to the Constitution announces, the purpose of the American government is to "form a more perfect union . . . and to secure the blessings of liberty . . . ." Government functions both to protect freedom and to promote it. Doing so requires that those qualities of citizens that enable them to act as free persons must be nourished and enhanced. Government must accept some responsibility for the public's capacity to understand both the technical and the moral significance of the decisions that have to be made.

Hence, public servants have a major educational responsibility. They cannot, and ought not try, to tell citizens what to think. But, they must make use of their considerable stature and expertise to frame questions so that public debate can be made coherent and intelligible. They must tease out the essential social and ethical issues from the welter of scientific data and legal formalisms in which those issues are enveloped.

Underlying the specific questions that guide individual policy debates is the strategic question that agency leaders must ask. Since different units of government inevitably share overlapping and intersecting responsibilities, no one agency can take the lead with regard to every issue over which it could conceivably claim jurisdiction. To a considerable degree, an agency can and must decide for itself how it should expend its energies. To do so wisely, agency leaders need to develop a complex view of their multiple responsibilities with regard to the specific problems they address, the agency they manage, and the constitutional order which they must preserve and protect. In the final chapter, we discuss how EPA should ask and answer this fundamental question.

As we demonstrate in great detail, the existing organizational and statutory framework inhibits EPA's ability to ask the right questions. By demanding that its officials ask such questions, we are also insisting that they devise means to overcome those restrictions. These are heavy burdens to place upon men and women who are already overextended. Because this duty is so critical, however, we will argue for a redefinition of the very function of agency leadership to see that it is carried out.

FORMAT OF THE BOOK

The book provides a chronicle of EPA from its founding in 1970 to the end of the Reagan Administration. Its centerpiece is a set of five specific cases from the Carter administration, each of which explores a different dimension of the agency's activities. The cases deliberately include considerable technical, administrative, and political detail. We require the reader to "get his hands dirty" to appreciate how difficult it is for anyone, including the participants, to see beyond the confusing welter of conflicting technical facts, legal requirements, and political exigencies to focus on central intellectual and ethical concerns.

"Revising the Ozone Standard" is about setting a standard to protect human health. It shows the problems that result when Congress instructs an agency to make such a decision without explicitly considering who is to be protected and at what cost. We discuss the relevant scientific studies in great detail, precisely to demonstrate that the crucial decisions that EPA made were not scientific but political. Although we do not delve as deeply into scientific issues in subsequent chapters, we had to do so at least once, to convince the reader that science alone cannot answer the deepest questions environmental health regulation poses.

"Writing the Resource Conservation and Recovery Act (RCRA) Regulations" focuses on an example of regulation writing. It illustrates two key points: the excruciating difficulty of integrating different professional and bureaucratic points of view, and the problems and opportunities created when Congress leaves it to the agency to make the most important judgments about what is to be accomplished. The need for an integrative organizational structure and a coherent strategic framework can best be appreciated by a close examination of what occurs in their absence. The case is also about the role of the courts in supervising regulation writing and about whether they can effectively force an agency to do such work in the face of its own structural and strategic deficits.

"Passing Superfund" is about EPA's legislative role. Executive branch agencies cannot control the legislative process, but they play a pivotal role in determining how Congress defines problems and in establishing the contours of congressional debate. We take the reader through the intricate maze of House and Senate subcommittee and committee deliberations to show that if the agency does not pose the right questions, those questions are not likely to emerge in the course of congressional consideration.

"Forging a Cancer Policy" is a study of a voluntary effort on the part of several agencies, including EPA, to devise a unified policy with regard to the regulation of cancer-causing chemicals. It was a serious experiment in overcoming agency parochialism, and the document that it produced represented an effort to educate the public about the risks that such chemicals pose. A detailed exegesis of that document is required to consider how successful that effort was and to describe the possibilities and limitations of this sort of integrative endeavor.

"Enforcing the Clean Air Act: The Steel Industry" is about both EPA's efforts to enforce its regulations vis-à-vis a specific industry and about how it became enmeshed in a broader effort by the federal government to improve an ailing industry's prospects. It shows how the agency's "domestic policy" (its approach to enforcement) influenced its "foreign policy" (its dealings with the White House, other agencies, labor, and the industry), and vice versa. It illustrates the difficult coordination problem that confronts managers when the boundaries of their organizations are narrower than those of the problems to be addressed.

WHY FOCUS ON ENVIRONMENTAL PROTECTION?

Almost any policy issue could serve to illustrate the themes we discuss, but environmental protection places them in particularly bold relief. It involves an extremely broad range of scientific information, including virtually all of the physical, biological, and social sciences. The questions of risk and uncertainty that it presents are especially subtle and vexing. Furthermore, the environment provides an excellent example of what happens when these problems merge with other profound philosophical and political dilemmas.

The tension between collective identity and individual choice is central to American political culture. From the beginning, "We the People" has co-existed uneasily with "The Invisible Hand." Environmental protection brings this tension to the surface. It forces a confrontation between political principles—a concern for public space, responsibility for future generations, and special help for those most at risk— and the market principle that requires that all goods, including environmental amenities, be allocated according to the expressed willingness of individuals to pay for them.[1]

Our relationship with the environment raises fundamental issues about who we are and what we care about. It challenges a basic tenet of industrial society, namely that mastery of nature is mankind's greatest project. It may force a choice between health and beauty on the one hand and prosperity on the other. It requires us to consider our relationships not only to fellow humans but to plants and animals as well.[2]

WHY FOCUS ON THE ENVIRONMENTAL PROTECTION AGENCY?

Many private organizations and all of the branches of government are active in environmental policy making. We could have focused on Congress, the courts, state agencies, or lobbying organizations. We choose instead to concentrate on the EPA.

First of all, an agency like EPA has substantial power, because of its expertise and its formal authority. Even when disputing its findings, all parties to a controversy often find themselves focusing on the information and analysis that EPA provides.

Moreover, agencies like EPA exercise much discretion in choosing which master to serve, and for what purposes. Constitutional law leaves their relationship to Congress and the President ambiguous. Statutes do not and cannot fully guide their behavior.[3] Thus they possess substantial power that needs to be understood, managed, and used to advance democratic purposes.

The Environmental Protection Agency is unique among environmental regulatory agencies in that it deals with both public health and resource management issues. Its comprehensive authority is reflected in its position in the Executive Branch. It is the only regulatory agency whose administrator reports directly to the President.

No single agency, however, can be understood in isolation. It is constantly influencing, and being influenced by, the courts, interest groups, the Congress, and other parts of the Executive Branch. They, too, have undergone significant changes in structure and outlook during the period under study. Thus, as we proceed, we examine such crucial innovations as the advent of the "new administrative law," the proliferation of congressional subcommittees, the rise of the Washington environmental community, and the establishment of central executive branch oversight of regulation, to see how these have affected the agency's performance.[4]

FOUR STANDARDS

We propose to judge the performance of the agency, and those who lead it, on the basis of four specific standards: fidelity to the technical merits, promoting civic education, responsiveness to the public, and building institutional capacity.

The Technical Merits

The history of environmental policy abounds with examples of the price to be paid for ignoring the technical merits. Repeatedly, policies have been adopted that were simply unworkable or whose announced goals could only be achieved at a higher cost than even avid proponents were prepared to pay.[5] To avoid such predicaments, government agencies must discern and respect the limits on policy choice imposed by the available engineering, scientific, and managerial understanding.

By "the technical merits," we mean the feasibility, effectiveness, and efficiency of proposed remedies. Feasibility means that the proposed solution can be put into place. Effectiveness means that the plan will produce the desired result. Efficiency means operating at minimum cost and conserving scarce public and private resources. Economists and decision analysts sometimes combine all these ideas into a multidimensional summary of the probabilities of achieving various consequences as a result of various levels of program cost and political effort. But for our purposes these rather more straightforward and simple ideas suffice.

The merits are often quite difficult to determine. Part of the difficulty is philosophical. Those who have different interests and perspectives tend to have different assessments of the merits if only because they value outcomes differently.

But the technical questions themselves may also be obscure. For example, regulations often contain standards that require industry to develop new technology.[6] Yet it is seldom clear whether or not waste sources will be able to develop and deploy the necessary innovations soon enough to meet the specified deadline and at a reasonable cost. Since the ecological and health benefits of a particular standard are also often hard to predict, determining whether any given proposal is defensible on the "merits" can be quite difficult.

These problems are not a justification for ignoring this criterion. It is often possible to demonstrate that a particular idea is infeasible, overly optimistic, too costly, or muddle headed in some other way. And, even when individual policy components are technically defensible, there remains the broader issue of strategic soundness that requires agency officials to show that the overall design of a policy constitutes a reasoned response in light of the available knowledge and the values it purports to advance.

Leaders of public agencies should serve as champions of the merits. When political pressures push them to over-promise, politicians need to be reminded that they can suffer a greater loss by failing to do the impossible than by making more limited commitments in light of inconvenient facts. Moreover, the public disillusionment that accompanies the failure of unsound initiatives is a cost to not only political leaders, but to the political capacity of the republic as well.

Civic Education

Government has the obligation to provide the civic education that strengthens the capacity of citizens for successful self-government. Civic education of this sort is, in part, about the technical merits. It is both possible and appropriate for the public to learn to distinguish policies that are coherent, reconcilable with the facts, and whose means are consistent with their ends, from those that are not. The authority of experts may pose a danger to democracy, but their expertise is crucial if citizens are to understand what is possible and what is mere wishful thinking. And, in so far as all participants in a dispute come to accept the same data and technical analyses, the nature of the disagreements that divide them can be clarified.

Civic education also concerns the ethical orientation that citizens adopt toward policy problems. Inevitably, public agencies are civic educators in this way as well. Policies and programs embody concrete lessons about the nature of civic responsibilities. They can encourage citizens to accept some degree of responsibility for a collective problem or to believe that someone else (perhaps "the government") can or will take care of it for them. The conversation provoked by agency proposals and actions influences citizens' views about the obligations they acknowledge as well as the rights they enjoy.

The link between the questions that agency officials ask and what the average citizen learns is complex and indirect. Most citizens do not participate in every (or even any) national policy debate. Each policy issue has its own set of interested publics, which can be conceived of as a set of concentric circles.[7] The innermost ring is composed of the relevant specialists in lead agencies and legislative committees. Surrounding them are staffers from other agencies (including oversight units like the Office of Management and Budget), members of interest groups, and other legislators. Farther removed are journalists, state and local bureaucrats, local opinion leaders, and activists. More distant still is the citizenry at large.

And yet, the mass public enjoys considerable influence. The electoral system, combined with the extraordinary capacity of the media to dramatize policy questions, makes the nation's elected leaders quite sensitive to the public will. Whatever the experts may say, policies are seldom enacted if they are demonstrably unpopular. Public opinion places severe constraints on what officials can and cannot do.

An agency like EPA has many opportunities to engage in public education. It testifies before Congress; it issues myriad documents that appear in the *Federal Register*; it gives briefings to the president and other executive officials; it holds press conferences; its leaders make speeches. Each time the agency is obliged to respond to interest group comments or to hold a public hearing regarding a proposed standard or regulation, it has the chance to explain and to clarify the core issues that a proposed action raises.

True, most of these efforts reach only the inner rings of the public. But, congressmen, state officials, interest group representatives, and journalists are, in turn, opinion leaders themselves.[8] What they think has a strong influence on those less deeply involved. Thus, the cumulative impact of the agency's educational efforts can be enormous if they are deployed in a coordinated way so as to increase public sophistication.

Therefore, one of an agency's most vital functions is to frame the questions that define those discussions. As we shall see, by defining the central question facing the agency as a technical one—how to provide safety—the EPA hindered meaningful political debate about critical environmental choices.

Responsiveness

Determining what the public wants is seldom obvious. Inevitably, an agency retains considerable discretion in deciding how much to respond and to whom.[9] In the environmental sphere, groups and individuals differ in the value they place on such goals as wilderness preservation, outdoor recreation, public health, and ecological balance. They also vary in the relative importance they accord to nonenvironmental concerns such as cost minimization, economic development, or distributive justice. Power is so fragmented in the United States that agency heads face a welter of conflicting signals from Congress, the courts, state and local leaders, other agencies, and from the White House. Nor is it easy to find out about the views of the general public. Public opinion surveys have the well-known difficulties of uninformed respondents and biased responses.[10]

One might argue that responsiveness can best be evaluated on the basis of how well an incumbent administration performs in the next election. Popularity at the polls is, after all, *prima facie* evidence that a government has responded to the public's wishes. This test is highly imperfect, however, when applied to a single agency. In any given election, much of the public may support a candidate despite his views on a particular subject, or it may simply ignore that subject altogether.[11]

These difficulties in judging responsiveness, however, do not free an agency, or its leaders, from this test. To do so satisfactorily, they need to conceive of responsiveness in strategic instead of tactical terms. The real issue is the overall coherence of an administration's program. Policies are "strategically responsive" if they help sustain and expand the long term coalition upon which the administration depends for its support, thereby fostering the cohesion and well-being of the political party that it represents.

If the educational role of an agency is largely about framing good questions, then its ability to be strategically responsive depends on its capacity to provide answers to those questions. We shall show that, because the EPA under the Carter Administration defined the questions it asked too narrowly and technically, it could not provide answers that helped to nurture and sustain the political coalition supporting the president.

Governmental Capacity

The success of a republic depends on the capacities of its institutions as well as those of its citizens. A government capable of performing the tasks already discussed must be more than a necessary evil. It must be an enterprise capable of sophisticated and wise action. Furthermore, to gain the trust of its citizens, government must ensure that they understand and appreciate its efforts.

Developing governmental capacity of this sort is the long run institutional counterpoint to civic education. For this to occur, civil servants need to be both technically and politically expert and perceived as such by citizens. Perpetuation of institutional memory, recruitment and retention of skilled personnel, and developing a capacity for honest and impartial judgment all require the attention of agency leaders, as does the communication of these strengths to the general public.[12]

THE ROOT OF THE PROBLEM

Government's frequent failure to measure up to these standards can be explained in a variety of ways. Liberals stress the venality of public officials and their vulnerability to "capture" by private (especially economic) interests. Conservatives stress the oppressiveness of government and the desire of public officials to pursue their own ideological and bureaucratic agendas. Liberals would hem bureaucrats in by specifying in great detail what they can and cannot do. Conservatives would require them to do less by relying on an alternative, and presumably less defective, method for solving problems, the market.[13]

We reject these prescriptions because we disagree with the assumptions on which they are based. Our interviews and observations of public officials have, for the most part, impressed us with their probity, their dedication to public service, and their desire to be responsive to the public's will. The few exceptions to this pattern, which occurred during Anne Gorsuch's term as EPA Administrator, serve only to increase our appreciation and admiration for the vast majority of EPA leaders and managers whom we studied. We find the source of the difficulty not in the motivation of public officials or in the obvious problems posed by organizational imperfections and limited data, but in something that lies behind all of these, the ways in which public managers think, reason, and decide.

Ideas are simpler than the reality they represent, often too simple. The mind economizes. Ordinarily, a few stylized principles and rough generalizations are enough to make sense of the world. Is the water "safe" to drink or not? Is that chemical a "carcinogen" and should it be banned from certain uses? Under pressure to make such choices, the mind seeks to avoid concepts that complicate the issue. Why add "maybe" to "yes" and "no" if the added complexity only leads to confusion and anxiety.[14]

The crude models and theories people ordinarily employ are useful precisely because they embody simple distinctions, and thereby serve to facilitate routine decisions. But, a choice is often required between alternative simplifications, categories, and models, none of which convey all that there is to know about a particular situation. Choosing among such alternative formulations requires some pragmatic considerations. What aspects of a situation are most important to explain? What variables do we most want to predict? Is it worse to make a few large errors or many smaller ones? Does a given way of looking at a problem highlight critical choices by leaving out extraneous detail; or obscure the situation by oversimplifying and omitting relevant options and consequences?

These features of thinking in turn complicate the organization and operation of government. The simplification characteristic of all analyses means that multiple accounts of a situation are possible, indeed likely. Real problems almost never match the domain and perspective of any particular specialist. Different professional and functional groups are likely to see the same problem differently and to disagree in ways they find it difficult to resolve.[15]

Advocacy groups take advantage of such disagreements to seek out and support those experts and arguments that advance their cause. Supposedly neutral expertise becomes the weaponry of partisan conflict. The result is at best confusion about the appropriate role (and limits) of technical knowledge, and, at worst, the widespread belief that experts are mere hired guns who have nothing to contribute to the policy debate. When that happens, both government and ordinary citizens find it difficult to learn about a problem since they do not know who, if anyone, can be trusted.

The imperfect fit between concepts and reality also enhances administrative discretion. In any particular case, there is almost always room for disagreement over what decision the rules require or over what rules the law implies.[16] As a result, legislators and rule makers must either try to specify laws or rules in great detail (to limit agency discretion as much as possible) or leave the language less explicit and detailed and place greater trust in the competence and probity of the implementors. The pressures that arise in controversial cases make an appropriate balance between vagueness and specificity difficult to attain.

The Manifestation of these Problems Within EPA

Because the agency is responsible for many different aspects of policymaking ranging from toxicity standard-setting to law enforcement, it employs a wide range of professionals including lawyers, economists, and engineers. The modes of thought characteristic of these professions express different conceptual simplifications. These differences often breed misunderstanding and conflict.[17] Lawyers, having read hundreds of appellate cases in law school, learn that there are two sides to every argument. Simple cases, after all, do not make it into the textbooks. Since defendant and plaintiff alike present cogent theories and precedents, lawyers learn that disagreements cannot be resolved by appealing to shared ideas. Hence, fair procedures (e.g., bargaining) may be the only way to resolve conflicts.

Engineers, on the other hand, are trained to solve problems, not resolve them. The formulae and rules on which they base their calculations are often arbitrary. Nonetheless, they have been trained to use them to provide numerical solutions. They come to believe that there are right answers to problems and that those can be arrived at by manipulating data according to a unique "best practice."

Unlike either lawyers or engineers, economists are trained to view all variables as continuous. Regardless of whether price, production, or consumption is at issue, choices are not "yes" or "no," but matters of amount or degree. Thus, economists instinctively see all issues as arenas for trade-off and outcomes that produce "a little of this and a little of that" are often judged desirable. Moreover, economists are seldom aware of the ethical assumptions built into their policy analytic tools. To them it is obvious that the correct test of a policy is its efficiency and that efficiency is defined by giving people as much of what they want as possible.

The EPA was and is organized in ways that reflect and reinforce these professional cleavages. For example, the Office of Policy and Management (OPM), whose name was later changed to the Office of Policy Planning and Evaluation, has been responsible for evaluating the cost-effectiveness of proposed rules and standards. Staffed by economists and policy analysts, its mission has fit in well with its members' own ideological and professional commitments.[18] In contrast, the lawyers in the Office of General Counsel (OGC) have been responsible for defending the agency against legal challenges. They have sought rules and standards that will withstand judicial scrutiny. While primarily concerned with insuring that EPA complies with procedural requirements, their desire for rules that seem "reasonable" has often made them allies of the economists in OPM.

Professional perspectives and bureaucratic incentives do not always coincide. The lawyers who work for the Enforcement Division often argue for very stringent rules to improve the agency's negotiating position with waste sources. The very infeasiblity of these rules gives polluters an added incentive to make a "reasonable" deal.[19]

Also, the deeply held differences in the society about environmental policy do not stop at the agency's doorway. While to some degree inevitable, the level of ideological conflict within the agency has at times risen to destructive levels when senior managers have been recruited who are closely identified with either the pro- or anti-environmental cause.

PLURALISM AND BEYOND

Both our definition of the problem and our analysis of its root causes run counter to pluralism, the mode of political analysis that has dominated the postwar period.[20]

Given the limited ability of analysis to "solve" problems, pluralists want to "muddle through" by a series of incremental political decisions. Legislative logrolling and vote trading function like the market, facilitating mutually beneficial trades that make all participants better off.[21]

Like economists, pluralists treat individual desires as fixed and given. Such desires determine the ends the state is to serve. The state in turn is but a handy organizing device, a traffic cop, helping the free market to function efficiently. Perhaps because it so complicates the analysis, pluralists seldom acknowledge that experiences change individuals, that is, the value someone places on preserving nature may depend on whether they have ever hiked in the wilderness.[22] Thus pluralists would reject our whole discussion of civic education, especially the ethical dimensions of such education, as at best irrelevant and at worst a recipe for tyranny.

This view privatizes all political action, reducing it to the uninhibited pursuit of selfish objectives. All motives and goals are equally legitimate or illegitimate. The public sector is merely an arena in which various interests are forged into winning or losing coalitions. The only questions are strategic: What can each group do to advance its goals? Any appeal to the "public interest" involves fuzzy thinking, manipulative self-promotion, or self-deception.[23]

Within this framework, government agencies have no special status. They are just additional players in the game. "Iron-triangles" —informal but stable alliances among congressional subcommittees, bureaus, and constituencies—are merely effective patterns of group defense or successful arrangements for securing joint benefits.[24]

Pluralist analysis offers no way of identifying "better" or "worse" outcomes, and pluralists have a hard time making recommendations or judgments about public policy questions. They might not like it when some group uses the government for its own ends, but it is hard to blame the successful for doing what pluralist theory itself urges them to do.

Yet in a modern democracy, political competition is rife with inequality. The resources needed to succeed in this competition—money, manpower, time, contacts, technical knowledge, political and media skills—are not evenly distributed.[25] In addition, small groups of individuals, with large stakes in a decision, find it easier to organize (and have more reason to do so) than larger groups of individuals with only small interests. As a result, the number and scope of government actions designed to confer benefits on well identified groups expands—especially for groups with the necessary resources. The costs of such policies are concealed in either a large and complex public budget, or higher consumer prices.[26]

Public concern with the expansion of government represents dissatisfaction with these pluralist dynamics. Farm price supports, oil depletion allowances, mass transit access for the handicapped, and veterans preference in state civil services are all accomplishments of a pluralist politics in which well-organized potential beneficiaries defeat broader but more diffuse opposition.

In desperation, some pluralists have offered procedural solutions to insure that decision processes are "fair." But, since the distribution of influence among participants in any political process will always be unequal, it is not clear where or how they have derived their standard of "fairness." Some advocate still more participatory processes. The cure for the ills of democracy is, in this account, even more representation.[27] Others have decided that current institutions give "too much" power to special interest groups and want to further formalize all decision processes to protect administrators from undue influence.[28] Still others want to perfect pluralism by the increased use of explicit negotiations among organized groups.[29] In all cases, one wonders how these writers know what to recommend since their theory offers no criteria for judging alternative processes or outcomes.

The problem is that in a pluralistic world, it is hard to make sense of the idea of the "merits" of an argument. Exaggerated claims and one-sided advocacy are merely defensible tactics. Counter arguments are to be overcome, not considered seriously. This optimistic irresponsibility presupposes—wrongly—that political and economic resources are unlimited and that in pursuing narrow objectives each party can with impunity ignore the long-term consequences of its own lack of restraint.

Pluralistic ideas are as old as the republic. They are enshrined in such concepts as federalism, separation of powers, and checks and balances. As the *Federalist* argues, accepting and harnessing diversity may be the only way to preserve liberty in a largely heterogeneous society.[30] Still, the question remains as to how, if at all, politics can compensate for, and move beyond the worst features of pluralistic institutions and practices.

Thomas Jefferson had hoped that political parties would serve to foster those important qualities like civic virtue, breadth of vision, and solidarity that the pluralism of the founders tended to erode.[31] Similar hopes have been expressed by defenders of political parties from V.O. Key to David Broder.[32] The steady and continuing decline of party identification and party loyalty, however, makes political parties unlikely places for new efforts to curb pluralism's worst tendencies.[33]

Our alternative is to concentrate on government itself. We look to the upper echelons of the bureaucracy to provide the leadership required to nourish and sustain the values of education, responsiveness and respect for the merits that we cherish. Specifically, we expect government to promote three critical processes that foster those values: deliberation, integration, and accountability.

Deliberation

Government based upon the consent of the governed must find ways to mobilize that consent, and to expand the knowledge available to the community. To moderate partial views and yet learn from them, to insure meritorious outcomes, and to serve the goal of civic education, government officials need to foster a continuing conversation about public policy and the values and ideals upon which such decisions depend. Good government needs deliberation because it needs capable citizens and the fruits of mutual instruction and mutual inquiry. By the same token, deliberation contributes to self-development, which is part of freedom in the modern republican ideal.[34]

Deliberation is a collective process for engaging in what Bernard Williams has called "reflective criticism."[35] People work back and forth among the norms and theories they claim to subscribe to, and the decisions they have actually made or are tempted to make. They seek to understand where their ideas conflict and where they are inconsistent with their actual practice. They explore those options that they find most compelling as their grasp of the situation improves. They accept that perfect consistency is unlikely but that pursuing it poses useful questions.

When people deliberate they explore what their various ethical ideas imply in a specific context. This acknowledges the possibility that individuals may be wiser than they know—that their decisions may embody ethical insights that their explicit theories have not fully captured. Participants ask themselves, and each other, which norms are relevant to the case at hand; what point of view should be adopted to make a particular decision, recognizing that different definitions of the question imply different answers.

Deliberation requires that those who participate take one another's positions and claims seriously. They must have enough in common to make the search for synthesis plausible. They must trust each other enough to offer an honest account of the facts they possess and the objectives they seek—even if the former does not always advance the latter. If they are to listen to and learn from one another, they cannot simply engage in posturing and strategic behavior.

Only through such experiences can citizens come to appreciate their role as citizens. In the process they will have to synthesize private concerns with their best common understanding of their civic interests and obligations. Only by struggling with the multiplicity of claims and perspectives can they develop a healthy mutual respect for, and enjoyment of, their own diversity, together with an acceptance of some common purposes. Along the way, citizens will also come to appreciate the complexity and difficulty of public decisions. As a result, they are more likely to understand and accept the policies that are chosen, even when their own views do not prevail.[36]

Deliberation is intended to lead to scientific as well as ethical enlightenment—which is what provides its link to the technical merits. It is not a contest between interests but an inquiry, which profits from the contributions of several minds. It expands available information, enlarges the set of options, and broadens understanding. At its best, it can even improve foresight by encouraging the full discussion of the possible results of alternative courses of action.

We understand that deliberation, as we have described it, is an ideal type—rarely encountered in its pure form. We are not seeking to reinvent either the Greek polis or the Burkean parliament. Still, we believe that such honest discussion is an ideal worth striving toward. Moreover, if the technical obscurity of key decisions means that experts must first deliberate among themselves before involving citizens and their representatives, then government must be structured to facilitate that activity as well.

Integration

A key difference between modern government and older forms is the complexity of its organizational structure and the high degree of specialization of its component parts. We have already alluded to the problems that this caused at the EPA. One cannot expect such diverse resources to be combined effectively simply by chance. Instead, the involvement of various kinds of experts must be fostered by conscious design. This endeavor we call integration.[37]

Integration is partly a matter of structure: the crafting of occasions and relationships so that relevant specialties come together to raise questions and to respond to each other's concerns. Structure alone, however, is insufficient. Integration also involves behavior and hence a concern for recruitment and reward patterns that foster the preferred behavior. An agency needs to attract people who have the capacity and desire to work with and learn from those with other points of view. Patterns of promotion must reflect these integrative commitments.[38]

Thus, within the Executive Branch, integration establishes the preconditions for deliberation. Also deliberation supports integration by demonstrating the palpable benefits to be obtained when discussions are freed from professional and bureaucratic impediments.

The problems of integration arise not only within agencies but also among them. As we will see, once EPA has decided what it wants to do about a problem like air pollution emissions from steel plants, its proposed policy still needs to be reconciled with the views of other units of government like the Commerce Department or the Council of Economic Advisors. The practical problem is that there is often, quite literally, no one other than the president with both the authority and the responsibility to undertake this task. Yet the president is typically too preoccupied with the affairs of state to be able to give sufficient attention to a relatively minor matter like plant emissions.'[39]

If there are no real joints along which problems can be neatly carved, the effort to divide them up will inevitably be imperfect.[40] Therefore, good organizational design is characterized as much by its flexibility as by its shape. Bureaucratic leaders must establish coordinating devices, ad hoc though they may be, that can foster deliberation of a sufficiently inclusive scope about the problem at hand. Integration, therefore, is not only a matter of organizational charts, but also of leadership.

Such bureaucratic leadership is mainly the responsibility of senior managers, those appointed by the president with the advice and consent of the Senate. This is not because they are the only important policy makers. Significant decisions are in fact made all along an agency's chain of command. Political appointees, however, have special significance because they set the terms and conditions under which the permanent civil service functions. They shape the organizational chart and create the special task forces and working groups. They also play a particularly significant role in an agency's relationship with the rest of government. This external function is critical with regard to the process of accountability as well as to integration.

Accountability

Nowhere do we more sharply part company with pluralism than in our view of bureaucratic accountability. In the pluralist mold, a bureaucratic leader is a guerilla captain, roaming the political landscape, adding troops, supplies, and territory by whatever means available. Asked what he wants, he would reply in Samuel Gompers' epic phrase, "more." Asked to whom he is responsible he would reply, "whoever can help me."

Despite the delights and rewards of such guerilla warfare, a public agency is also a part of the government and must act on that understanding. Its leaders are accountable for acting to "preserve, protect, and defend" those constitutional values that allow democratic government to function and that perpetuate it.[41] Agency heads and their appointees are also accountable to a particular president. Neither job can be done by people who see themselves only as entrepreneurs or guerillas. An agency's leadership should try to ascertain how its mission fits with that of the overall administration program and articulate its activities with the rest of the president's team.

But an agency is not simply a cog in the executive engine. It is also accountable to statutes and to the Congress and the courts. Agencies have obligations to remain faithful to the law even when it puts them at odds with presidential directives. They also have an obligation to enrich congressional and judicial debate concerning the matters under their purview. Neither the Congress, the courts, nor the public at large possesses the expertise, and the time, to make sense of the myriad decisions that officials make every day. They are dependent on the agency's leadership to provide them with a clear and coherent picture of what the agency is doing and why.

Without that guidance, both executive and legislative overseers will resort to surrogate measures of performance. One such is the agency's level of activity: the number of enforcement actions brought or the number of dumps closed. The other measure is probity, the presence of officials who do not favor particular private parties. Neither of these standards is irrelevant. The Gorsuch years at the EPA illustrate the perils that befall an agency when it fails to live up to them. But zeal and honesty do not themselves guarantee reason and wisdom.

For others to be able to hold it accountable, the agency's leaders must provide a cogent account of the considerations that lie behind their actions. They need to focus their attention on key questions and how they propose to answer them. To achieve accountability, therefore, agencies must develop and communicate an explicit strategy.

Strategy

Strategy is the articulation of targets, and plans to reach those targets, in light of available resources and constraints.[42] Strategic planning is done not in the service of plan writing, but to facilitate the making of real choices. Strategy serves as a source of perspective and proportion. It enables one to differentiate between big problems and small ones; to separate significant issues from trivial ones; to conserve resources and coordinate behavior. It gives subordinates a clear sense of what their superiors are really trying to achieve.

As the case studies show, developing a clear problem definition is a necessary step in developing a strategy. This means characterizing which elements of the problem are most important and explaining why other aspects ought to be taken less seriously. Only such explicitness makes it possible to determine whether the means chosen for addressing the problem are adequate and whether they are being deployed effectively.

Strategy fosters deliberation. By providing an agenda it enables the parties to any policy discussion to focus their attention on a sufficiently limited and clearly defined set of questions to make progress possible. It can also make integration easier. A clear problem definition reveals which elements of the problem are to be taken most seriously and, therefore, provides a basis for establishing effective organizational relationships.[43]

The criteria and principles we have offered may seem overly elaborate and abstract. But, the relevent test is whether these ideas can inform our understanding of the actual cases we have studied. To put these cases in an appropriate historical perspective, we must first review the development of EPA from its founding under President Nixon to the advent of the Carter Administration.

NOTES

1. On the tension between collective and market principles, see Kenneth J. Arrow, "Values and Collective Decision Making," in Peter Laslett and W. G. Runciman, eds., *Philosophy, Politics and Society* Third Series (Oxford: Basil Blackwell, 1969), 215–232; Michael Walzer, *Spheres of Justice: A Defense of Pluralism and Equality* (New York: Basic Books, 1983); Fred Hirsch, *The Social Limits to Growth* (Cambridge MA: Harvard University Press, 1976), especially Chapter 6. See also Deborah A. Stone, *Policy Paradox and Political Reason* (Glenview, IL: Scott Foresman, 1988), especially 13–26, and Marc Landy and Henry Plotkin "Limits of the Market Metaphor," *Society* (May/June 1982), 8–17,

2. The most cogent and thoughtful consideration of the difficult choices raised by man's relationship to the environment is contained in Joseph Sax, *Mountains Without Handrails: Reflections on the National Parks* (Ann Arbor, MI: University of Michigan Press, 1980) and John McPhee, *Encounters with the Archdruid* (New York: Farrar, Straus, Giroux, 1971). For a radical critique of conventional environmentalism, see Bill Devall and George Session, *Deep Ecology* (Salt Lake City, UT: Gibbs M. Smith Inc., 1985). On the question of the ethical relations between man and animals, see Paul W. Taylor, *Respect for Nature: A Theory of Environmental Ethics* (Princeton: Princeton University Press, 1986), Chapters One and Five.

3. The classical discussion of this problem is contained in James Landis, *The Administrative Process* (New Haven, CT: Yale University Press, 1938), especially 47–88. On the role of statutory ambiguity with regard to environmental issues, see Angus MacIntyre, "Administrative Initiative and Theories of Implementation: Federal Pesticide Policy," in

Helen M. Ingram and R. Kenneth Godwin, eds., *Public Policy and the Natural Environment* (Greenwich, CT: JAI Press, 1985), 231–232.

4. On the impact of the "New Administrative Law," see R. Shep Melnick, *Regulation and the Courts: The Case of the Clean Air Act* (Washington, D.C.: The Brookings Institution, 1983) and Richard B. Stewart, "The Reformation of American Administrative Law," *Harvard Law Review* 88 (1975):1667. Also see Kenneth Culp Davis, *Administrative Law of the Seventies* (Rochester, NY: Lawyers Co-operative Publishing, 1976). On developments in Congress, see Dennis Hale, ed., *The United States Congress* (New Brunswick, NJ: Transaction Books, 1984); Arthur Maass, *Congress and the Common Good* (New York: Basic Books, 1983). On changing patterns of organizational life in Washington, see Hugh Heclo, "Issue Networks and the Executive Establishment," in Anthony King, ed., *The New American Political System* (Washington D.C.: The American Enterprise Institute, 1978), 87–124.

5. A detailed account of this problem as applied to water quality control can be found in Bruce Ackerman, et al., *The Uncertain Search for Environmental Quality* (New York: MacMillan, 1974). As applied to New Source Performance Standards under the Clean Air Act, see Bruce Ackerman and William Hassler, *Clean Coal/Dirty Air* (New Haven, CT: Yale University Press, 1981). As applied to the air emission standard for benzene, see Albert Nichols, *Targeting Economic Incentives for Environmental Protection* (Cambridge, MA: MIT Press, 1978), 127–158.

6. For a discussion of this problem, see Donald Dewees, "The Costs and Technology of Pollution Abatement," in Ann F. Friedlaender, ed., *Approaches to Controlling Air Pollution* (Cambridge, MA: MIT Press, 1978), 291–334.

7. See Hugh Heclo,"Issue Networks and the Executive Establishment," in Anthony King, ed., *The New American Political System*(Washington: The American Enterprise Institute 1978), 87–124. See also Kay Lehman Schlozman and John Tierney, *Organized Interests and American Democracy* (New York: Harper and Row, 1986), 62–87.

8. See Lewis Anthony Dexter, *The Sociology and Politics of Congress*(Chicago:Rand McNally, 1969), especially Part Two.

9. Raymond A. Bauer, Ithiel de Sola Pool and Lewis Anthony Dexter, *American Business and Public Policy: The Politics of Foreign Trade*, 2nd edition (New York: Aldine de Gruyter, 1972); see also Francis E. Rourke, *Bureaucracy Politics and Public Policy* (Boston: Little Brown, 1984).

10. For an excellent discussion of these and other difficulties, see V.O. Key Jr., *Public Opinion and American Democracy* (New York: Alfred Knopf, 1961), chapters 4,9. See also Dennis F. Thompson, *The Democratic Citizen: Social Science and Democratic Theory in the 20th Century* (Cambridge, England: Cambridge University Press, 1970).

11. On the problem of mandate uncertainty, see Kay Lehman Schlozman and Sidney Verba, "Sending Them a Message-Getting a Reply: Presidential Elections and Democratic Accountability," in Kay Lehman Schlozman, ed., *Elections in America* (Boston: Allen and Unwin, 1987), 3–26.

12. See Hugh Heclo, "OMB and the Presidency - The Problem of Neutral Competence," *The Public Interest* (Winter 1975), 80–98. That courts do not by the very nature of their role have this institutional capacity is the argument of Donald Horowitz, *The Courts and Social Policy* (Washington, D.C.: The Brookings Institution, 1977).

13. The capture thesis was initially proposed by Samuel Huntington, "The Marasmus of the I.C.C.," *Yale Law Journal* 61 (April 1952), 467–509. See also Marver Bernstein, *Regulating Business by Independent Commission* (Princeton, NJ: Princeton University Press, 1955). Ralph Nader and his colleagues provide the classic illustration of the liberal view. See, for example, Mark J. Green with Beverly C. Moore Jr. and Bruce Wasserstrom, *The Closed Enterprise System: A Ralph Nader Study Group Report on*

Antitrust Enforcement (New York: Grossman Publishers, 1972). For a conservative view, see George Stigler, "The Theory of Economic Regulation," *Bell Journal of Economics and Management Science* 2(1971), 3–21; E.S. Savas, *Privatization: The Key to Better Government* (Chatham, NJ: Chatham House, 1987), 1–30; Charles Wolf Jr., "A Theory of Non-market Failure," *The Public Interest* 55 (Spring 1977), 114–133.

14. For an elaboration of this discussion, see Marc J. Roberts, "On the Nature and Condition of Social Science," *Daedalus* (Summer 1974), 47–64. Also Stuart Hampshire, *Thought and Action* (New York: Viking Press, 1959); George Lakoff and Mark Johnson, *Metaphors We Live By* (Chicago: The University of Chicago Press, 1980); and Stephen Jay Gould, "The Streak of Streaks," *The New York Review of Books* vol. XXXV, no. 13 (August 18, 1988), 8–12.

15. Marc J. Roberts, Stephen R. Thomas and Michael J. Dowling, "Mapping Scientific Disputes That Affect Public Policymaking," *Science, Technology and Human Values*, Vol. 9 (Winter 1984), 112–122; John D. Graham, Laura C. Green, and Marc J. Roberts, *In Search of Safety: Chemicals and Cancer Risk* (Cambridge, MA: Harvard University Press, 1988), esp, Chapter 7.

16. See Melnick, *Regulation and the Courts*, 11–13. Also Douglas Yates, *Bureaucratic Democracy* (Cambridge, MA: Harvard University Press, 1982); Gary C. Bryner, *Bureaucratic Discretion* (New York: Pergamon Books, 1987).

17. Marc J. Roberts and Jeremy S. Bluhm, *The Choices of Power: Utilities Face the Environmental Challenge* (Cambridge, MA: Harvard University Press, 1981), 35; also Melnick, *Regulation and the Courts*, 38–43, 320–328.

18. See Chapter Four (RCRA).

19. See Chapter Five (Steel).

20. Seminal pluralist texts include David Truman, *The Governmental Process: Political Interests and Public Opinion* (New York: Alfred A. Knopf, 1951); Robert Dahl, *Who Governs?* (New Haven, CT: Yale University Press, 1961); and Aaron Wildavsky, *The Politics of the Budgetary Process*, 1st ed. (Boston: Little Brown, 1964). In recent years, public choice theory has systematized many of the assumptions of pluralism. The Introduction to one recent collection of papers in public choice says,

> checking democratic pathologies is an even more difficult task than Madison and Hamilton and Jay envisioned. Nonetheless the central thrust of public choice theory is congruent with that of *The Federalist*: people are essentially the same when they act publicly as when they act privately; self-interest is dominant throughout human affairs.

James D. Gwartney and Richard E. Wagner, "Public Choice and the Conduct of Representative Government," in Gwartney and Wagner, eds. *Public Choice and Constitutional Economics* (Greenwich, CT:JAI Press, 1988), p. 26.

Regardless of what may or may not be in *The Federalist*, this "central thrust" is just wrong. Self-interest is capable of producing highly variable results, and it is not always the dominant motive in politics. The authors of *The Federalist* wanted self-interest to moderate "passions" and "opinions."

21. Charles E. Lindblom, "The Science of Muddling Through," *Public Administration Review* 19 (Spring 1959):79–88. See also Anthony Downs, *Inside Bureaucracy* (Boston: Little Brown, 1967).

22. See John C. Harsanyi, "Welfare Economics of Variable Tastes," *Review of Economic Studies* 21 (1953–1954): 204–213. For enormously stimulating further development of this view see Albert O. Hirschman, *Shifting Involvements: Private Interest and Public Action* (Princeton: Princeton University Press, 1982), where several technical papers in psychology and philosophy are cited; for a political scientist's use of the same ideas, see

Aaron Wildavsky, "Choosing Preferences By Constructing Institutions: A Cultural Theory of Preference Formation," *American Political Science Review* Vol. 81 (March 1987): 3–21.

23. Truman, *The Governmental Process*, 50–51.

24. For early discussions of "subgovernments," see Douglas Cater, *Power in Washington* (New York: Random House, 1964). The similar idea of "policy whirlpools" is in Ernest Griffiths, *Impasse of Democracy* (New York: Harrison - Hilton, 1939). For a good contemporary account, see Randall Riply and Grace A. Franklin, *Congress the Bureaucracy and Public Policy*, 4th ed. (Homewood, IL: Dorsey Press, 1987).

25. E.E. Schattschneider, *The Semi Sovereign People* (New York: Holt Rinehart and Winston, 1960), 20–46, 97–113. For an excellent case study documenting this phenomena, see J. Clarence Davies III, *Neighborhood Groups and Urban Renewal* (New York: Columbia University Press, 1966).

26. James Q. Wilson, *Political Organizations* (New York: Basic Books, 1973), 330–333.

27. Harvey C. Mansfield, Jr., "Modern and Medieval Representation," in J. Roland Pennock and John Chapman, eds., *Representation* Nomos XI (New York, 1968); Dennis F. Thompson, *John Stuart Mill and Representative Government* (Princeton: Princeton University Press, 1976), Chapter 3; Stephen R. Thomas, "Publics, Markets and Expertise," (unpublished Ph.D. dissertation, Harvard University, 1980), Chapter 3.

28. Kenneth C. Davis, *Discretionary Justice: A Preliminary Inquiry*, (Urbana, IL: University of Illinois Press, 1971) v, 3–4.

29. Lawrence Susskind, *Breaking the Impasse: Consensual Approaches to Resolving Public Disputes* (New York: Basic Books, 1987). For a skeptical critique of this view, see Thomas, "Publics, Markets and Expertise," Chapter 2.

30. Alexander Hamilton, James Madison and John Jay, *The Federalist*, ed. Clinton Rossiter (New York: New American Library, 1961), see especially papers 10 and 51. There are many editions of these papers.

31. On this point in Jefferson, see Wilson C. McWilliams, *The Idea of Fraternity in American Politics*, 200–224; and Harvey C. Mansfield, Jr., "Thomas Jefferson," in Morton J. Frisch and Richard G. Stevens, eds., *American Political Thought* (New York: Charles Scribner's Sons, 1971).

32. David Broder, *The Party's Over* (New York: Harper and Row, 1972); V.O. Key Jr., *Politics, Parties and Pressure Groups*, 2nd ed. (New York: Thomas Y. Crowell Company, 1947), 666–675, 698–700. See also J. Roland Pennock, "Towards a More Responsible Two Party System," *American Political Science Review* 44 (Sept. 1950): Supplement.

33. Walter Dean Burnham, "The Changing Shape of the American Political Universe," *American Political Science Review* 59 (1965):7–28. See also his essay, "Elections as Democratic Institutions," in Schlozman, ed., *Elections in America*, 27–60, and Robert Kuttner, *The Life of the Party: Democratic Prospects in 1988 and Beyond* (New York: Viking Penguin, 1987).

34. Thomas, "Publics, Markets and Expertise," Chapter 2, "Deliberation;" Joseph Tussman, *Obligation and The Body Politic* (New York: Oxford University Press, 1960); Samuel H. Beer, "The Strengths of Liberal Democracy," in William S. Livingston, ed., *A Prospect of Liberal Democracy* (Austin: University of Texas Press, 1979); Thompson, *John Stuart Mill and Representative Government*, Chapters 2 and 4; Benjamin Barber, *Strong Democracy: Participatory Politics for a New Age* (Berkeley and Los Angeles: University of California Press, 1984), Chapters 8 and 9.

35. Bernard Williams, *Ethics and the Limits of Philosophy* (Cambridge, MA: Harvard University Press, 1985), 116. See also 93–119.

36. Deliberation is not the same as cooperation, but repeated opportunities to ex-

change ideas and opinions are likely to increase empathy with and sympathy for the position of one's erstwhile antagonist. See Irving Janis, *Victims of Groupthink* (Boston: Houghton Mifflin, 1972).

37. The classic work on this problem is Paul R. Lawrence and Jay W. Lorsch, *Organization and Environment* (Homewood, IL: Richard D. Irwin, 1969). See also Roberts and Bluhm, *The Choices of Power*, 355–376.

38. See Herbert Kaufman, *The Forest Ranger: A Study in Administrative Behavior* (Baltimore, MD: Johns Hopkins Press, 1960), 176–182

39. Hugh Heclo, "The Executive Office of the President," in Marc Landy, ed., *Modern Presidents and the Presidency* (Lexington, MA.: Lexington Books, 1985), 79. On the president's difficulties managing the Executive Branch, see also Colin Campbell SJ, *Managing the Presidency: Carter, Reagan and the Search for Executive Harmony* (Pittsburgh, PA: University of Pittsburgh Press, 1986); Edward Paul Fuchs, *Presidents, Management and Regulation* (Englewood Cliffs, N.J.: Prentice Hall, 1988); and Stephen Hess, *Organizing the Presidency* (Washington DC: The Brookings Institution, 1976).

40. On the impossibility of designing a single set of boundaries to perfectly encompass all problems, see Marc J. Roberts, "Organizing Water Pollution Control: The Scope and Structure of River Basin Authorities," *Public Policy* (Winter 1981), 89–92, 98–113; and Robert Dahl, "The City in the Future of Democracy," *The American Political Science Review* 61 (Dec. 1967): 958–960.

41. For a penetrating discussion of this question, see Donald J. Maletz, "The Place of Constitutionalism in the Education of Public Administrators," paper presented at the annual meeting of the American Political Science Association, Chicago, Illinois, Sept. 1987. See also John Rohr, *To Run a Constitution: the Legitimacy of the Administrative State* (Lawrence, KS: University of Kansas Press, 1986); Herbert Storing, "American Statesmanship: Old and New," in Robert A. Goldwin, ed., *Bureaucrats, Policy Analysts, Statesmen: Who Leads* (Washington D.C.: American Enterprise Institute, 1980), 88–113.

42. See Kenneth Andrews, *The Concept of Corporate Strategy* (Homewood, IL: Dow Jones-Irwin, 1971).

43. Lawrence and Lorsch's *Organization and the Environment* discusses how structure should be adjusted to strategy. It is also clear that the reverse occurs, i.e structure causes strategy to adjust. See also Joseph L. Bower, *Managing the Resource Allocation Process* (Boston: Harvard Graduate School of Business Administration, Division of Research, 1970). For a historical overview of this relationship as applied to corporations, see Alfred D. Chandler, *Strategy and Structure* (Cambridge, MA: Harvard University Press, 1962).

2

The Origins and Development of the Environmental Protection Agency

BACKROUND

In this chapter, we review the history of the Environmental Protection Agency and the larger social and political setting in which the agency was born. We do this because the origins and the early years of an organization often leave a lasting impact on its structure, and on the ideas, goals, and attitudes of its members. After describing some of the economic and political processes that helped produce the environmental movement, we trace the evolution of congressional politics and the development of environmental legislation. We then review the short but tumultuous history of EPA from its founding in 1970 to Douglas Costle's arrival in the spring of 1977, at which point our detailed case narratives begin.

Our story really begins with the great movement to the suburbs after World War II. Hoards of upwardly mobile white collar workers left crowded cities for localities with clean air, gardens, and grass. However, the reality of suburban life—smog, traffic jams and strip development—too often left this bucolic fantasy unrealized. At the same time some rural folk watched with dismay as their small towns became urbanized.[1]

The population was becoming younger, more secure financially, and better educated. Between 1950 and 1974 the percentage of adults with some college education rose from 13.4 to 25.2%.[2] This was coupled with a streak of unprecedented prosperity. Affluence, leisure, mobility, and a greater understanding of physical and biological science combined to create a new awareness of, and interest in, the natural world.

The New Deal and the war had taught Americans that government could be used to achieve social goals. When the new suburbanites found that uncontrolled growth produced results they disliked, they turned to public policy to achieve their aims. The overwhelming public support demonstrated for environmental legislation in the 1960s signaled the arrival of a new political constituency, one more concerned with "quality of life" than "bread and butter."[3]

The translation of widespread but unfocused public concern into specific policies and programs was shaped by the political institutions of the time. The characteristics of these institutions decisively affected the form and the content of the public debate about environmental questions, and the results that were obtained. The changes that occurred in political parties, in the media, and in the environmental community were of particular importance.

Political Parties

The electoral potential of the new suburban constituency was not lost on either political party, both of whose core constituencies were dwindling. Democrats confronted a decline in union membership and in the number of small farmers and a decrease in party loyalty among traditionally Democratic groups. Republicans were suffering from the continued depopulation of the smaller cities and towns and rural areas. Each party saw its salvation in the suburbs. There resided the young, well-educated voters who were to become the mainstays of environmentalism.[4] The political leverage of suburbia was further increased by the Supreme Court's reapportionment rulings of the early 1960s.[5] States were forced to decrease the representation of the depopulated countryside and decaying inner cities and recognize the enormous relocation that was taking place.

Responding to all this caused both parties great political strain. Rural Democrats deeply appreciated the public works projects that had brought electricity to their homes, paved roads to their doorstep, and saved their land from the dual ravages of flood and drought. They were upset at the new criticism of such projects, as were the construction unions whose members built these same dams and roads. Other large unions like the Auto Workers and the Steel Workers likewise were dismayed to find their industries at the top of the list of pollution sources.[6]

Republicans too were torn. Environmentalism risked antagonizing their traditional friends in the business community. Fiscal conservatives and opponents of government intrusion were also upset by demands for increased public and private spending on pollution control.[7]

Leaders in both parties sought to frame environmental programs to avoid these cleavages. A multibillion dollar grant program for the construction of local sewage treatment plants allowed congressional Democrats to make pollution control palatable to the construction unions. Bold statutory preambles, which trumpeted the elimination of all pollution, were often joined to cumbersome enforcement mechanisms designed to appease both labor and management in target industries. The resulting incoherence in the design of environmental programs was therefore due in part to the ambivalent motivations that produced them.[8]

The Media

The second important development was the growing role of television news. During the 1960s and 1970s both network and local news air time increased dramatically. Potentially very profitable, the news is substantially cheaper to produce than entertainment programs, and widely viewed. Local competition is especially intense because the local news is often the first show viewers watch each evening.[9] This gives stations an opportunity to capture viewers for the entire evening. It has become industry gospel that a good news segment must have (1) visual interest, (2) a strong story line, and (3) viewer identification with its outcome.[10]

Environmental stories meet all three requirements. Oil covered birds, belching smoke stacks, rusting storage drums, and inspection crews in "moonsuits" are all visually compelling. Environmental stories have the particular advantage that often the crew can set up at its leisure, obtain the desired angles, and have plenty of vivid footage to show at the six and the eleven PM newscasts.[11]

The story to be told is familiar yet poignant. The victims, like the typical viewer, are middle class homeowners. A dangerous polluter threatens their home, livelihood, or family. The perpetrator, a large and faceless corporation, ignores their protests to pursue vast profits. David takes on Goliath while feckless bureaucrats and venal politicians stand idly by.

This is not to say that media journalists concoct environmental stories. They do, however, have powerful incentives to accentuate their pictorial and dramatic qualities. As a result, viewers repeatedly encounter such disputes as life threatening contests between good and evil.

Furthermore, the need for news organizations to "get both sides" has contributed to a demoralizing portrait of public agencies. In a contest between good and evil, they are often more easily seen as being evil. And whenever they are quoted, someone with a contrary opinion gets equal time. Thus, agencies are deprived of whatever impartiality or special expertise they might possess in the public's mind.

The Environmental Movement

The third important development involved the evolution of the environmental movement. Some environmental organizations, The Sierra Club (1892), the Audubon Society (1905), the Wilderness Society (1935), and the National Wildlife Federation (1936), have existed for decades and have large numbers of members whose dues support their activities. But in the early 1970s they were joined by newer organizations like the Environmental Defense Fund (EDF) and the Natural Resources Defense Council (NRDC). These had far fewer members and relied on grants from foundations for their sustenance. From the outset their main goal was to influence governmental policy.[12]

Despite their differences, the various organizations have worked well together. Since they each have only limited resources and confront an enormous range of issues, they are virtually forced to specialize and cooperate. Each organization takes the lead on a few issues. Coalitions develop with the leadership provided by the one or two groups whose staff members are expert on that question.[13] Alliances are also often sought with nonenvironmental groups. For example, the Clean Air Coalition, founded in 1973, included the United Steel Workers, the League of Women Voters, and the American Lung Association, as well as environmental groups. The resulting organization worked hard and effectively to mobilize grass roots support to pressure legislators not normally in the environmental camp.[14]

Thus, on any one issue, a relatively few individuals define problems, determine priorities, and craft the overall strategy. The Washington staffers who exercise this power are by no means a microcosm of either the population at large or even the environmental community. Most of them were lawyers who came to political maturity in the 1960s; their political world view was formed by those stormy times. A small, tight-knit community in constant contact with each other, they are bound by ties of loyalty, friendship, and common experience.[15]

This group was well prepared to exploit the opportunities that the development of television news offered. Raised on the civil rights revolution, the media-oriented tactics of that extraordinary effort were their political birthright. They appreciated the publicity value of good visuals, clear symbols, and human drama, and they were skilled at providing these to a often sympathetic press corps.[16]

Environmentalists make good copy. They enable the media to report highly inflammatory charges without violating its own canon of impartiality. To take Alfred Marcus' apt example, a reporter cannot himself characterize the Federal Trade Commission as a "lethargic pawn of lobbyists." But there is no problem reporting that this charge had been made by a Ralph Nader study group. Thus, environmental activists are rewarded for extremism. The more vivid the language they use, the more melodramatically they criticize industry and government, the more air time and headlines they receive.[17]

The media impact of environmentalists is also increased by their role as a source of technical information that nonexpert journalists find difficult to acquire. The press has come to distrust industry, and environmentalists are often the only alternative source of information.

Having become politically aware during the presidency of Lyndon B. Johnson and having reached political maturity during the administration of Richard M. Nixon, many of these activists shared a distrust of executive authority. Their view of the presidency and of the executive branch in general was deeply shaped by what they considered to be the usurpations and prevarications of these two men. New Deal reformers had looked upon the federal bureaucracy as a champion of the public interest. To the new generation the Forest Service, the Bureau of Reclamation, and the highway program were the perpetrators and abetters of ecological rapine.

By contrast, the courts and the Congress were viewed much more favorably. As the Ralph Nader Congress Project wrote in 1972,

> Congress has its problems, but the only way to restore balance to the government is for it to stand up for the rights it retains and to fight for the return of those that have been taken from it.[18]

Whatever its defects, the decentralized structure and varied membership of Congress rendered it accessible to any group willing to work hard to identify and contact likeminded members and/or their staffs.[19] In contrast to the executive branch, which appeared to be growing ever more isolated, the proliferation of subcommittees and the weakening of the leadership of Congress made it easier for almost any group to obtain a hearing there.

The courts had likewise moved in directions that environmentalists favored.

The Supreme Court, under the leadership of Earl Warren, had demonstrated a willingness to protect the weak that many environmentalists regarded as heroic. The relaxation in the rules of standing and the other innovations comprising the "new administrative law" gave environmental groups increased access to the federal bench. Indeed, judges wrote much of the "new" law in environmental cases. The D.C. Circuit was notoriously friendly to environmentalists' views and several judges made much of their innovative reputation in dealing with such issues. This sympathetic ear encouraged the environmental movement to press for an expanded judicial role. Moreover, its own success made the legal strategy ever more attractive.[20]

Furthermore, there were plenty of young attorneys eager to sign on to the ecological cause. Since the beginning of the New Deal, the Justice Department and other federal agencies had employed many freshly minted, reform-minded law graduates. With the election of Richard Nixon, young attorneys eager to use their skills for liberal ends preferred to sue the government rather than represent it.[21]

CONGRESS AND THE ENVIRONMENT: MUSKIE'S ROLE

This preoccupation with judicial remedies only increased the importance of Congress. Changing modes of judicial interpretation presented new opportunities to those who knew how to manipulate the nonstatutory aspects of the legislative process: hearings records, committee reports, colloquies on the floor, preambles to legislation, and so on. This development has been keenly and extensively analyzed by R. Shep Melnick. He has described numerous examples of the courts' willingness to divine the "intent" of a Congress that was clearly divided by referring to the actions of specific committee members. For example:

> If there was any congressional intent to add PSD (prevention of significant deterioration) to the provisions for attaining the national air quality standards established by the (clean air) act, then that intent existed only in the minds of the staff members who inserted a few pregnant phrases into the Senate report and the committee leaders who subsequently berated the EPA for refusing to address the Sierra Club's concerns.[22]

Environmental lobbyists became adept at working with cooperative staffers and members to insert specific phrases in the record that could then serve as "hooks" on which to hang future litigation. The Congress as a whole had no means of controlling these activities. The willingness of courts to elevate these devices to quasi-statutory status thus served to increase greatly the power of the committee and subcommittee chairmen and their staffs who were in a position to fashion them.

Furthermore, once the court had found in favor of the environmentalists, the inertial bias of Congress came to their aid. Their enemies were then forced to undertake the arduous and disaster-prone process of mobilizing the Congress to pass new legislation.

Because of the highly technical nature of these issues, most members of Congress deferred to those who specialized in them. One member in particular,

Senator Edmund Muskie, succeeded in establishing himself as the nation's preeminent designer of environmental policy.

Muskie's role in the environmental field began inauspiciously. As a freshman senator from Maine in 1958, he offended Senate Majority Leader Johnson and was relegated to a set of undesirable committee assignments, including the Public Works Committee. Viewed as a pork barrel affair, Public Works was avoided by senators eager to be seen as leaders and statesmen. It ranked eleventh of fifteen Senate committees in a survey of member preferences.[23]

Public Works' jurisdiction over water pollution derived from its responsibility for flood control and the dredging of rivers and harbors. In 1955, it took on air pollution by the simple expedient of holding hearings on the subject and reporting out a bill. In 1963 Muskie was appointed chairman of a newly created environment subcommittee, a post he retained until his departure from the Senate seventeen years later.[24]

Muskie's influence came from several sources.[25] One was his ability to assemble a competent and aggressive staff. He also made full and skillful use of his institutional position, developing a detailed knowledge of pollution issues and getting to know environmental officials at the state and local levels. In addition, his own longevity contrasted with the high turnover among other members of the committee. Between 1963 and 1970 twelve members left Public Works for other more attractive assignments. Thus, the senator from Maine chaired a subcommittee whose staff was devoted exclusively to pollution matters and whose other members were far less knowledgeable about the subject than he was. The very undesirability of Public Works thus, ironically, worked to his advantage.

Muskie's central role in the Senate had no parallel in the House, where pollution control issues were assigned to various committees.[26] Air pollution became the province of the Commerce Committee whose jurisdiction extended to questions of "public health and quarantine."[27] Water pollution went to House Public Works, which had responsibility for flood control, dredging, and other water-related construction. Neither created a special pollution subcommittee.

In the Commerce Committee, in particular, pollution issues were quite subordinate. The House Commerce Committee is very prestigious, and next to Ways and Means, it handles the largest amount of domestic legislation. With a staff whose size has not been commensurate with its enormous workload, Commerce could only deal with the most highly visible and contentious issues.[28]

During the 1960s the committee focused on health care and securities regulation. Paul Rogers, Chairman of the Health Care Subcommittee, achieved a preeminence in that field that paralleled Muskie's role in pollution. This same subcommittee had jurisdiction over air pollution. Only in the 1970s, as the health import of air pollution aroused greater attention, did Rogers and his subcommittee begin to play a more active role.[29]

The situation in House Public Works was similar. John Blatnik, Chairman of the Subcommittee on Rivers and Harbors, gradually became a force on water pollution matters. His subcommittee however, dealt with a wide variety of issues of great concern to the rest of the committee, and to the House as a whole. He did not enjoy the luxury of concentrating his attention or his staff on pollution questions as Muskie did.[30]

Muskie authored a series of pathbreaking water and air quality statutes during the 1960's that had an enormous impact upon both environmental policies and the institutions established to implement them. In 1963 he proposed amendments to the existing Water Pollution Control Act that transferred authority for water pollution control from the Public Health Service to a newly established Federal Water Pollution Control Administration within HEW.[31] The bill gave the Secretary of HEW power to establish water quality standards for interstate waterways if the standards established by the states failed to meet with his approval. The bill passed the Senate but failed in the House. Muskie reintroduced it in the following session and it ultimately became the Water Quality Act of 1965. The following year, Muskie authored the 1966 Clean Water Restoration Act, which gave the states 3.5 billion dollars in federal grant money to be spent for sewage construction (Muskie's original bill had called for six billion).[32]

In 1967 Muskie, who had first pushed through air pollution legislation in 1963, placed his imprimatur upon a far reaching air quality bill. The bill required the states to establish air quality standards on the basis of scientific studies to be carried out by the federal government. The states would also bear primary responsibility for enforcing those standards. Both the standards and the enforcement plans were subject to review and approval by the federal government. The act also required the federal government to set emission standards for automobiles.[33]

The Johnson Administration had pressed for greater involvement of regional bodies in setting standards and in planning. Its air pollution proposal also included a provision for national uniform emissions limits. Muskie, a former governor, had successfully resisted these efforts.[34] The legislation he sponsored reflected his belief that while federal assistance and involvement was necessary, the states should be primarily responsible for both the formulation and implementation of pollution control policy.[35]

THE 1972 ELECTION

As the 1960s drew to a close, Edmund Muskie's positions underwent a profound metamorphosis. The senator had emerged from his vice presidential candidacy in 1968 as the leading contender for the 1972 Democratic nomination. As Nixon sought to strengthen his environmental credentials, Muskie responded by advocating strongly pro environmental positions that he had opposed just months before.[36]

The first Democrat elected statewide in Maine for many decades, Muskie had reflected his state's moderate-to conservative outlook. The environmental statutes he authored during the 1960s expressed this moderation and his concern for state prerogatives. As he prepared to campaign for the presidential nomination, however, he found that those statutes, especially the 1967 air law, were being derided as ineffective.[37]

The 1967 Air Quality Act was to be reauthorized in 1970. Progress had been disappointing. As of 1970, not a single state implementation plan had been approved by the federal government.[38] More seriously, the act's reliance on the states led some to doubt as to whether it could ever succeed. Critics feared that states would compete for new industry by keeping standards permissive and enforcement lax.[39]

Muskie felt that it was too soon to judge his handiwork a failure. He continued to oppose nationally uniform standards for either emissions or for ambient air quality. During hearings in March and early April of 1970 he defended the 1967 Act, and blamed the Nixon Administration for failing to implement it adequately.[40]

Much to his surprise, Muskie found himself subjected to a three-pronged attack. First, the House displayed unaccustomed initiative. Paul Rogers' Subcommittee held extensive hearings on air pollution that spring and produced a record that was intensely critical of the 1967 Act. The subcommittee then took the unprecedented step of drafting legislation more stringent and ambitious than Muskie's proposal.[41]

Even more devastating was the report in May of a Ralph Nader study group, *Vanishing Air*. It judged the 1967 Act a failure and ascribed Muskie's refusal to endorse a complete overhaul to his "selling out" to industrial interests. It particularly chastised Muskie for his failure to support national uniform emissions limits.[42] In its closing sentences the task force equated his behavior with that of his rival, the president:

> In 1970, for the first time, Americans rallied around the cause of preserving the environment. But their enormous enthusiasm has yet to find direction or true leadership. The two men with the greatest obligation to chart new passages—Richard Nixon and Edmund Muskie—instead dusted off old maps, and are now attempting, each in his own way, to steer the same course which has brought us to our present peril.[43]

Most damaging of all, the White House itself proposed more ambitious legislation than did the Senate Subcommittee. The administration urged that the Secretary of HEW be empowered to establish national health-based ambient air quality standards and to create national emissions limits for new sources of pollution.[44]

As a legislative craftsman from Maine, Muskie had taken an incremental approach to pollution problems. He had sought to balance improved air quality against other political and economic goals,[45] especially since his home state was desperately in need of more employment.[46] Enforcement under the 1967 Act was admittedly slow and cumbersome, and further legislation might well prove necessary. But it was best to proceed slowly and not threaten the web of political alliances he had painstakingly created in the Senate and at home.

Once Muskie decided to run for the presidency, however, he faced new incentives and pressures. For example, his top aide, Don Nicoll, had become executive secretary of the Maine Democratic Party in 1954, the same year that Muskie was elected governor. Together they had worked to revolutionize the politics of that previously rockribbed Republican state. Unfortunately, Nicoll had neither the experience nor the capacity to delegate that were necessary to run a national campaign. After months of procrastination Muskie replaced him.[47]

The air quality controversy required him to make a similar choice between past commitments and new demands.[48] He had to respond to the public's new interest in quick, visible results even if that meant repudiating previously held positions.

Muskie proposed a revised bill that contained the most stringent federal pollution control program that had ever been attempted. It went far beyond both the House and the Nixon Administration versions. It called for both emissions limits on all new sources and federal national air quality standards to be set without consideration of cost. Automobile emissions were to be reduced by 90 percent by 1975. States retained primary responsibility for devising plans for implementing these requirements, but these plan's too had to be approved by the federal government.[49]

The enormous weight of Muskie's leadership was put behind four ideas that were to prove central to later debates over environmental policy: (1) nationally uniform ambient standards, (2) no balancing between health risks and economic costs, (3) rigid deadlines to be adhered to regardless of economic and technological obstacles, and (4) uniform emissions limits for new sources, even in areas without current pollution problems.

Muskie ,then, shifted from a proponent of broad executive discretion, at both the state and federal levels, to an advocate of a narrow "mission" orientation for EPA. He began to characterize the EPA as "quasi-independent," and wrote laws giving the EPA Administrator—as opposed to the president—the power to make key decisions. He then tried to use his congressional power base to insist that EPA respond to Congress, not the White House, and, in particular, that it heed the nonstatutory instructions it received from the subcommittee.[50] Muskie adhered to these views for the rest of his political career.[51]

THE VIEW FROM THE WHITE HOUSE: ORGANIZATIONAL REFORM

Nixon wanted the support of environmentally oriented voters, but he also wanted to insure the compatibility of environmental policy with other objectives such as economic growth. This led him to a series of apparently contradictory decisions. On the one hand, he signed into law landmark legislation like the National Environmental Protection Act (NEPA) and the Clean Air Amendments of 1970, and he created the Environmental Protection Agency by executive order. On the other hand, he presided over a series of efforts to keep environmental regulators in check.

In May 1969, he appointed a cabinet committee called the Environmental Quality Council, with his science advisor, Lee DuBridge, as executive secretary. The committee was charged with preparing a comprehensive institutional strategy to address environmental questions.[52]

When the Council failed to deliver the needed plan, the president asked his top domestic policy advisor, John Ehrlichman, to assemble a task force to do the job. Ehrlichman, a former land use lawyer in Seattle, was well versed in environmental matters. He appointed John Whitaker of the White House staff to direct the task force. It included several men who would eventually perform high level duties within EPA, including Alvin Alm, later Deputy Administrator, John Quarles, the first General Counsel and Roger Strelow, the first Assistant Administrator for Air and Waste Management.[53]

In February 1970, the task force produced a preliminary report that recommended the establishment of a new Department of Environment and Natural Resources (DENR). The new department would replace the existing Department of Interior, which had recently acquired control of water pollution activities from HEW. The new department would absorb the Forest Service, Soil Conservation Service and the pesticide program from the Agriculture, the Army Corps of Engineers, and Air Pollution Control from HEW. The Bureau of Indian Affairs would leave Interior for HEW. With resource development and environmental regulation functions in the same department, politically troublesome trade-offs could then nominally be made several steps away from the Oval Office.[54]

Nixon indicated his approval of the DENR concept and delegated responsibility for its development to an advisory commission on government reorganization, which he had created in 1969. Chaired by Roy Ash, former head of Litton Industries, that group came to be known as the Ash Council. Ash had ambitious ideas about government reorganization. He wanted to merge all of the domestic side of the executive branch into four "super-departments:" Community Development; Economic Affairs; Human Resources; and Natural Resources.[55]

Ash assigned responsibility for designing the new Natural Resources Department to Amory Bradford, a former *New York Times* executive and official in the War on Poverty. Bradford's staff included Douglas M. Costle, (whose later tenure as EPA Administrator we will explore extensively) and J. Clarence Davies, a political scientist from Princeton.[56]

This staff, known as the environmental protection group, opposed the DENR plan. Congress, they argued, would not accept the enormous disruption of committee jurisdictions that creating DENR would produce. Even if Congress should acquiesce, the group believed that the new department would be unable to digest such a diverse collection of responsibilities and personnel. Like the existing Department of Interior, it would become little more than a holding company composed of more or less autonomous agencies, not an integrated structure. The president's objective of finding a way to trade off between development and environmental considerations would not in fact be met.

More fundamentally, however, the environmental group did not accept the essential premise of the DENR scheme. Its members did not believe that resource development and environmental quality objectives should be played off against one another within an agency in which development concerns were likely to dominate. Moreover, because it would be impossible to merge all the agencies whose perspectives needed to be taken into account, they viewed the scheme as incomplete and misguided.[57]

Instead, the group proposed to consolidate pollution control programs in a new and separate agency, reporting directly to the president. They offered three reasons for making this change. First, pollution problems were so great that simply providing more resources was not enough. A more visible organization with direct access to the president was needed. Second, a clear focus for environmental advocacy would help ensure that such objectives got attention within the executive branch. Finally, to develop a comprehensive and systematic approach to pollution, a single agency had to be responsible for all aspects of the problem.[58]

The environmental group won Bradford's support but, on April 8, 1970, Ash turned down its recommendation. He believed strongly that competing viewpoints had to be reconciled within departments, not separated out into different "mission" agencies. The following day he convened a meeting to discuss the DENR proposal with representatives of the affected cabinet departments.[59]

Unsurprisingly, the attitudes of the different departments reflected how they would fare under the proposed reorganization. Secretary Hickel of Interior enthusiastically endorsed the plan. The rest (except Secretary Volpe of Transportation who was largely unaffected) opposed it. For example, Maurice Stans of Commerce argued that resource development, air quality, and water quality did not belong in the same department. Air pollution was a health problem, he said, and, therefore, belonged in HEW; water pollution involved recreation and belonged in Interior; those programs dealing with development ought to be in the government's development agency, the Department of Commerce.[60] Similarly, Secretary Hardin of Agriculture argued that the nation's farmers would refuse to allow crucial matters concerning forests, soil conservation, and pesticide use to be removed from the department they loved and trusted.[61]

President Nixon could not fail to be impressed by the depth and breadth of cabinet opposition to the DENR plan. In addition, he harbored deep reservations about the administrative capacity and political judgment of the man who would inherit DENR, Walter Hickel. Hickel had endorsed Earth Day with a zeal the president deemed excessive. Worse, he had written a letter to the president sharply criticizing the White House response to the killings at Kent State. That letter had somehow leaked to the press and Hickel himself was suspected of leaking it. While reluctant to risk firing Hickel, Nixon was not enthusiastic about significantly increasing his authority.[62]

Nixon then withdrew his support for the DENR plan and, for a time, it appeared as if no reorganization would take place. The environmental group's plan for a separate environmental agency had surfaced during the cabinet meetings. But it had been opposed by those—most notably the Secretaries of Agriculture and HEW—who would have lost functions to the new agency.

In the wake of the DENR defeat, however, two important shifts occurred. The HEW accepted the creation of a separate agency, although it would be called upon to give up three of its bureaus. Furthermore, Ash decided to support the proposal as an improvement over the status quo.[63]

From the president's point of view, a separate agency had two very attractive features. First, it was a highly visible and innovative action. Second, it represented a compromise between those who wanted to totally redesign the executive branch and those who wanted to change nothing. Ironically, then, the creation of a powerful environmental advocacy agency was a compromise devised for parties virtually none of whom was an environmental advocate.[64] The important consequences stemming from this decision will be considered repeatedly throughout this book.

The reorganization plan creating the EPA was submitted to Congress on July 9, 1970, and since neither house chose to oppose it within sixty days, it went into effect on September 9, 1970.[65] The order called for the transfer to the new agency of the Federal Water Quality Administration and the Office of Research on Effects of Pesticides on Wildlife and Fish, from the Department of the Interior; the Bureau of Water Hygiene, the Bureau of Solid Waste Management, the National Air Pollution Control Administration, the Bureau of Radiological Health and the Office of Pesticides Research, from HEW; the Pesticides Regulation Division from the Department of Agriculture; the Division of Radiation Standards from the Atomic Energy Commission; and the Interagency Federal Radiation Council.

THE EARLY YEARS OF THE ENVIRONMENTAL PROTECTION AGENCY: DIFFERING EXPECTATIONS

The White House, on the one hand, and environmentalists and their congressional allies, on the other, disagreed sharply about the mission of EPA and its relationship to Congress and to the rest of the Executive Branch. The president and his aides expected the leader of EPA to be a balancer and integrator, to pursue environmental protection in ways that were compatible with industrial expansion and resource development. The advocacy community, in contrast, wanted EPA to champion environmental values against counterpressures from elsewhere in government. To insure congressional supremacy in environmental policy making, they sought to write statutes that imposed strict limits on the executive.

This issue surfaced during the first set of oversight hearings held by Muskie's subcommittee to review EPA's implementation of the Clean Air Act. Richard Ayres, attorney for the National Resources Defense Council, blamed the White House and the Office of Management and Budget (OMB) for undermining environmental policy:

> The White House Office of Management and Budget is reviewing in secrecy every major action of the Environmental Protection Agency. The public is completely excluded from this review, but the most anti-environmental Federal agencies, such as the Commerce Department and the Federal Power Commission appear to have full access to it. These agencies, acting as spokesmen for industrial interests, have effective power to veto EPA's actions . . . [66]

Six years later, testifying against efforts of another president to impose external review on EPA, Ayres put forward a fullfledged theory of executive-legislative relations that sought to justify EPA's special isolation from presidential control:

> Congress has delegated powers along a continuum. At one end, the President has vast discretion, with little review, . . . at the other, where sound policymaking requires expert knowledge, where decisions need to be insulated from political interference, and where the exercise of judicial review requires safeguarding the integrity and fairness of the record on which the agency acts, his control over the agencies is far more circumscribed.[67]
>
> The writing of regulations by quasi-independent agencies such as EPA, whose

powers are explicitly delegated to them by the Congress, whose judgments require great technical knowledge and whose actions are subject to judicial review, falls near the latter end of this continuum.[68]

Thus, despite the lack of any statutory authority, Ayres wanted EPA to be treated as a quasi-independent agency—like the Interstate Commerce Committee (ICC) and the Federal Communications Commission (FCC)—rather than like a normal federal department.

The agency that confronted these differing expectations, was initially staffed primarily by bureaucrats transferred from other federal departments. Mostly scientists and engineers, they had long labored in the bowels of large departments. Although suffering from budgetary and status deficiencies, they enjoyed the absence of scrutiny that this obscurity offered. They hoped that in the new EPA their status and budgets would increase while their work remained undisturbed. In addition, they brought with them the concepts, attitudes, and skills that had served their former agencies. For example, the pesticide group from the Department of Agriculture had long been more interested in promoting agricultural productivity than in protecting human health and the environment. Over the years, these units had often developed comfortable working relationships with their state and local counterparts. Since early environmental laws emphasized planning, grant giving, and technical assistance, their members did not have to adopt the adversarial posture required by regulatory enforcement. By both training and conviction, they were not prepared to shift from being aid and advice givers to aggressive violation hunters.[69]

The very substantial burden of dealing with these disparate expectations was placed on unlikely shoulders—an orthodox Republican politician from Indiana. A decade before, as a State Assistant Attorney General, William Ruckelshaus had prosecuted water pollution violators. But such matters played little role in his subsequent career as a state legislator, unsuccessful candidate for the U.S. Senate, and U.S. Assistant Attorney General.

Ruckelshaus' selection to be the first EPA administrator appears to have rested on two factors. First, he benefited from an active lobbying campaign by his old friend Jerry Hansler, a career officer in the U.S. Public Health Service. Assigned to the Indiana Health Department in the early 1960's, Hansler provided the expert testimony that enabled Ruckelshaus to successfully prosecute several water pollution cases. In the process they became close friends. When EPA was created, Hansler was in charge of federal air pollution regulation in the New York region, and was well placed to lobby for Ruckelshaus within the federal pollution control establishment. He also happened to be Senator Sam Ervin's (D,NC) son-in-law.

Hansler lined up a list of prominent environmental figures to endorse Ruckelshaus. This helped ensure him serious consideration. His ultimate selection, however, was due to the backing of his boss. Perhaps the single most powerful figure in the administration after Nixon himself, Attorney General John Mitchell's strong support was sufficient to gain Ruckelshaus the appointment.[70]

Ruckelshaus perceived a need to "hit the ground running," to convince the public that the Nixon administration was serious about environmental protection. The most immediate question was how to meld the various bureaus EPA had inherited into a coherent organization. Nixon's executive order, stressing comprehensive environmental management, pointed in the direction of a thorough restructuring. That was the view of Alain Enthoven, a former senior Defense Department official who was acting as a White House consultant. Based in part on the Ash Council staff recommendations, Enthoven proposed that EPA be organized functionally (e.g., research, standards setting, enforcement), thereby abolishing the agency's legacy of separate units for water, pesticides, and so forth.[71]

Douglas M. Costle, future EPA administrator, was the Ash Council staffer in charge of transition issues. He chose not to recommend Enthoven's plan to Ruckelshaus. Costle feared that the massive changes Enthoven proposed would create a great deal of friction and chaos. Only after the Agency had proved itself, Costle believed, could the administrator risk the broad restructuring Enthoven advocated.

Instead, Costle proposed a three-stage approach. Initially, each of the program areas would be left intact. In a second phase, new functional divisions would be added. Finally, the program offices would be abolished and merged into the new functional units.[72]

Ruckelshaus went through stages one and two of Costle's plan. Initially he established five commissioners, one for each of the existing programs. In April 1971, he created a revised organizational plan that corresponded closely to Costle's stage two proposal. The Agency was divided into five divisions, each headed by an assistant administrator. Three of these, Planning and Management, Enforcement and General Counsel, and Research and Monitoring, were organized along functional lines. The other two, Air and Water, and Pesticides, Radiation and Solid Waste, retained a program orientation. No effort was made, then or later, to implement the final stage of Costle's proposal. As we mentioned in the previous chapter, and as we will see extensively in the cases, differences in professional training and functional responsibilities, and crosscutting jurisdictions caused these various units to coexist uneasily with one another.[73]

Ruckelshaus decided to give first priority to enforcement. Other directions were certainly possible, even attractive. The woeful state of environmental science and pollution control technology provided a strong rationale for devoting the lion's share of agency resources to research and development. Alternatively, Nixon's stated commitment to comprehensive environmental management could have justified an emphasis on regional planning. Ruckelshaus, a lawyer, chose neither. He had succeeded as a law enforcement officer both in his home state and in Washington. As a Republican, a Nixon appointee, and a protege of John Mitchell, he knew that he was viewed with suspicion by the Democratic majority in Congress, the anti-Nixon press, and the environmental community. To establish his personal credibility he felt he had to immediately demonstrate his good faith and commitment to environmental objectives.[74]

Alternate strategies would only produce results in the distant future. They were also more difficult since it would take considerable time and effort to effectively assimilate the various units that had come from other agencies. In contrast, with a legal staff already assembled, Ruckelshaus could begin to litigate immediately. The bigger the targets, the more credibility he could amass. As John Quarles, his General Counsel, later phrased it:

> Ruckelshaus believed in the strength of public opinion and public support. . . . He did not seek support for his actions in the established structures of political power. He turned instead directly to the press and to public opinion, . . . The results were impressive, especially during the period of public clamor for environmental reform.[75]

During its first sixty days, EPA brought five times as many enforcement actions as the agencies it inherited had brought during any similar period. Atlanta, Cleveland, and Detroit were sued for illegal sewage discharges; the Reserve Mining Corporation was sued for dumping taconite filings into Lake Superior, Armco Steel was sued for polluting the Houston Ship Channel, U.S. Plywood-Champion Papers for polluting the Ohio River, and ITT Rayonier for dumping pulp waste products into Puget Sound. (The preponderance of water pollution cases reflects the fact that it was easier to sue under existing water quality legislation than under the pre-1970 air quality statute.)[76]

As the case of Union Carbide illustrates, however, lack of enforcement authority did not deter Ruckelshaus. One of that company's plants seemed to be primarily responsible for the particulate matter that blackened the skies in nearby Parkersburg, West Virginia, and Marietta, Ohio. For years, state and federal officials had been unable to move the company to begin a cleanup. Ruckelshaus had EPA send Carbide a lengthy letter detailing the history of the case. The Agency threatened to initiate legal action within ten days unless the company agreed to comply with the recommendations of a previous interstate conference that had considered the problem. Unsure of his formal legal authority, Ruckelshaus immediately released the letter to the press. On the day of the deadline, the company announced its willingness to comply.[77]

Ruckelshaus must have known that his confrontational approach would horrify the White House. He gambled that Muskie's electoral threat would keep the White House wolves at bay until after the 1972 election. Despite Muskie's unexpected failure to capture the Democratic nomination, the gamble worked. EPA's audacious efforts to haul polluters into court created such a positive image for the agency in general and Ruckelshaus in particular that Nixon reappointed him after the election.[78]

While he did establish the Agency's credibility, Ruckelshaus' decision to forego a thorough reorganization had long-term effects. The inherited bureaus and divisions continued to seek support from their traditional allies outside the Agency. The addition of various functional units, as we will see later on, resulted in an organization full of conflicting objectives that has made integrated management extremely difficult. It is also arguable that the early commitment to enforcement led to a shortchanging of other functions, such as research and development and planning. As a result, EPA has had to struggle to develop the scientific capability that would render its work technically credible.

The enforcement strategy had other costs as well. In the eyes of many conservatives, and those sensitive to business perspectives, it demonstrated that EPA was nothing more than an environmental advocate. Ironically, that is exactly what some in the environmental protection group and in Congress had sought to create. This, in turn, led to the first of many efforts by the White House to find ways to control and manage EPA.

THE QUALITY OF LIFE REVIEW

On May 21, 1971, OMB Director George Schultz announced the establishment of a procedure entitled "Quality of Life Review." All proposed EPA regulations were to be submitted for scrutiny by other relevant agencies, with the review process to be coordinated by OMB.[79] This was to ensure that economic development and fiscal concerns received due consideration in the process of writing regulations. In many cases, this review process delayed the promulgation of regulations for several months.

Environmentalists claimed that the process violated the law by, for example, allowing the OMB and not the EPA administrator to make the final decision on the content of air quality regulations. To support this charge they cited the many revisions that were made in EPA's regulations governing state implementation plans (SIPs) after their quality of life review.[80] Ruckelshaus maintained that while he paid close attention to the recommendations of other agencies, the final decision in all cases was his alone.[81]

The history of these reviews (like the later experience of the "Regulatory Analysis Review Group" under President Carter, which we consider in several of the cases) raises troublesome questions. In particular, who is "the White House" and who can responsibly invoke the authority that flows from the elected head of state? The White House after all is not a single person, but rather an ill-defined collection of groups and individuals all striving to influence policy. During the Nixon years, the president's preoccupation with foreign policy, and later with Watergate, meant that others were emboldened to speak in his name on relatively less important matters, like the formulation and enforcement of environmental policy.

The initial skirmish between the White House and the agency occurred in the fall of 1971. It concerned the case that EPA brought against Armco Steel for polluting the Houston Ship Channel.[82] After a favorable court decision, EPA enforcement chief John Quarles was summoned to the White House by presidential aide Peter Flanigan. Flanigan showed Quarles a letter to the president from William Verity, president of Armco and a powerful figure in the Ohio Republican Party. Verity claimed that the decision would result in the closing of important segments of the Armco facility and the loss of about 300 jobs. Flanigan contended that the loss of 300 jobs ran counter to the president's policy and that EPA had to remedy the situation.

Flanigan also asked the Justice Department to join Armco in seeking a sixty-day stay of the court order so that the company and the government could try to negotiate a settlement.[83] On October 8, however, John Whitaker, the White House aide responsible for environmental policy, convened a meeting to determine the official administration position on the matter. He ruled that Flanigan's action did not represent White House policy and that no stay should be sought until the government had time to establish its negotiating position vis-à-vis Armco.

Flanigan, a successful investment banker, was far more influential in White House circles than Whitaker. Whitaker's willingness to countermand him suggests that Flanigan had not in fact been acting for the president. Such behavior by the White House staff helped undermine whatever responsibility EPA's leadership might have felt to follow White House orders.

THE TRAIN INTERVAL

Ruckelshaus left EPA in April of 1973 to replace L. Patrick Gray as FBI Director after the latter had become implicated in the Watergate scandal.[84] He was succeeded at EPA by Russell Train. A career public servant and conservationist, Train had served as a staff member on two congressional committees, joined the Treasury Department and had then been appointed a U.S. Tax Court Judge. In 1965 he became the president of the Conservation Foundation. In 1969 he was appointed Under Secretary of the Interior and, a year later, was made Chairman of the Council on Environmental Quality.[85]

A lifelong Washingtonian, Train enjoyed close personal relations with many members of the Congress and the career civil service. His personal stature and relationships were a decided asset for the Nixon administration during Watergate and in turn helped give him significant independence from the rest of the Executive branch. When he insisted on having written confirmation that the EPA administrator had final authority over the substance of all EPA regulations, the White House provided it. Similarly, in March of 1974, Train refused to support an administration proposal that he claimed would gut the Clean Air Act.[86]

Train remained in office until after Gerald Ford's defeat by Jimmy Carter in 1976. The initiatives undertaken during his regime reinforced the image of EPA as an aggressive environmental advocate. The untidy mix of functional and program offices was retained. For the purposes of this study, the most important new development during these years was EPA's increased dependence on the Congress.

Train's independence limited his influence within the Nixon administration. As a result, he found it all the more necessary to cultivate ties with Congress. As a former senior EPA official put it:

> Train survived because Richard Nixon did not. He drew his support from the Air and Water Pollution Subcommittee of the Senate Public Works Committee in resisting the Nixon White House's campaign to dismantle environmental legislation. What had been to Ruckelshaus a voice of encouragement was to Train a lifeline.[87]

All of this enhanced the power and influence of the Muskie Subcommittee. The subcommittee and its staff came more and more to view its responsibilities in quasi-executive terms. The caretaker status of President Ford's administration limited his ability to disrupt this relationship or to force EPA to abide by White House, as opposed to congressional dictates. These warring views about who had the right to oversee environmental policy, and for what ends, formed a crucial and troubling aspect of the legacy received by Train's successor, Douglas M. Costle.

THE NEW TEAM

Douglas M. Costle, EPA administrator throughout the Carter years, was in many ways archetypical of that administration. He had had little or no prior involvement with the Carter campaign or with Carter personally. Nor was he well connected to the party organization or to some element of the party coalition. In fact, he was barely a Democrat at all, having served in Republican administrations at both the national and state level.

A native of Seattle, who went to Harvard and the University of Chicago Law School, Costle first worked in OMB in the Nixon administration. We have seen how, as a staffer for the Ash Council, he played a major role in designing EPA. He then became the commissioner for Environmental Affairs in Connecticut under a Republican governor. With the advent of a Democratic regime in Connecticut, Costle returned to Washington to work in the newly created Congressional Budget Office. His record as a successful administrator, combined with his involvement in the creation of EPA, made him attractive to an administration interested in administrative reform and enhanced efficiency. The Carter people, political mavericks themselves, were not put off by Costle's lack of orthodox Democratic credentials.[88]

The story of Costle's selection is typical of the transition process. Various lists were circulated and various candidates lobbied for. During the late stages of the process, Costle himself was working on transition matters—but on the structure of the Department of Energy. Senator Muskie wanted his longtime subcommittee staff director Leon Billings to be the EPA administrator. Afraid of the power this might give the senator from Maine, the transition team heeded the advice of one of its members, an ex-McKinsey and Company consultant named William Drayton, who had worked with Costle in Connecticut. Despite the fact that the Democratic governor of Connecticut was not one of his greatest admirers, the transition team settled on this relatively unknown candidate.[89]

As we have seen, the EPA Costle inherited had come to be viewed, and to view itself, as an environmental advocate. Its strong sense of purpose and its hostile relations with Nixon had indebted it to its friends in Congress and in the environmental community, and made it highly resistant to pressure from the White House. Internally, the uneasy coexistence of programmatic and functional units invited turf conflicts and political "free lancing" on the part of subordinates. Costle was not obliged to perpetuate this legacy, but he recognized that any revisions would meet both with internal and external opposition and would, therefore, require considerable effort and political capital.

Recognizing that much of his time would be taken up with external work, he had initially planned to pick a strong administrator for the number two job in the

Agency, that of deputy administrator.[90] Soon after his own appointment, however, he was asked by Hamilton Jordan, Carter's top aide, to consider Barbara Blum for that post. Blum, a Georgian, had been a local environmental activist before becoming a key campaign organizer. She had no previous experience in agency management.

Costle had to weigh Blum's lack of administrative qualifications against her strong personal ties to the White House inner circle. Although he had not been ordered to take her, a refusal would obviously disappoint Carter's "Georgia Mafia" and strain his relationship with them. Costle, therefore, decided to appoint Blum, and to use her as a political emissary to the White House. This meant he had to find another source of help in running the Agency.

The Office of Planning and Management (OPM) was the obvious alternative. That job, however, went to transition staffer William Drayton, who, before helping Costle land the EPA job, had helped him institute several important regulatory innovations in Connecticut. Drayton's great strength was not in operational matters but policy analysis and conceptual innovation, as well as a legendary capacity for hard work. Again, Costle's desire to build a general management team was frustrated by other considerations.

Similarly, to placate Muskie, and the rest of the Senate subcommittee, Costle chose a former committee staffer, Thomas Jorling, as the assistant administrator for Water Programs. Although assigned to the committee minority, Jorling had played an active role in drafting environmental statutes and mobilizing a bipartisan committee consensus behind them. After leaving the committee, Jorling moved to Williams College to direct its environmental center. His strong commitment to environmental improvement and his zest for battle, however, made the lure of a key post at EPA irresistible.

Other major appointments also went to those with an environmental identification. To head the other large program component of EPA, the Air Office, Costle relied on Muskie's advice and chose David Hawkins. Hawkins was a staff attorney for the Natural Resources Defense Council and one of the environmental community's chief spokesman on air quality issues. As head of the Office of Congressional Relations Costle picked Charles Warren, former chief legislative aide to Senator Jacob Javits (R, NY). Although Warren's own legislative responsibilities had not been centered on environmental matters, his wife, Jacqueline, was a leading environmental attorney. Thus, he too was closely identified with the environmental community.

A NEW STRATEGY

During the campaign, Carter had pledged to make government more efficient and responsive. Costle felt that to keep EPA in the good graces of the White House, he had to actively promote and adopt some of the administration's reform schemes. During the transition period, he and Drayton worked on the White House's proposed Civil Service Reform Act. He lobbied for it in Congress and worked to implement it at EPA. The EPA was also the first federal agency to adopt Carter's zero based budgeting scheme. Throughout his tenure, Costle, with Drayton's active assistance, would strive to make EPA a model of "good government."[91]

In Costle's view, however, the EPA needed to do more than embrace managerial reform. Although opinion polls continued to reflect widespread public support for pollution control, the Washington community increasingly believed that this support was diminishing.[92] At the same time, there was growing recognition of the large costs of pollution abatement. To justify such expenditures at a time of slow economic growth seemed to require a more compelling goal than ecological purity. Preventing disease, especially cancer, offered just such a justification. Although most of EPA's regulations were based on health considerations, in the public's mind the agency's mission was more involved with protecting nature. Costle became determined to convince the public that EPA was first and foremost a public health agency, not a guardian of birds and bunnies.[93]

The public health mission also provided a response to the threat to the Agency posed by executive reorganization. There was widespread speculation about reviving the DENR idea and merging the EPA into the Department of the Interior. If that happened, Costle would find himself reporting to Secretary of the Interior Cecil Andrus, who was close to Carter and the administration's lead spokesman on environmental matters.[94] Indeed, Andrus had been involved in the choice of Costle, and when they met before Costle's appointment, the two had established a good rapport that would blossom into a close friendship.

Andrus gave Costle private assurances that he would oppose an EPA-DOI merger. He was well aware of the role Costle had played in convincing the Ash Council to recommend the creation of EPA and he agreed with those arguments.[95] In fact, when Carter's plan for a Natural Resources Department was announced, EPA was not included. The administration's interest in organizational neatness, together with Andrus' preeminence, however, made it clear that EPA could not expect to expand its ecological mandate.

The environmental health arena, by contrast, was wide open, and within it EPA possessed certain advantages. It was the largest of the regulatory agencies concerned with environmental health and the only one that combined a significant research effort with its regulatory duties.

Costle lost no time in implementing the public health strategy. He toured the country giving a speech entitled "The Chemical Revolution," in which he outlined the threat posed by the multiplicity of environmental toxins and discussed how the threat should be met. He was quoted in *Science* and the *National Journal* announcing that "EPA is a preventive health agency as well as an environmental agency."[96]

In a speech delivered to the American Chemical Society on September 11, 1978, he reiterated the Agency's intention to shift its regulatory focus from conventional pollutants to toxic substances. He justified the need for EPA to adopt a more aggressive posture toward environmental disease by stating that the Agency "cannot wait for proof positive in the form of dead bodies" to regulate suspected carcinogens.[97]

The clearest evidence of this new approach surfaced at a briefing outlining the agency's 1979 budget plan. Congressional reform of the budgetary process created a two-year budget cycle that meant that EPA's 1979 budget was really the first one that Costle could place his own imprint upon. After once again announcing that EPA had become a public health agency, he proceeded to mention "health" six more times in his brief talk. He was then succeeded on the podium by William Drayton, who also stressed that agency priorities had been shifted towards public health; toxic substances, pesticides, drinking water, and health research were slated to receive additional funds; "Older air, water and solid waste programs have been refocused on toxics and hazardous materials." He also noted that both funding and personnel for research on ecological effects of pollution had been reduced to expand research on human health effects.[98]

This shift in strategy bore fruit quickly. It enabled EPA to acquire new resources at a time of strict austerity. Whereas the overall 1979 budget for environment and natural resources increased by less than one percent (from 12.1 to 12.2 billion), EPA was able to increase its operating budget by twenty five percent. One third of that increase was slated to be spent on programs devoted to controlling pollutants posing a public health risk.[99] The sentiments expressed by Costle and Drayton were echoed in the very first sentence of the natural resources and environment section of the OMB's Budget in Brief, which stated that "our national needs for natural resources and the environment are to protect public health by assuring a clean environment."[100]

In summary, Costle made several decisions which shaped the context in which the cases that follow evolved. He left the structure of the Agency intact and maintained its identity as an advocacy agency. He formed his leadership team from people with strong loyalties to outside constituencies, and, most importantly, he redefined the central mission of the agency: protection of public health replaced the maintenance of ecological balance as EPA's central objective.

NOTES

1. For a comprehensive account of suburbanization and its impact, see Kenneth T. Jackson, *Crabgrass Frontier: The Suburbanization of America* (New York: Oxford University Press, 1985).

2. Andrew S. McFarland, *Public Interest Lobbies: Decision Making on Energy* (Washington, DC: The American Enterprise Institute for Public Policy Research, 1976), 6.

3. See McFarland, *Public Interest Lobbies*, 4–24; Jeffrey M. Berry, *Lobbying for the People* (Princeton, NJ: Princeton University Press, 1977), 18–44; and J. Clarence Davies III and Barbara S. Davies, *The Politics of Pollution*, 2nd ed. (Indianapolis, IN: Bobbs-Merrill, 1975), 80–86.

4. Robert Cameron Mitchell, "Public Opinion and Environmental Politics in the 1970's and 1980's," in Norman J. Vig and Michael E. Kraft, eds., *Environmental Policy in the 1980's: Reagan's New Agenda* (Washington D,C,: CQ Press, 1984), 51–74.

5. See especially Reynolds v. Sims 377 U.S. 533; 84 S. Ct. 12 L. Ed. 2d 506 (1964).

6. Norman Ornstein and Shirley Elder, *Interest Groups: Lobbying and Policymaking* (Washington D,C,: CQ Press, 1978), 155–186; John McPhee, *Encounters with the Archdruid* (New York: Farrar, Giroux and Straus, 1971), 151–245.

7. The attitudinal gaps between environmentalists and businessmen are explored by Kathy Bloomgarden in her article "Managing the Environment: The Public's View," *Public Opinion,* 6 (Feb./Mar. 1983):47–51.

8. On the political advantages of statutory ambiguity, see Bruce A. Ackerman and

William T. Hassler, *Clean Coal/Dirty Air* (New Haven, CT: Yale University Press, 1981), 35–38.

9. Edward J. Epstein, *News From Nowhere: Television and the News* (New York: Random House, 1973), 88, 98.

10. Edwin Diamond, *The Tin Kazoo: Television, Politics, and the News* (Cambridge, MA: MIT Press, 1975), 87–109; Epstein, *News from Nowhere*, 4–5.

11. Diamond, *The Tin Kazoo*, 93–94; Epstein, *News From Nowhere*, 147.

12. For thumbnail sketches of the major old and new environmental organizations, including membership figures, see *The National Journal* (13 Dec. 1980): 2118–9; Davies and Davies, *The Politics of Pollution*, 87–95.

13. Berry, *Lobbying for the People*, 205; McFarland, *Public Interest Lobbies*, 18, 71.

14. For a discussion of the Clean Air Coalition, see Ornstein and Elder, *Interest Groups: Lobbying and Policy Making*, 155–186.

15. Michael J. Malbin, *Unelected Representatives: Congressional Staffs and the Future of Republican Government* (New York: Basic Books, 1980), 19–24; David Broder, *Changing of the Guard: Power and Leadership in America* (New York: Penguin, 1980), 225–253.

16. Broder, *Changing of the Guard*, 225–253.

17. Alfred A. Marcus, *Promise and Performance: Choosing and Implementing Environmental Policy* (Westport, CT: Greenwood Press, 1980), 63.

18. Mark J. Green, James M. Fallows, and David R. Zwick, *Who Runs Congress?* (New York: Bantam, 1972), 130.

19. Green, et al., *Who Runs Congress?*, 5.

20. R. Shep Melnick, *Regulation and the Courts: The Case of the Clean Air Act* (Washington D,C,: The Brookings Institution, 1983).

21. Broder, *The Changing of the Guard*, 229–230.

22. Melnick, *Regulation and the Courts*, 374.

23. Bernard Asbell, *The Senate Nobody Knows* (Baltimore, MD: Johns Hopkins University Press, 1978), 120; Charles O. Jones, *Clean Air: The Policies and Politics of Pollution Control* (Pittsburgh, PA: University of Pittsburgh Press, 1975), 56.

24. Asbell, *The Senate Nobody Knows*, 76.

25. Jones, *Clean Air*, 57.

26. Jones, *Clean Air*, 176.

27. Jones, *Clean Air*, 55.

28. Melnick, *Regulation and the Courts*, 33.

29. Davies and Davies, *The Politics of Pollution*, 66–67.

30. Davies and Davies, *The Politics of Pollution*, 33.

31. Davies and Davies, *The Politics of Pollution*, 34.

32. Melnick, *Regulation and the Courts*, 26–27.

33. John C. Esposito, et al., *Vanishing Air* (New York: Pantheon, 1970), 270–271.

34. Theodore Lippman, Jr. and Donald C. Hansen, *Muskie* (New York: WW Norton and Co., 1971), 144.

35. Jones, *Clean Air*, 179.

36. Lippman and Hansen, *Muskie*, 228.

37. Davies and Davies, *The Politics of Pollution*, 52–53.

38. Esposito, et al., *Vanishing Air*, 259, 298.

39. Jones, *Clean Air*, 192–193.

40. Jones, *Clean Air*, 202.

41. Esposito, et al., *Vanishing Air*, 306–307.

42. Esposito, et al., *Vanishing Air*, 309–310.

43. Esposito, et al., *Vanishing Air*, 304; Jones, *Clean Air*, 204.

44. Lippman and Hansen, *Muskie*, 148.

45. Lippman and Hansen, *Muskie*, 145–146.
46. David Nevin, *Muskie of Maine* (New York: Random House, 1972), 120–137.
47. Marcus, *Promise and Performance*, 553–584.
48. Jones, *Clean Air*, 194–214.
49. Melnick, *Regulation and the Courts*, 28, 97–101, 253–254.
50. Robert L. Sansom, *The New American Dream Machine: Toward a Simpler Lifestyle in an Environmental Age* (Garden City, NY: Anchor Press, 1976), 9,26,30.
51. John C. Whitaker, *Striking a Balance: Environment and Natural Resources in the Nixon-Ford Years* (Washington, DC: American Enterprise Institute for Public Policy Research, 1976), 27.
52. Whitaker, *Striking a Balance*, 29–30.
53. Whitaker, *Striking a Balance*, 31.
54. Marcus, *Promise and Performance*, 32.
55. Marcus, *Promise and Performance*, 33.
56. Marcus, *Promise and Performance*, 34–36.
57. Marcus, *Promise and Performance*, 36–37.
58. Marcus, *Promise and Performance*, 38.
59. Marcus, *Promise and Performance*, 40–41.
60. Whitaker, *Striking a Balance*, 54.
61. John Quarles, *Cleaning Up America: An Insider's View of the Environmental Protection Agency* (Boston, MA: Houghton Mifflin, 1976), 16–18.
62. Marcus, *Promise and Performance*, 32.
63. Quarles, *Cleaning Up America*, 20.
64. Marcus, *Promise and Performance*, 56.
65. Senate Committee on Public Works, Subcommittee on Air and Water Pollution, *Implementation of the Clean Air Act Amendments of 1970—Part I*, 92nd Congress, 2nd Session, Hearings February 16,17,18, and 23, 1972, Committee Serial No. 92-H31, 4. See also Marcus, *Promise and Performance*, 45.
66. Senate Committee on Environment and Public Works, Subcommittee on Environmental Pollution, *Executive Branch Review of Environmental Regulations*, 96th Congress, 1st Session, Hearings February 26 and 27, 1979, Committee Serial No. 96-H4, 30.
67. *Executive Branch Review of Environmental Regulations*, 31.
68. On the EPA bureaucratic inheritance, see Marcus, *Promise and Performance*, 45–53.
69. Quarles, *Cleaning Up America*, 47–48.
70. The story of Ruckelshaus' appointment is told in Quarles, *Cleaning Up America*, 21–24.
71. Marcus, *Promise and Performance*, 102.
72. Marcus, *Promise and Performance*, 103–104.
73. Marcus, *Promise and Performance*, 105.
74. Quarles, *Cleaning Up America*, 39; Marcus, *Promise and Performance*, 85–99.
75. Quarles, *Cleaning Up America*, 36.
76. Sansom, *the New Amercian Dream Machine*, 24–25,43; Quarles, *Cleaning Up America*, 117–118.
77. Quarles, *Cleaning Up America*, 37–57.
78. Quarles, *Cleaning Up America*, 42–43.
79. Marcus, *Promise and Performance*, 125; Quarles, *Cleaning Up America*, 118.
80. See testimony of Richard Ayres of the Natural Resources Defense Council, *Implementation of the Clean Air Act Amendments of 1970*, 3–24.
81. Ruckelshaus' testimony, *Implementation of the Clean Air Act Amendments of 1970*, 243.

82. See Quarles statement, *Executive Branch Review of Environmental Regulations*, 3–4.

83. This incident is related in Quarles, *Cleaning Up America*, 59–76.

84. Quarles, *Cleaning Up America*, 190.

85. Quarles, *Cleaning Up America*, 199.

86. Quarles, *Cleaning Up America*, 119; and Sansom, *The New American Dream Machine*, 48.

87. Sansom, *The New American Dream Machine*, 25.

88. Douglas M. Costle, former EPA Administrator, interview with Harvard School of Public Health and Kennedy School faculty, Cambridge, MA, 2 Mar. 1981.

89. Costle interview, 2 Mar. 1981.

90. Costle describes the process of picking his management team during this extensive interview.

91. Costle interview, 2 Mar. 1981.

92. *BNA Environmental Reporter*, 1978, 1614.

93. *Science* 202 (10 Nov. 1978):598.

94. *National Journal*, 12 Mar. 1977, 382–384; *National Journal*, 15 Nov 1977, 1616.

95. Costle interview, 2 Mar. 1981.

96. *National Journal*, 17 Dec. 1977, 1976.

97. *Science* 202 (10 Nov. 1978):598; *National Journal*, 28 Jan. 1978, 140.

98. *BNA Environmental Reporter*, 15 Sep. 1978, 914.

99. *National Journal*, Jan. 1978, 140.

100. *National Journal*, Jan. 1978, 140.

Part II

3

Revising the Ozone Standard

With Valle Nazar

Under the Clean Air Act, it is the Environmental Protection Agency's responsibility to set what are called air quality standards, specific numerical values that describe the condition of the "ambient" air around us.[1] These standards are to be set at levels "the attainment and maintenance of which . . . allowing an adequate margin of safety, are requisite to protect the public health."[2] Achieving these standards is the goal that guides subsequent pollution control efforts. The Act tells EPA to set standards for all compounds that may "reasonably be anticipated to endanger public health and welfare."[3]

In this chapter we examine EPA's first attempt to revise such a standard, namely the 1979 effort that dealt with ozone. This decision involved questions that were both subtle and difficult. What constituted a "health effect?" What made a margin of safety "adequate?" Such issues involve values as well as facts, politics as well as science. The extent to which EPA did not acknowledge and respond to this duality, and the confusion and misunderstanding that resulted, are two of our continuing themes.

The case also illustrates the limits on congressional capacity to constrain executive discretion even in this comparatively simple decision. In the Clean Air Act, the Congress tried to provide explicit guidance to an administration it did not fully trust. Despite these efforts, the world was and remains more complex than the words the Congress used to describe it. For example, individuals vary widely in their "sensitivity" to air pollution, but there is no sudden break or discontinuity in that sensitivity. Hence, delineating the "most sensitive group"—a key distinction in legislative history—is not straightforward.[4] Who then should the standard be set to protect? In this and many other respects, Congress left the agency with significant decisions to make.

Within EPA, there was substantial internal conflict over this decision, conflict that did little to clarify the technical or philosophical issues at stake. This combat was repeated when EPA tried to justify its decision to the rest of the government. No one short of the president himself was ever in a position to ask, "All things considered, what ozone standard is in the best interest of the United States?"

Ozone is one member of a class of compounds called photochemical oxidants. These are produced in the atmosphere when various "precursors" react with each other in the presence of ultraviolet light (whose role in these reactions accounts for the term photochemical).[5] These precursors (hydrocarbons and nitrogen oxides) come from natural sources and from human activities, particularly the combustion of fossil fuels (including that in automobile engines).

In 1971, to comply with the 1970 Clean Air Act Amendments, EPA had hastily promulgated a standard of 0.08 parts per million (ppm), not to be exceeded for more than one hour more frequently than once a year. Although the standard was described as applying to photochemical oxidants in general, the regulations specified a monitoring technique that measured only ozone. The standard was justified primarily by reference to a single epidemiological study, by Schoettlin and Landau,[6] which reported that asthmatics had more attacks on days when ozone levels were above 0.10 ppm. The 1971 standard was promulgated without either challenge or court review.

This did not last long. The photochemical oxidant standard was the main justification for the highly controversial controls that were being imposed on automobile emissions. Thus the ambient standard itself became controversial. In 1974 Congress asked the National Academy of Sciences to study the standard. The Committee appointed by the Academy reported that it could find no compelling reason for change. This was despite finding that the technical basis for the standard was "clearly inadequate" and based on "limited evidence."[7]

Meanwhile, in response to industry criticism, EPA reassessed the Schoettlin and Landau study and found that the exposure levels reported to be 0.10 ppm had actually been 0.25 ppm. Apparently there had been calibration difficulties with some of the measurement devices used in the study.

In 1976, an Interagency Task Force on motor vehicle air pollution reported that the 1971 standard was proving to be unattainable. Natural background levels in some areas were often close to 0.08 ppm, and sometimes even higher.[8] In such locations, draconian controls on stationary sources, including limits on new construction and on motor vehicle use (such as travel and parking bans), might become necessary. In December of 1976, facing industry pressure and aware that the existing level was becoming technically and politically indefensible, EPA announced that it had initiated a review of the standard.[9]

THE NEW ADMINISTRATION

That review turned out to be one of the first tasks of the new administration. EPA Administrator Douglas Costle had several general objectives in mind as he undertook this job, including enhancing the Agency's scientific credibility.[10] As he put it, "We have to consolidate our authority . . . we need to improve the scientific basis of our regulation."[11] Yet at the time neither EPA's own laboratories nor its grant-supported research were held in particularly high regard. Scientists had come from a variety of agencies when EPA was created, and not all of these scientific units were first rate. In addition, external awards were often made without competitive bidding or peer review to a few favored investigators. Costle wanted to convince both scientists and the general public that EPA could itself undertake, and use, good scientific work, and he wanted to involve leading scientists in the agency's decision making.[12]

In addition, and unlike his predecessors Ruckelshaus and Train, Costle lacked a Washington political base.[13] Costle was thus regarded by some as likely to be less independent of the White House.[14] Yet, to be effective, Costle had to establish his independent authority, as well as his "reasonableness." He had to convince industry and environmentalists that they both would have to and could deal with him.

Furthermore, the Carter administration was committed to conflicting policy objectives. The president had campaigned on an environmentalist platform with environmentalist support, yet mounting economic problems were leading some to ask whether the country could afford a very tough policy on clean air. Carter had also run on a platform of efficiency in government and in April he announced his intention to cut the "inflationary costs of regulation."[15] A few months later he created the Regulatory Analysis Review Group (RARG) to analyze and comment on the cost and inflationary implications of selected new regulations.[16] As we will see, RARG became a major actor in the ozone decision and Costle had to find a way to deal with these pressures.

The scientific background work on the standard was done within EPA's Office of Research and Development (ORD). When Costle's team was fully assembled it was under assistant administrator Steve Gage. Recommending an actual number was the responsibility of the Office of Air Quality Planning and Standards (OAQPS) located at Research Triangle Park in North Carolina. Headed by deputy assistant administrator Walter Barber, it reported to David Hawkins, the assistant administrator for air programs. Although there was provision for an elaborate internal review process, setting the numeric value for the standard was ultimately and unambiguously Costle's decision.

DEVELOPING THE CRITERIA DOCUMENT

The first major step was the development of a "criteria document" by ORD. The *Criteria Document* was required by law to:

> accurately reflect the latest scientific knowledge useful in indicating the kind and extent of all identifiable effects on public health and welfare, which may be expected from the presence of such pollutant in the ambient air, in varying quantities . . .[17]

The team preparing the *Criteria Document* was headed by two senior people from ORD, Lester Grant of its Environmental Criteria and Assessment Office, and Gordon Hueter, associate director of its Health Effects Research Laboratory. It began work in early 1977, searching the literature and contacting various outside experts.[18]

On March 3, 1977, EPA asked its Science Advisory Board (SAB) to play a role in the process. Established by the agency in 1974, the Board was a loosely structured group of external scientists whose function was to provide advice to EPA whenever asked. (The Board's role was formalized by legislation toward the end of 1978.)[19]

Acting on Costle's behalf, Richard Dowd, executive director of the SAB, asked James Whittenberger, a professor of physiology at the Harvard School of Public Health, to be the chair of a special subcommittee of the SAB to review the *Criteria Document*. Whittenberger selected six other scientists for the Subcommittee and three experts to act as consultants.[20] They were told that a draft would be ready for review sometime in late spring.

Meanwhile, in the first two months of 1977, Joseph Padgett, Director of the Strategies and Air Standards Division in OAQPS, had created yet another external scientific group to consider ozone. Padgett arranged for an advisory panel of clinicians and scientists knowledgeable about photochemical oxidants to help OAQPS interpret the *Criteria Document* and the available data on ozone.[21]

He took this initiative because OAQPS had found its previous interactions with ORD less than satisfactory. The OAQPS believed that ORD, as a research agency, regarded preparing the *Criteria Document* as an unwanted burden. Indeed the *Criteria Document* was not even supposed to consider what specific standard the evidence would justify. Conscious of its scientific limitations, OAQPS feared it would be left on its own to interpret the voluminous clinical and epidemiological data.[22] Padgett wanted access to the outside advice he felt OAQPS would need to make a prompt and informed recommendation.[23]

Carl Shy of the University of North Carolina was asked to lead what was called the Advisory Panel on Health Effects of Photochemical Oxidants (hereafter, the "Shy Panel").[24] A former EPA employee and at the time a grantee, Shy would later speak on behalf of the American Lung Association at EPA's first public hearing on the ozone standard. He would strongly advocate retention of the existing 0.08 ppm standard, as he had done on previous occasions.[25]

Under Shy's leadership, the panel did not wait for the *Criteria Document*. Instead it met on June 7 and 8, 1977, in North Carolina. Three days later the first draft of its report was available to its members. The draft stated that a variety of physiological effects

> . . . may be induced by short term exposures to ozone in the range of 0.15 to 0.25 ppm. An unreplicated study of airway resistance in humans and repeated studies of susceptibility to experimental respiratory infection in animals further suggest the possibility of adverse effects in some persons at short term ozone exposure of 0.10 ppm.

It concluded that:

> There is no compelling reason to suggest a change from . . . the existing standard . . .[26]

The development of the *Criteria Document* was also going forward and by September 1977 the first external review draft was available.[27] When the SAB subcommittee and interested parties met on November 10 and 11, Gordon Hueter shared with the subcommittee a copy of the unfinished report of the Shy Panel. According to Hueter, Whittenberger and other members recommended specific changes in that report to make it a useful part of the "Summary and Conclusions" chapter.[28] Hueter made the recommended revisions and subsequent drafts of parts of the "Summary and Conclusions" chapter relating to human clinical and epidemiological evidence were quite similar to parts of the Shy Panel Report.

In his opening remarks at the meeting, Whittenberger reminded subcommittee members that EPA had asked them to consider the draft criteria document's treatment of various issues. Two of these reveal a fundamental conceptual problem that was to plague the whole process.[29]

> What are the critical health effects and what are the susceptible segments of the population?
>
> What is the medical significance of eye irritation, student athletic performance, general discomfort, increased asthma, and other observed effects?

Despite (or perhaps as a result of) these instructions, there was some question at the meeting about the function of the *Criteria Document* and the role of the SAB. Where was the boundary between evaluating evidence and policy judgment? The narrow view was that the *Criteria Document* was simply to describe the effects that could be expected as a result of different levels of air pollution. But as Whittenberger's questions suggest, the distinction was easier to state than to implement. Deciding what effects were "critical" or which had "medical significance" was not a purely technical matter. Sheldon Murphy, a consultant to the Subcommittee, commented that,

> maybe what we have to address is the question of acceptable (risk) . . . Mr. Costle is not going to make this decision on his own. Somebody will have to give him guidance as to how to use this data.[30]

George Chapman from ORD, and one of the primary authors of the *Criteria Document*, disagreed:

> . . . a criteria document should basically be a recounting of evidence and with interpretation that arises from that evidence.[31]

Whittenberger countered that the margin of safety and the population at risk seemed part of the interpretation, and Murphy added, "if you don't (provide interpretation) . . . then any kind of effect becomes equal. . . ."[32] Joseph Padgett responded that from the agency's viewpoint the *Criteria Document* should not deal with health judgments or with computations of populations at risk. "That is part of the process of converting the information to a standard," he said.[33]

Padgett's statement did not resolve the issue. Later, in a discussion of what role to give economic information, Whittenberger remarked that,

> in spite of what the legal restraints are, it seems to me that eventually we are going to have to take economics much more into consideration.[34]

As this discussion illustrates, from the very beginning the SAB subcommittee found it difficult to think about questions like what was a "significant" health effect, or what segment of the population should be considered the "sensitive group," without also considering the social and economic consequences of such decisions.[35] In general the SAB was not entirely pleased with the draft and at the end of the two day meeting left the EPA staff with many suggested changes.

A few weeks later, EPA announced in the *Federal Register* that it would hold a public meeting on January 30 to receive comments from interested parties before taking any formal agency position on the revised standard. The notice indicated that,

> Preliminary reviews of the health evidence presented in the draft revised criteria document suggest that the range of possible standard levels lies between 0.08 and 0.15 ppm and that the choice within that range will be based on the level and nature of health risks which can be tolerated.[36]

According to Whittenberger, the SAB was not informed of this meeting: this was the beginning of an "uncoupling" between the *Criteria Document* development and the standard-setting processes.[37] Equally as interesting was the wording of the notice. How was the EPA to determine what "health risks" "can be tolerated?" The announcement offered no suggestions on this point.

In January 1978, the Shy Panel report was published, and there were heated reactions both inside and outside the Agency.[38] Stephen Gage protested that because Dr. Shy's viewpoint was well known, it was going to be difficult for EPA to portray the report to outsiders as a balanced analysis. Richard Dowd and Frank Press, Science Advisor to the president, both took the matter to Costle.[39] He in turn was furious that a process that was not part of the agency's carefully developed formal procedures had occurred. That could threaten both the apparent integrity and legal defensibility of the agency's rule making.[40] As Costle put it,

> We are going to be very insistent that whatever decision such an expert body makes, that it be supported by the record, . . . and that it has been arrived at through a fair and open process . . . [D.C. Circuit Court Judge] Skelly Wright would want to be sure somebody's not getting an extra inning.[41]

Nevertheless, the Shy Panel report had already been used by the SAB in revising the *Criteria Document*, and it was to be used again by EPA in substantiating its ultimate proposed standard in the *Federal Register*.

January also saw the release of another critical study in the standard setting process, a quantitative risk assessment of ozone.[42] A panel of experts was asked about the probability that sensitive individuals would suffer various health effects at various ozone levels. The panel was asked to consider a person who was more sensitive than ninety-nine percent of the members of some sensitive group, but less sensitive than one percent of that group. However, the definition of that "sensitive group" was left to each scientist, and their definitions varied. In all, nine experts were consulted, four of whom were also members of the Shy panel.[43]

This document was later severely criticized by another SAB subcommittee. That subcommittee objected to the report's failure to explain how the experts selected their estimates and to their anonymity, which meant that individuals could not be held responsible for their judgments. It found the methodology to be "obscure" and "impenetrable" to all but mathematical specialists.[44] Despite these objections and despite the lack of peer review, the study was used to justify the ultimate standard.[45]

When the SAB subcommittee convened to review a second draft on February

23, these disagreements immediately surfaced. Whittenberger, who had known about the Shy Panel for months, opened the meeting by angrily expressing his disapproval of the panel's very existence.[46] He complained that OAQPS was using the Shy Panel report to set a standard, rather than waiting for an approved *Criteria Document* and the advice of the SAB. He criticized Padgett for putting together an "unauthorized" body and infringing on the role of the SAB. Worse, Whittenberger also contended that the Shy Panel was "stacked" and that Shy had been associated with EPA studies in which data had been "overinterpreted" to support a view that a large margin of safety was required to protect public health.[47]

Due to time constraints, the subcommittee agreed that only the "Summary and Conclusions" chapter of the draft needed to be reviewed by all members once it had been further revised. The other chapters, when redone, would be sent only to selected members. This occurred during March and April. Although they still had comments and reservations, six of the eleven board members voted to accept the *Criteria Document*. Three members said that they were unable to approve or disapprove, while two, including Whittenberger, voted to reject the document. Whittenberger commented, in writing, that the document's authors had "responded insufficiently to many criticisms made by members of the subcommittee" and that "the health risk assessments are largely speculative, incomplete and heavily dependent on studies of questionable value."[48]

These criticisms caused yet another version to be prepared and circulated. During the first two weeks of April, Mike Jones of OAQPS polled the subcommittee members by telephone. All responded with endorsements or statements of "no major objections" remaining.[49] Whittenberger wrote suggestions in the margins of the thrice-revised summary chapter and returned the copy to Chapman with a handwritten note:

> I think there are substantial improvements to the presentation and interpretation of the evidence. Any remaining complaints I have are minor.[50]

Whittenberger later expressed his dismay that Stephen Gage "passed this [note] off as the Chairman's approval" of the draft *Criteria Document*.[51] Gage recalls that it was Dick Dowd who used Whittenberger's note to indicate to Costle that the document had been approved.[52]

THE SCIENTIFIC ISSUES

What were the issues that the members of the SAB found so difficult and controversial? The story of the ozone standard would be incomprehensible without some understanding of the nature and limits of the available scientific evidence.

The most fundamental problem has to do with the way in which humans respond to air pollution. The Clean Air Act directs the administrator to pick a standard that "protects the public health." This presumes that there is a "safe level" below which adverse health effects do not occur. But suppose that there are always some adverse effects—more when the air is dirtier and less when it is cleaner—but always some impact. No level of ozone concentration other than zero, therefore, would satisfy the Act.

There are several reasons to suspect that this might be the case with ozone. First, very low levels do appear to bother some particularly sensitive individuals (e.g., some of those with serious asthma). Second, low levels produce effects that are small but measurable with sensitive instruments, such as changes in blood biochemistry or respiratory patterns. Finally, at low pollution levels, the probability of an effect on any one person may be very small, but not zero. Together these possibilities imply that a decision had to be made about which effects, and to whom, are significant enough to constitute an impairment of "public health."

This question of definition was a policy question. Calling something an "adverse health effect" implied taking certain actions to avoid the effects in question. The members of the SAB found such judgments difficult to make precisely because they were told to ignore consequences and only think about the science. Yet terms like *sensitive group*, *health*, and *adequate margin of safety* are not self-defining. The science of the situation could not, by itself, produce a decision. No matter how intelligent and sincere they are, scientists with different values were likely to disagree about what constituted an "adverse health effect" that should be prevented.

Some observers clearly understood at the time that there were no sharp discontinuities between "safety" and "health." A National Academy of Science report concluded:

> In no case is there evidence that the threshold levels have a clear physiological meaning . . . [The] amount of health damage varies with . . . concentrations of the pollutant, and with no sharp lower limit . . . Even if there were sharp threshold levels for individual persons, the levels would certainly not be the same for different persons, or even for the same person in different states of health . . . at any concentration, no matter how small, health effects may occur.[53]

The significance of this conclusion, however, was either unappreciated or avoided for various strategic reasons.

This dilemma was made much more difficult by the scarcity of information. There were few studies, and most of those had not been designed to contribute to regulatory decision making. Given that the question EPA was asking was inherently difficult to answer, individual studies tended to be ambiguous and subject to divergent interpretations. When there is no clear preponderance of evidence, a scientist can always defer a decision until more data are collected. EPA did not have that luxury. The available evidence consisted of four groups of studies. The first of these involved experimental observations on human subjects.

Human Clinical Studies

In the human clinical studies a small number of subjects breathed air containing controlled concentrations of ozone (and/or other pollutants) for specified periods of time. Various measurements were made of blood characteristics and breathing patterns and were compared with data from control observations where subjects were in similar situations with no ozone exposure.

Because these studies were time consuming to conduct, and because volunteers who would agree to be exposed to airborne chemicals were hard to find, samples tended to be both small and nonrandom. Thus, it was not clear how the results of ozone chamber studies should be interpreted in terms of where on the continuum of ozone sensitivity the experimental subjects lay and how representative they were of the population in general, asthmatics in general, or the most sensitive one per cent of asthmatics. If they represented asthmatics in general, then a random sample of 500 to 1,000 would be needed to determine the experience of the most sensitive one per cent of that group.

With such small sample sizes, it is also difficult to know what to make of the statistical significance, or lack thereof, of any results. Formal methods are designed to protect experimenters from mistaking chance differences for real ones. Yet, when there are only a few cases, this burden of proof is difficult to meet. And in this case, EPA might well be more concerned with discovering any effect of ozone that actually occurs, thereby lowering the risk of a false negative, even if that increased the risk of mistaking a chance effect for a real one (a false positive).

Two studies were pivotal. DeLucia and Adams exposed six healthy men, using breathing tubes, to 0.15 ppm or 0.30 ppm of ozone for one hour, either at rest or exercising at different rates.[54] At 0.30 ppm and the highest exercise rate (later characterized as jogging a 6 minute mile), there were statistically significant effects on respiratory patterns. Subjects breathed shallower and faster, and they showed some decrease in lung capacity and breathing flow rates as compared with their experience in air without ozone. At 0.15 ppm the same effects were recorded, but were noticeably smaller and not statistically significant. These averages are hard to interpret because, as DeLucia and Adams noted, there were large differences among individuals. Two subjects were largely unaffected, whereas two others (one an asthmatic in childhood) showed increases in respiratory frequency of fifty to one-hundred per cent and had to complete the task at a lower work rate. However, we have no way of knowing just how sensitive DeLucia and Adams' two sensitive subjects were. At the highest work rate "most subjects" at both 0.15 ppm and 0.30 ppm reported symptoms such as headaches, wheezing, congestion, or coughing.

The final draft of the "Summary and Conclusions" chapter put great emphasis on this study:

> Though it remains unreplicated, the study of DeLucia and Adams raises the distinct possibility that an ozone concentration of [0.30 ppm] exerts a temporary effect on the lung function of healthy subjects exercising fairly strenuously. The same investigators have raised the question as to whether ozone concentrations as low as [0.15 ppm] exert effects in a portion of healthy subjects exercising strenuously.[55]

The second study cited in the final rule making was by von Nieding and colleagues in Germany.[56] Eleven healthy young adult men were exposed to ozone, nitrogen dioxide, and sulfur dioxide, alone or in combination, for two hours in an exposure chamber while performing "easy" work. The experimenters measured both oxygen in the blood and flow resistance in the respiratory path. Seven of eleven subjects exposed to 0.10 ppm of ozone showed lower blood oxygen levels and higher respiratory flow resistance than control subjects.

The first draft of the *Criteria Document* produced by the EPA staff pointed out that the techniques used in von Nieding's study were nonstandard, but nonetheless emphasized these results:

> The results of the von Nieding studies serve to reinforce the conclusion that changes in mechanical function of the lung may well occur in some subjects at ozone concentrations less than 0.25 ppm for two hours, and that there may be some risk of inducing functional changes at levels in the range of 0.15 to 0.25 ppm.[57]

At the first review meeting of the SAB subcommittee, an American Petroleum Institute (API) representative criticized the study's unusual protocol for measuring airway resistance.[58] Several members of the SAB, including its Chairman, were also opposed to relying so heavily on a piece of work about which they had serious technical misgivings. The final *Criteria Document* reflects these views and describes von Nieding's findings in one sentence, adding that, "until confirmed with generally accepted methods, these studies must be interpreted with caution."[59] Nonetheless, along with DeLucia and Adams, it is one of the two clinical studies cited in the *Final Rulemaking* that reported effects below 0.2 ppm[60]

Equally controversial were two studies that found minimal or no effects from ozone at substantially higher levels. One of these, by Linn and colleagues, did not appear until after the *Criteria Document* was completed.[61] This study examined twenty- two asthmatics with "minimal asthma to moderately severe chronic airway obstruction," although those with "marked disability" were excluded. (Given the way these subjects were chosen, it is not possible to determine whether this sample matched EPA's definition of the sensitive group.)

The volunteers were exposed to ozone concentrations of approximately 0.20 ppm for two hours, with secondary stresses of heat and intermittent exercise. The investigators reported "no meaningful" changes in lung function, but did see a "slight" increase in subjectively reported respiratory symptoms, as well as blood biochemistry changes that were small but statistically significant.[62]

Hackney and co-workers exposed several sets of adult male volunteers to ozone at 0.25, 0.37, and 0.50 ppm.[63] Although some subjects showed some effects at the higher exposure levels, all subjects, including four with some history of respiratory problems, showed no effects at 0.25 ppm, even while exercising.

The API eventually cited the Hackney study when it advocated a standard of 0.25 ppm. It argued that a study showing "no effects" should be given at least as much weight as a study showing positive effects that might be spurious.[64] The sample sizes in Hackney's studies, however, are so small that they say very little about whether effects will occur to the most sensitive one per cent of all asthmatics.

What conclusions should EPA have drawn from these results? It seems that some blood biochemistry changes do occur at ozone exposure levels below 0.20 ppm. But is that to be considered an "adverse health effect" when the changes are small and disappear one hour after exposure? Furthermore, some individuals may experience changes in lung function at levels below 0.20 ppm, especially when exercising. But are these adverse health effects? How significant are reports of discomfort when the subjects are also exercising vigorously? How do we combine the results reported by DeLucia and Adams with the finding that various asthmatics in other studies showed little or no response, even to higher exposure levels? Furthermore, even if we knew the distribution of responses among asthmatics of varying sensitivity, how do we decide which individuals are "too sensitive" to protect?

In addition, none of those studies tell us about the longer term effects of continuous or repeated exposures. Perhaps such chronic effects occur at even lower levels. Since long-term experiments are not feasible, exploring that question required epidemiological studies.

EPIDEMIOLOGICAL STUDIES

Unlike clinical trials, epidemiological studies measure the effects of exposure in the real world. But nature does not always cooperate by conducting well-designed experiments. Often, environmental variables, like temperature or other air pollutants, will increase or decrease in parallel to changes in ozone concentrations. In such cases, it is difficult to attribute any observable health effects to one particular pollutant. Furthermore, recorded variations in ozone levels tend to be relatively small, and variations in response among individuals relatively great. This implies a need for large samples, especially when the issue is the effects of ozone at low concentrations, where effects are likely to be small and hence particularly hard to detect.

It is also difficult to know exactly what "end points" to use in epidemiological studies, that is, how to measure "effects." Possible measures include giving subjects breathing tests as in the laboratory studies, asking them to report symptoms such as headaches or asthma attacks, looking for changes in their use of health care, and exploring indirect indicators such as diminished athletic performance. Each has its difficulties. All measures based on self-reporting risk unreliability. Breathing tests, although objective, do not directly measure disease. Highly indirect measures, like running times or doctor visits, tend to be only weakly related to environmental conditions because they are influenced by so many other factors and, arguably, reflect adverse health effects only very indirectly.

For all these reasons, it would have been very difficult for epidemiological studies to pinpoint the "threshold" in the dose-response relationship to ozone. The two epidemiological studies that were at the head of the list in the final *Federal Register* notice demonstrate the difficulty. One by Kagawa and Toyama[65] and the other by Kagawa, Toyama, and Makaza[66] followed a single group of twenty Japanese school children over two years. Data from the first six months are reported in the first study, and data from the last eighteen months in the second study. The children were given respiratory function tests every Wednesday afternoon and air pollution was monitored on the roof of their school.

In these studies, correlation coefficients were computed to see if each child's measured breathing patterns tended to change, week by week, in a way that paralleled changes in various pollution measures. Some subjects' breathing in each study showed some positive correlation with some of the pollution measures, although the effect diminished—and in warm weather in the second study, disappeared—when temperature was taken into account.

Unfortunately, this study says almost nothing about the shape of the dose-response curve in the area of interest. Even if there were no effects at all below 0.15 or 0.20 ppm we still could have significant correlations if there was a positive relationship at higher exposures. Without the underlying data, there is no way to tell what the correlations signify. In addition, when anyone computes dozens of correlations, at least some of them are likely to be "significant" simply as the result of chance.

The third Japanese epidemiological study compared self-reported symptoms of discomfort among school children, on days when ambient ozone levels were below 0.10 ppm to those when it was above 0.15 ppm (maximum hourly average).[67] Again, this analysis does little to identify the threshold, if there is one. Moreover, as several subcommittee members stressed, ozone levels were reported daily in the Japanese media, and these self reports could have been influenced by subjects' awareness of what they were "supposed to" feel.

The subcommittee discussion of these studies was heavily influenced by the views of one of its consultants, James McCarroll. McCarroll said that he had visited the Japanese scientists and had found it "extremely difficult to understand what it is they are doing, what they are measuring and what they are actually finding." McCarroll suggested: "I don't know how you can put this into a criteria document in any polite way but I think some caveat has to be given that these data are not generally accepted by the scientific community or by anyone else."[68] The *Criteria Document* ultimately put it this way:

> Japanese and U.S. investigators as yet have great difficulty on exchanging specific scientific concepts. Until the quality of scientific communications between these two groups of investigators increases, the ability to interpret the Japanese studies and to apply them to situations in the United States will remain severely limited.[69]

Attitudes toward these studies tend to reflect an individual's whole approach to the ozone problem. Consider, for example, the views of David Hawkins, who, we noted, came to EPA from an environmental group. Some years after the decision he expressed regret that these epidemiological studies, as well as others on animal infectivity, had not been taken more seriously. "No one I talked to had read them," he said, "even though they counseled us not to rely on them." In his view, the subcommittee had, at times, been "quite undisciplined" in its willingness, based on second-hand information, to "badmouth studies," perhaps because scientists tend to be skeptical of results until they are clearly established.[70]

Several U.S. epidemiologic studies were also conceivably relevant. Wayne and colleagues found an inverse correlation between the oxidant level one hour before high school cross-country races in California and the extent to which runners' times decreased week by week.[71] The authors did present the full scatter diagram of their results, something the Japanese investigators had failed to do, and the SAB subcommittee concluded in the *Criteria Document* that there was "no obvious relationship below 0.15 ppm" in the study. More critically, it is difficult to perceive what this explains about the presence of an adverse health effect.

The Schoettlin and Landau study of 137 asthmatic patients, the basis of the original standard, was reevaluated in light of the new information that had been developed about their measurement techniques. It was now taken to demonstrate an increased frequency of acute asthma episodes, when ambient ozone concentrations were 0.25 ppm and above.[72]

Durham found visits to student health centers were explainable in part by a set of four statistical constructs that summarized sixteen meteorological and pollution variables.[73] This study, too, was mentioned in the *Final Rulemaking*. The fact that so many variables were combined, however, meant that the study cannot be used to find the threshold in the ozone dose-response relationship.

Several other studies tended to support the existence of health effects (e.g., asthma attacks) at ozone levels above 0.30 ppm. As we will see, however, these were not of much relevance since the argument was about standards at roughly half that level.

Animal Studies

Animal studies raised still other problems. Are animal responses similar to human responses? Even if they are qualitatively similar, how should differences among species in the amount of air breathed, physiology, size, and so forth affect the translation of a given dose and its effects in a rodent to an equivalent dose and its effects in humans?

The first set of animal studies looked at the extent to which ozone exposures made animals more susceptible to infection. Several studies showed that, following ozone exposures as low as 0.01 ppm, some animals exhibited an increased susceptibility to bacterial infections.[74] In the view of the *Criteria Document*:

> These findings have definite human health implications, although different exposure levels may be associated with such effects in humans. These reactions in mice represent effects on basic biologic responses to infectious agents . . . Humans possess equivalent defense systems.[75]

The EPA, however, had no accepted method for extrapolating from animals to humans when it came to such studies, so the exposures in the animal studies could not be used to directly determine the level at which human effects would likely be observed.

A second set of animal studies focused on what are called "radiomimetic" effects, that is, effects of ozone that mimic the cell-damaging action of low level ionizing radiation.

Brinkman and co-workers found that when humans breathe air containing 0.20 to 0.25 ppm ozone, the shapes of some cells become distorted, at least as long as the exposure continues. They also showed that the same levels of ozone damaged certain classes of cells from rabbits, and that mice exposed to 0.10 to 0.20 ppm ozone for seven hours a day for three weeks had increased neonatal mortality and a reduced number of litters.[76]

Zelac found that ozone at 0.20 ppm for five hours increased the frequency of chromosomal breaks in certain cells of hamsters.[77] In a highly controversial analysis, Zelac extrapolated to the human case, and argued that exposures of 0.10 ppm (over a forty hour week) would produce breakage frequencies six orders of magnitude greater than those resulting from one week at permitted occupational radiation exposures.

Radiomimetic effects from ozone would, if they exist, be very important.[78] First of all, they would imply that ozone might cause cancer since agents that damage chromosomes often have that property. Second, such effects might occur, albeit with ever lower probability, at any positive ozone level. If so, no ambient ozone level would be "safe" in the sense of producing no physiological effects.

The *Criteria Document* staff at ORD had reviewed these studies and concluded that they should be mentioned in passing, but that they did not provide sufficient evidence of carcinogenicity to warrant regulatory action. At the first SAB subcommittee meeting one member commented that any reference to cancer generally raised a "red flag" in peoples' minds. Several participants at the meeting urged that if the question of carcinogenicity were to be raised at all, it should be treated thoroughly. At the same time, some subcommittee members felt that the draft put too much emphasis on some of these studies, particularly Zelac's, which other investigators had not been able to replicate and that these studies should not be mentioned at all in the "Summary and Conclusions" chapter.[79] That chapter in the final *Criteria Document* ultimately contained a brief section on these studies but did not use the words "carcinogenic" or "radiomimetic."[80]

Integrating the results of diverse results was hardly straightforward. The first draft of the "Summary and Conclusions" chapter of the *Criteria Document* cited only a few studies, placing great emphasis on von Nieding's work. The draft also discussed a "demonstrated effects level" of 0.15 to 0.25 ppm. Apparently as a result of the first SAB subcommittee meeting, all mention of such a level was removed from the document, although it reappeared in the agency's justification for the final rule, which the SAB did not review. The subcommittee's severe criticism of the von Nieding study and the Japanese epidemiological studies also seems to have led to their reduced prominence in the *Criteria Document*, although they too were cited in the ultimate rulemaking.

The study by DeLucia and Adams was not mentioned in either the first draft of the *Criteria Document* or the Shy Panel report. It was not until December 1977 that an outside scientist happened to mention it to an EPA staffer, who in turn brought it to Gordon Hueter's attention.[81]

To appreciate the difficulty facing the EPA administrator in setting the standard, one must keep in mind what the available scientific information did and did not reveal. The epidemiological work, given the difficulties of measuring exposures and effects and the problems of simultaneous movement of many environmental variables, was of little help in identifying precisely what effects, if any, were likely to appear at a given exposure level. The animal studies that suggested radiomimetic or immunosuppressant effects were controversial, and in some cases unreplicated. It was not clear how to extrapolate effects from animals to humans. Furthermore, it was not certain just how sensitive the asthmatics in the clinical trials really were. The most sensitive were unlikely to have been included. There was also the possibility that still other relatively unstudied groups, such as children and the elderly, might prove even more sensitive.

Regarding the human clinical studies, very high ozone levels (above 0.25 to 0.30 ppm) were clearly injurious to some relatively healthy individuals. Below 0.20 ppm some sensitive persons (i.e., some, but not all asthmatics) experienced transitory and statistically insignificant (but consistent) impacts on lung function and some subjective discomfort, especially when performing strenuous exercise. There also were some small and apparently transitory effects on blood chemistry at such levels.

Despite these difficulties, the administrator had to decide just how important it was to protect various kinds of sensitive persons from different nonpermanent decrements in lung function, or decide that there was some acceptable frequency of such events and the discomfort and activity limits they implied. He also had to determine what weight he should give to the few animal studies that suggested there was absolutely *no* safe level of ozone. Clearly he faced a confusing situation in which both technical judgment and political values had to play a major role in the way he chose to exercise his discretion.

THE PROPOSED RULE

In June of 1978, as a result of an internal compromise, EPA proposed an ozone standard of 0.10 ppm, not to be exceeded more than once per year. On one side David Hawkins, who believed that there was no threshold in the dose-response function, had advocated retention of the 0.08 ppm standard. Hawkins felt that the proponents of increases had to prove that they would be safe. Thus, the poor data that was available implied, in his view, maintaining the status quo. Recognizing the likelihood that he would eventually be overruled, Hawkins nevertheless felt an obligation to present the environmental point of view so that Costle could hear all the arguments.[82]

In the middle, Walter Barber, OAQPS deputy assistant administrator, believed that the effects of low levels of ozone were reversible. The debate focused on the margin of safety that was appropriate below the threshold level which was considered to be about .15 ppm. When Barber suggested a 0.10 ppm revised standard in a draft OAQPS recommendation, Hawkins persuaded him to change the recommendation to 0.08 ppm.[83]

At the opposite end of the spectrum, OPM wanted to go even higher. It was impressed by the high cost of achieving dubious gains for only a small minority of citizens.[84]

Ultimately Costle himself decided to propose 0.10 ppm. He realized that he might have to relax the standard even further, but felt he would be hard pressed to go lower if the level in the preliminary notice was set too high. Furthermore, if he were going to raise the standard above 0.10 ppm, Costle believed that it might be easier to do so in two steps.[85] Inside the agency everyone realized that the real battle would come over the final standard.

At the press conference that announced the standard,[86] Steve Gage said that health effects at levels below 0.15 ppm appeared completely reversible, whereas irreversible effects might occur at ozone levels above 0.25 ppm, and even higher levels might trigger diseases such as heart attacks (in fact, little is known about the effects of ozone, given the lack of long-term epidemiological studies).

At the same briefing, David Hawkins tried to head off any criticism from the environmental community by arguing that a 0.10 ppm standard represented only "a marginal change" and would not make much "practical difference." As a press release in Costle's name put it, "because the attainment problem in most urban areas is so severe, the proposed relaxation of the standard is not expected to change the level of control requirements in the near term." Hawkins also said that under the new standard, approximately ten to fifteen of the nation's Air Quality Control Regions might still not reach the ambient standard by 1987, rather than an estimated thirty to sixty regions if the standard remained at 0.08 ppm. Much of the change would come in rural areas that would now be in compliance.

The EPA's *Advance Notice of Proposed Rulemaking* appeared in the *Federal Register* the week after the press conference.[87] Besides changing the standard, it also changed the designation from photochemical oxidant to ozone, which is what the approved measurement actually gauged. Thus, the new standard would be easier to enforce. In addition, EPA apparently felt that it would now be able to dismiss arguments from city governments that their mix of photochemical oxidants was unique and required a special standard.[88]

On the other side, several environmental groups were dismayed by this change. They argued that other oxidants were much more dangerous than ozone. They were concerned that the regulation of these other compounds might never be restored.

Their concern was shared by several members of the SAB subcommittee. James Pitts, an atmospheric chemist and a member of that group testified at an August 24 hearing in Los Angeles that an "ozone" standard would inadequately characterize complex photochemical oxidant pollution and its attendant hazards. He also argued that it would allow industry to demand a far more relaxed standard than 0.10 or 0.12 ppm because pure ozone did not seem to produce noticeable effects at low levels.[89]

In justifying the 0.10 ppm level, the agency went some way toward accepting the 1974 National Academy view that "thresholds" were not a helpful way to pose the regulatory problem. The *Notice of Proposed Rulemaking* said in part:

> The concept of a "threshold" may not be an appropriate term for describing the impact of ozone on human health. Since "thresholds" will depend on who is studied and what is measured, it is unlikely that scientific evidence for a specific effects threshold can be satisfactorily derived for protecting public health . . .[90]

And later:

> . . . no clear threshold can be identified for health effects due to ozone . . . Selecting a standard . . . does not imply some discrete or fixed margin of safety that is applied to a known "threshold."[91]

How then is the standard to be selected? The discussion of this point in the *Notice* is not very informative. It characterizes the choice of a standard as ". . . a judgment of prudent public health practice," but says very little about what that involves. It does acknowledge that, although the administrator cannot take costs into account:

> . . . controlling ozone to very low levels . . . will have significant impact on social and economic activity. It is thus important that the standard not be any more stringent than protection of public health demands.[92]

In justifying the proposed level, the *Notice* argues that the animal infectivity studies cannot be extrapolated to humans and that the studies of radiomimetic effects "should not be deemed important because of the problems of replication and relevance." Instead, "aggravation of asthma," "reduction in pulmonary function," and "chest discomfort and irritation of the respiratory tract" are the effects to be avoided. The work of DeLucia and Adams was characterized as "the study demonstrating the lowest effect level in man," which was taken to be 0.15 ppm. The notice goes on to say that interviews with the Shy Panel and experts consulted in the risk analysis "are reasonably consistent" in providing "lowest effect level estimates . . . ranging from 0.15 to 0.18 ppm." It concludes:

> Based on this data and . . . the criteria document, it can be concluded that health effects have been demonstrated at ozone levels of 0.15 ppm.[93]

Given the animal infectivity studies and the Japanese epidemiological studies that "it would not be wise to totally disregard," the agency chose a .10 ppm standard in part because the lower the standard, the less likely it becomes that the "level at which health effects are virtually certain (0.15 ppm") will in fact be exceeded during adverse meteorological conditions.[94]

This exposition does not go into all of the relevant facts. The Shy Panel report says at one point, "a variety of adverse effects are likely to occur in some segments of the population from short term exposures of 0.15 to 0.25 ppm."[95] But this is not as definite a statement as the conclusion that the *Notice* uses it to support. Still more questionable was the presentation of the special risk analysis. The unusual part of that report was that the experts provided a whole probability distribution to describe the likelihood that effects would occur at or below various levels. The *Notice*, however, only reported the medians of those distributions. Then the medians for each effect were averaged to produce a value apparently consistent with the Shy Panel.

This average of the medians is most certainly not the level at which the experts thought effects were "virtually certain," yet that is how the *Notice* characterizes these numbers. In fact, in only one of nine cases did an expert estimate that the probability of a health effect occurring at or below 0.15 ppm was greater than sixty per cent. In that case the expert in question characterized the sensitive population as asthmatics, infrequently exposed to ozone, exercising at high altitudes.

In summary, from an educational point of view, the *Preliminary Notice* had both good and bad points. The idea of a threshold was openly questioned. In addition, the "impact on social and economic activity" was given a role to play in standard setting. On the other hand, EPA made no effort to explain just what was in the black box of "prudent public health judgment." Nor did EPA explicitly pose the question of how to set a standard when there is a no-threshold dose-response relationship. Finally, parts of the science, particularly the difficulties of the epidemiological studies and the risk analysis, are just not adequately presented. In general, much of the *Notice* seems designed to lay the groundwork for defense of the agency in a law suit, as opposed to telling a confused citizen what was actually going on.

Given all these questions, however, and the substantial economic stakes involved, one could anticipate that the June 22 publication represented more of a call to arms than a set of truce terms. The various parties responded accordingly.

During July and August of 1978, EPA held four public hearings on the standard. At these meetings the agency encountered the expected range of views. Robert Rauch, a staff attorney with the Environmental Defense Fund (EDF), cited the studies of radiomimetic effects. He concluded that it was "obvious" that the 0.08 ppm standard had provided "virtually no margin of safety from mutagenic hazards." He also raised the "no threshold" argument:

> Given the fact that scientists have been unable to discover a threshold below which carcinogens do not have an adverse effect, there is absolutely no justification for relaxing the existing standard of 0.08 ppm.[96]

Carl Shy, speaking on behalf of the American Lung Association, asserted that health effects had "certainly" been demonstrated at 0.10 ppm.[97]

The State and Territorial Air Pollution Programs Association (STAPPA) and the Virginia State Air Pollution Board supported a 0.12 ppm standard. The Virginia representative said their recommendation "represents the consensus of these people in the field who are closest to and most familiar with the problem, yet are less biased and unduly influenced by the opposing forces of environmentalists and private industry."[98]

At the other extreme, when the API representative was pressed by the EPA hearing panel for a number, he endorsed a standard of 0.23 to 0.28 ppm. This was justified, he said, by the lack of firm evidence of significant adverse health effects near the proposed standard.[99]

OZONE IS RARGED

After publication of the proposed standard, those inside the government who did not share EPA's perspective began to try to influence the decision. The main focus of concern was the Regulatory Analysis Review Group (RARG), which President Carter had established by Executive Order. In a way that clearly paralleled the "Quality of Life" review process under Nixon, all federal agencies were to conduct an economic impact analyses of each proposed regulation that was expected to have a "substantial impact" (i.e., cost more than 100 million dollars a year). The RARG was to selectively review those analyses. The group included representatives of many cabinet departments and other executive branch agencies, but the actual work was done by economists from the Council on Wage and Price Stability (CWPS) and the Council of Economic Advisors (CEA). Charles Schultze, Chairman of CEA, was in charge of the operation.[100]

Near the end of the public comment period, Bob Litan, a young RARG economist and attorney, gave a speech on the benefits of regulatory reform. A former classmate in the audience, John Hahn, was working for a law firm that represented API. He claimed to Litan that the proposed ozone standard was by far the most costly regulation the administration had ever devised, and sent him some supporting analyses. Stunned by the huge cost estimates, Litan took the issue to Charles Schultze. Schultze was also disconcerted and asked EPA to extend the public comment period to give RARG time to respond to the proposed rule.[101]

The RARG was seen by its own members as an organization with a mission. Its staff economists espoused the common professional view that allocative efficiency was a, if not the, central virtue of all public policy. As representatives of the economic viewpoint, a rule that would impose huge costs for little or no gain was, to the RARGers, an affront both to public morality and to their sense of professional responsibility.[102]

In addition, the president's own ambiguity about his priorities and purposes made it difficult for the RARGers to behave as his representatives. As one of them said later, "Since I never knew what decision the president would make, or would have made if an issue ever got to him, I had no choice but to pursue my own vision of what was good."[103]

But RARG had to proceed carefully given its own recent and highly embarrassing defeat on the OSHA cotton dust standard. As with ozone, the cotton dust standard had been delayed to allow the economists to review it. After that review, Charles Schultze, with the president's consent, sent a memo to Labor Secretary Ray Marshall instructing him to relax the proposed standard. Marshall responded angrily and key congressmen and labor groups also pressured the president. After meeting Marshall, Carter reversed himself and accepted the Department of Labor's proposal with two relatively minor changes.[104] As an advisory process with no power except its claim on the President's ear, RARG could not afford another humiliating public repudiation. Indeed David Hawkins suspected that one reason RARG took on ozone was to show that they could still have an impact.[105]

RARG did have some allies within EPA, namely several young economists in OPM who shared some of RARG's concerns. They had developed some close working relationships, especially with Larry White of the CEA staff who did most of the actual work on the RARG analysis. Indeed, according to Litan, EPA was the source of most of the cost data used by RARG.[106] Costle had a different view:

> I would say that probably three out of every four CWPS comments on our rule making were cribbed right from industry briefs . . . partly because it suited their economic biases about these issues, and their own perception that they were custodians and keepers of the regulatory reform flame.[107]

The RARG report was submitted to EPA as a formal comment from Barry Bosworth, Chairman of CWPS.[108] The report claimed that EPA had sought to create an artificial distinction:

> From the continuum of effects, EPA has focused on a particular point as defining the onset of health effects . . . Next, EPA has tried to determine the ozone concentration level at which this . . . occurs . . . [109]

RARG criticized EPA for making choices for which there was no adequate rationale. For example, EPA arbitrarily chose the 99th percentile individual among the sensitive group as the person to be protected, without offering any justification or explanation. RARG also criticized EPA's assessment of literature, especially the reliance on DeLucia and Adams, and commented that "virtually all of the remaining clinical studies cited in the *Criteria Document* describe health effects at ozone concentrations of 0.25 ppm or above."[110]

The report also criticized EPA for underestimating costs. Where EPA had estimated 6.9 to 9.5 billion dollars per year, RARG expected annual costs of 14.3 to 18.8 billion dollars. Such costs depended on how state implementation plans (SIPs) would be revised and enforced. RARG believed that EPA failed to count the costs of some very expensive measures that would be necessary to meet a 0.10 ppm standard. The EPA countered that these steps were so costly that it was unlikely they would ever be implemented, whatever the standard.[111]

Given that a threshold did not exist, RARG suggested focusing on the marginal cost of reducing the aggregate amount of exposure above some given level (in person hours). At some point, RARG argued, the marginal cost per reduced person-hour of unhealthy exposure would begin to increase sharply.[112] Such a point should determine the standard. This approach, RARG argued, was a rational step toward equalizing the marginal costs of protecting health across different health and safety programs.[113]

Since the report emphasized the low marginal cost per reduced person-hour of unhealthy exposure in going from a standard of 0.18 ppm to 0.16 ppm (between $170 and $440) relative to the cost of going from 0.12 to 0.10 ppm (between $2100 and $4100), EPA interpreted RARG as recommending a standard of 0.16 ppm.[114]

On November 7, 1978, EPA officials briefed Richard Neustadt, of the White House domestic policy staff, on the agency's response to the RARG report. The basic response was contained in a memorandum from Walter Barber to David Hawkins. Barber first faulted RARG for having focused on economics:

> Nowhere does the Act authorize such balancing of costs and benefits across pollutant and media lines, . . . The Act does not authorize the administrator to abandon air pollution control simply because the marginal cost is less for a comparable health unit in another public health or safety program.[115]

The approach taken by RARG, Barber argued, had implications well beyond ozone, as it would undermine EPA's interpretation of the statute and the justification for a whole series of other regulatory actions.

Barber also had trouble with RARG's approach to the scientific evidence. RARG "completely dismisses the entire body of medical evidence regarding adverse effects . . . with the exception of irrefutable clinical studies showing effects at concentrations of approximately 0.37 ppm." RARG excluded consideration of "uncertain" (unreplicated) effects, and, "virtually the entire body of scientific data showing effects at levels down to 0.10 ppm, including those demonstrating synergistic behavior with other pollutants." The memo concluded that any approach, including RARG's, must be wrong if it could characterize air quality as acceptable in cities such as Atlanta, where citizens complained strongly of smog problems.

Given the force of the claim that the Clean Air Act required EPA to ignore costs, the RARGers began to shift their attack to the scientific basis for the standard.[116] Frank Press, the president's science advisor, was being lobbied by both sides to get involved in the controversy. Press delegated the job to his staff member Gilbert Omenn.[117] Omenn telephoned the scientific investigators whose work was cited in the *Criteria Document* and the science advisors EPA had retained.[118] When he finished, Omenn felt that it was hard to deny that clinical effects were demonstrated at 0.20 ppm, at least in some exercising individuals. Once one took into account the need to protect the sensitive segment of the population, possible synergisms with other air pollutants, the animal studies, and uncertainties in the data base, an acceptable range for the standard, he felt at the time, ran from 0.10 to 0.16 ppm.[119] This did not do much to resolve the dispute, however, since most parties were making proposals in this range.

Nor did the other comments on the proposed standard help EPA very much. By the time the extended comment period closed on October 16, 1978, EPA had received 168 written comments, most of which opposed 0.10 ppm. State and local air pollution control agencies generally endorsed 0.12 ppm or higher. Most industrial groups urged 0.15 ppm or higher. Environmental groups in contrast generally encouraged EPA to retain 0.08 ppm.[120]

The comments raised critical questions regarding EPA's basis for selecting a standard. Those commenting wanted to know what "adverse health effects" the agency sought to prevent, and questioned the significance of reversible effects. Disagreements were expressed with EPA's interpretation of the various studies. Some comments emphasized that the effects reported by DeLucia and Adams at 0.15 ppm were not statistically significant at the level of confidence sought by the investigators. Others argued that not enough weight was given to negative studies at still higher exposures and that too much was made of some of the Japanese epidemiological work. There were comments that EPA was unnecessarily stringent in selecting the sensitive population, and that the standard would be less stringent if EPA accounted for the portion of time that persons were indoors and, therefore, not exposed to higher ambient concentrations. There were also questions about EPA's unwillingness to consider costs, on the risk analysis, and on the lack of formal SAB approval.[121]

On the other side, environmentalists raised the possibility of carcinogenic effects and the evidence of the animal infectivity studies. They argued that the statute focused only on health, not on costs, and that even sensitive individuals were to be protected. They went on to argue that since no one could prove that even 0.08 ppm was safe, there was certainly no basis for raising the standard.[122]

Clearly this one was not over yet.

COSTLE DECIDES

Late in 1978, Costle became personally involved in ozone, essentially because of the developing dispute between Bill Drayton and David Hawkins over the issue. He felt that the previous standard had no clear scientific basis. Furthermore, the problem seemed to him to be a good candidate for formal analysis. After reviewing all the data, OPM endorsed a standard around 0.15 ppm.[123]

Hawkins, as we have seen, urged retaining 0.08 ppm. In his view, cost-effectiveness reasoning here was both legally indefensible and irrelevant. Any money saved by relaxing the standard was not likely, in fact, to go toward some other, more effective public health program.[124]

Their disagreement came to a head on December 12. Drayton's office wrote a letter that went out over Costle's signature to James McIntyre, Director of OMB.[125] In discussing EPA's plans, the letter said that by changing the ozone standard the nation could expect "to save over one billion dollars each year." David Hawkins reacted angrily, criticizing Drayton for "confusing people outside the agency as to the basis for our decision."[126] The language used, he said, suggested that cost was a factor in EPA's decision making, which conflicted with the administrator's statutory duties. He told Drayton that all future communications that described the Air Program's activities should go through Hawkins' office for review.

As he managed the growing controversy, Costle tried to keep Drayton off the record and uninvolved. He wanted to insure that EPA could not be fairly accused of using statutorily proscribed cost-based arguments in setting the new standard.[127]

Other EPA managers were also concerned with ozone. Richard Dowd, secretary to the SAB, wanted the standard set as high as possible without producing health effects. Control of ozone was so expensive that he was reluctant to urge a decision as conservative as those he had supported in other cases. He tried to represent a compromise viewpoint between Hawkins and Drayton, although his position was much closer to Drayton's. The poor data simply would not, he felt, support Hawkins' position. Dowd believed that Costle was aware of his own scientific limitations and thus relied heavily on the advice of both him and Steve Gage.[128] Gage, too, had concluded that 0.08 ppm standard was unjustified. He preferred 0.15 ppm.[129]

Costle made it clear from the start that he wanted to make his own judgment about the science. He knew that the original standard was based on little more than one study that had since been found deficient. Given industry's complaints, leaving the standard unchanged would, he believed, damage the credibility of the agency. To accomplish this, Costle closeted himself for almost two full days with Gage, Dowd, Hawkins, Barber, Roy Gamse of OPM, and Drayton.

The original standard, they all agreed, had no clear scientific justification. Even Hawkins dropped his burden of proof argument for 0.08 ppm when he considered the possibility of external attacks on the agency's credibility.[130] In the view of many in this group, however, the Japanese and von Nieding studies did not offer clear-cut proof of effects at levels below 0.15 ppm. Steve Gage recalls feeling that there was a "pretty clear" threshold around 0.15 ppm given the DeLucia and Adams studies. Above 0.15 ppm the evidence was more clear cut.[131]

Describing this experience later, Costle said: "Once you begin to review the studies themselves . . . you wind up narrowing the basis for the decision. The minute that you begin looking at concrete data and evidence, it's not quite as vague as the statute might suggest."

Costle left the meeting, as he often did, without revealing his position. Discussing his decision later, Costle said that:[132]

> It was a political loser no matter what you did . . . The minute you picked a number, . . . everybody can argue that it can't be that number, or it could just as easily be another number . . . [It] was a value judgment . . .

Whatever number he picked, the administrator wanted to put it in the best possible light:

> That is . . . how could I characterize the basis for my judgment in such a way that would have respectability, given the antagonism of the opposing forces? I kept asking myself, 'Is there high ground for EPA in this?'

On December 21, a group from EPA including Hawkins and Barber briefed seven or eight White House staffers.[133] According to EPA, the purpose of the briefing session was not to come to a decision, but to communicate EPA's thinking. Much of the meeting was spent reviewing EPA's determination of 0.15 ppm as the "probable effects level" for sensitive segments of the population. Some from the White House argued that it was not for EPA to prevent the reversible physiological effects seen at low levels of ozone exposure, especially when the cost per prevented exposure-hour was so high at those levels. The meeting also discussed the relationship of the standard to state implementation plans (SIPs), and considered what steps, and costs, would be needed to reach various standards.[134]

After this first briefing, Larry White from CEA met several times with Roy Gamse and Deborah Taylor, of the Office of Planning and Management, to discuss the costs of reaching different standards. The EPA continued to insist, meanwhile, that the Clean Air Act forbade consideration of costs in setting the standard.[135]

On the first day of 1979 Costle asked Dowd to telephone selected scientists for their views on the standard. Over the next week Dowd called Whittenberger and Pitts from the SAB subcommittee, as well as two of the subcommittee's consultants. Whittenberger selected 0.12 ppm. Pitts said he would like to see 0.10 ppm. Most or all of those consulted chose levels between 0.10 and 0.15 ppm. Costle later described the results of this effort as follows:

> It did help to establish a range. Then you could take the Clean Air Act criteria and that would tend to drive the decision in the direction of one end of the range as opposed to the other. Put another way, once you had a range, the Clean Air Act was clearly biased toward the cautionary end of the range.[136]

The process also had "political value." Costle felt that if he could "ever pin them down on what they thought the right number ought to be, it would ultimately lend more credibility to the decision." For if he could do that,

> . . . it would then be possible for me to go up to Capital Hill and defend a decision by saying I not only reviewed the advice of the EPA scientists on this but I went outside EPA and reviewed the advice of reputable independent scientists . . . [137]

As Costle moved to a final decision, he became increasingly worried about White House involvement. He was concerned both for the president's image and for the integrity of the rule making process. On Monday, January 8, with his set of experts' opinions in hand and a meeting scheduled the next day with Schultze, Alfred Kahn, Frank Press, and Stuart Eizenstat, Costle prepared an internal memo to Hawkins to formalize his decision: "As I told you Friday, I have concluded that the primary ozone standard should be set at 0.12 ppm."[138] The memo requested that Hawkins set in motion the process of drafting a final *Federal Register* package.

THE WHITE HOUSE BECOMES INVOLVED

Costle's decision still had to be merged with the concerns of the RARG and the views of the rest of the government. But it is not clear that there was anyone below the level of the president who could play that integrative role. One possibility was Stuart Eizenstat, chief of the domestic council policy staff. A long-time Carter aide and political, as well as policy, advisor, Eizenstat was concerned that the nation was still watching for Carter's promised tough antipollution stance. It would not help to have the administration appear to be at war with itself. Eizenstat also believed, however, in the substantive and political importance of inflation-fighting. He wanted to find a way to avoid the developing confrontation.[139]

His goal received a serious setback when he received a memo from Barbara Blum, on Costle's behalf, saying that there was "no way the administrator, in good conscience, can provide justification for a higher level (than 0.12 ppm)." Schultze was furious, since once Costle's decision was in writing, any subsequent change might be open to legal challenge.[140]

At the January 9 meeting, Schultze still pushed for 0.16 ppm or at least 0.14 ppm. He argued that since there was no threshold, determining which adverse health effects were to be prevented should include consideration of the costs and benefits. Besides, he argued, the data on health effects at low exposures were very shaky.[141]

Since the economists could not claim to be experts on the science, the views of Frank Press and Omenn were critical. As noted previously, the latter had settled on the range of 0.10 to 0.16 ppm, and Press never offered a specific alternative.[142]

Costle held firm. Schultze, he felt, had always disliked the Clean Air Act's failure to consider costs. Eizenstat's views, too, seemed to him to be based in part on a distaste for the statute itself. They could argue about the statute as much as they wanted, Costle felt, but they could not fault him for following it. He let the White House staff know that he was approaching the standard from a "lower-end," point of view: ". . . I felt very much compelled by the way the Act is structured in giving much more weight to some of the more cautionary studies than some of the more conservative scientists would."[143]

In addition, Costle did not believe that the very high cost estimates offered by the RARGers. First, they were based on self-interested figures from API and various firms in the petroleum and automotive industries. The analysis was also highly conjectural, since the costs would depend on what states wrote into their SIPs, and then actually enforced. Costle believed that there would be flexibility in the implementation process if costs got too far out of hand.[144]

Although Dick Dowd continued to telephone ozone scientists,[145] Costle postponed a previously scheduled press conference on the standard in order "to review more economic data."

At a series of meetings between various White House and EPA staff, a compromise was proposed: set the standard at 0.12 ppm but allow four exceedences a year instead of only one. This would imply the same control requirements, and cost savings, as a standard of 0.14 ppm but would not conflict with the Blum memo or Costle's January 5 memo. In addition, it would mean very little difference in public health. Eizenstat was very taken with this idea and went personally to see Costle to convince him to accept it.[146]

To Eizenstat's distress, Costle was unmoved. Costle thought the compromise would be seen as a transparent subterfuge and successfully attacked as such. On January 23, Costle sent a memo to Eizenstat with a review of the relevant health data, statutory provisions, and legislative history. The memo stated that all of the polled ozone experts would find a 0.12 ppm standard acceptable, and some would object to a 0.14 ppm standard as unsafe.[147] Whatever the opinion of of Eizenstat and Schultze, President Carter chose not to overrule Costle's decision to set the standard at 0.12 ppm.[148]

THE FINAL RULEMAKING

On January 26, 1979, Costle held a press conference to announce that the revised standard would be set at 0.12 ppm. He explained his reasoning this way:

> The new standard is principally designed to provide protection for the nation's five million to ten million asthmatics and others with chronic respiratory diseases, many of whom reside in urban areas where high ozone levels are observed . . . However, even healthy individuals engaged in vigorous outdoor activities, such as construction

> work, freight hauling, and athletics, may well experience discomfort and the first indications of adverse health effects as a result of exposure to ozone concentrations ranging from 0.15 ppm to 0.30 ppm.[149]

The reaction was predictable. *The Wall Street Journal* attacked the rule as "A $100 Billion Cough Remedy," urged 0.25 ppm, and suggested that Congress retract EPA's standard-setting authority.[150] The API petitioned the Court of Appeals, claiming, "It is far more stringent than medical evidence shows is necessary to protect public health."[151] The ruling would, said API, add to inflation, cause unemployment, and reduce economic growth. Albert Frye, from the Business Roundtable, predicted general hostility to mandatory inspection and maintenance (I&M) programs for automobiles. "I think the public would be upset with I&M. Up to now the public hasn't had to pay directly for pollution costs."[152] On the other hand, he said that businesses should give Costle credit for "taking a step most environmentalists won't like." The Environmental Defense Fund and the Natural Resources Defense Council also considered litigation. According to Robert Rauch, of EDF, the "evidence suggests that at 0.12 we will see health effects."[153]

Costle was questioned at the press conference about whether pressure from the White House economists had influenced his decision. He commented briefly that he had discussed the standard with the economists, but only after he had decided and then only to explain his decision.[154]

The *Federal Register Notice of Final Rulemaking* contained some subtle but important shifts in argument from the proposal.[155] Instead of attacking the idea of a threshold it said:

> Although the concept of an adverse health effect threshold has utility in setting ambient air quality standards . . . [a threshold] cannot be identified with certainty.

In this light the agency introduced a new term, the "probable effects level," to refer to the level that in its best judgment is "most likely to be the adverse health effects threshold concentration."

Hence, all previous suggestions that there was no threshold vanished. All that remained was the assertion that we cannot know for sure where that threshold is. Indeed, EPA argued, uncertainty as to "the" level is what justified the "margin of safety" requirement in the Act.[156]

The virtue of this claim, from EPA's viewpoint, was that it allowed the agency to defend the choice of a standard as a scientific task. The central premise of the counter argument—that in the absence of a clear threshold one needs to make a social policy judgment about who to protect against what effects — is simply assumed away.

On the basis of the various studies, the *Final Notice* said, "EPA remains convinced that at levels in the range of 0.15–0.25 ppm adverse health effects will almost certainly be experienced by significant numbers of sensitive persons."[157] As in the *Proposal,* the animal infectivity and epidemiological studies are cited to ". . . suggest the real possibility of significant human adverse health effects below 0.15 ppm."[158]

Thus, EPA argued, the standard must be below 0.15 ppm. Criticisms of the animal and epidemiological studies, however, convinced EPA "that they do not dictate as wide a margin of safety as was established in the proposal." Hence 0.12 ppm, EPA said, "is sufficiently prudent."[159]

The *Final Notice*, like earlier statements, was written as a legal brief for the rule. It is even less informative and self-critical than the earlier document. In sum, it would be very hard for someone to use it to understand either the science or the policy issues. There was no analysis of how to define the "sensitive group," or consideration of the difficulties of deciding which "effects" should count as adverse health effects.

The advocacy approach also appeared in the comments on specific studies. Each study was discussed in a way designed to emphasize its positive findings. For example, von Nieding's results "cannot be ignored" even though "they are unconfirmed and must be interpreted cautiously."[160] The problems in using the Japanese epidemiological papers to determine a threshold are not fully explained. Instead, the *Notice* merely says that although "it is very difficult to interpret their results . . . EPA must consider" these studies because they represent "uncertainties which medical research has not yet resolved."[161]

The White House economists were dismayed that they had not had a greater effect. Although their opponents credited them with responsibility for the increase in the standard, they felt defeated. They felt Costle had not been swayed by either their cost figures or their health arguments.

Bob Litan, however, had to admire the compromise Costle had struck with the final rule. After all, Costle had twice raised the standard, and it was ultimately a fifty percent relaxation compared with the 1971 standard. "I think he pleased Carter," Litan offered.[162]

Environmentalists, too, were distressed by the revised standard. Although Rauch admired Costle for finding middle ground where the courts would uphold him, he felt that the administrator had not been the environmental advocate Jimmy Carter had wanted.[163] Costle's conciliatory nature, he believed, had undermined Congress' intent that EPA be an advocate, rather than a mediator of competing interests.[164]

CONGRESS NEITHER ADVISES NOR CONSENTS

On the CBS news program, *Face the Nation*, Senator Edmund S. Muskie voiced his concern that EPA had based its decision on economic factors, thereby violating the law that he had largely written.[165] He called Costle, Schultze, and Kahn before his subcommittee and questioned them vigorously about the White House role.

Schultze and Kahn held that, as advisors to the president, they had the legal right to discuss a decision with EPA officials at any point in the process. They were acting, Schultze insisted, not as outsiders, but as members of the administration.[166]

On the substantive issues, they argued that prudent public health practice included consideration of costs. Indeed, Schultze explicitly contended that in the absence of a threshold, there simply was no other way to proceed.

In response, Senator Muskie emphatically argued that EPA was an independent agency, created to be an environmental advocate. The administration's regulatory review process, he said, was "slanted against environmental regulation." Insofar as RARG participated in discussions after the close of the comment period, the economic point of view was gaining a disproportionate amount of weight in the decision process.

Senator Muskie also angrily reminded the economists that, as a major author of the Clean Air Act, he was quite aware of the intent of Congress:

> The statute clearly prohibits the use of economic considerations in the setting of health standards . . . There was no other way to do it . . . We couldn't use a technology handle . . . we had to find the threshold . . . We have learned since that there is no real threshold. Even with practically minimal emissions there are health effects to someone. . . . It may be an oversimplification. But the validity of the health standard as a health standard ought to be subject to constant review by the Congress because it is the Congress that established it and if the standard is . . . unnecessarily harsh, from a health point of view and only from a health point of view . . . the Congress should have a chance to modify it . . . it is the heart of the Clean Air Act.[167]

Thus, Muskie was simultaneously granting Schultze's empirical arguments and rejecting their implications. He was willing to accept that there was no threshold, but insisted that the "simplification" that there was one had to be maintained because there was no other way to proceed. Just how this was to be done, however, remained obscure at best.

When Costle testified, he argued that most of his meetings with the White House staff were just briefings. "My decisions have simply not been changed by the new argumentation."[168] He noted that if new facts or arguments had been presented after the close of the comment period, then the docket would have had to have been re-opened to allow for additional public comment:

> The final decision was my best personal judgment . . . based on our best knowledge of what those health effects were, . . . and a very conservative reading of those studies. I erred on the side of preserving the public health, if at all.[169]

Other senators questioned Costle, too, some from different viewpoints. Chafee, of Rhode Island, suggested that Costle could have decided on just about any level for the new standard and found evidence to rationalize it. Costle agreed: ". . . that is the nature of my job, to review the arguments and make the best judgment call I can."[170] Chafee went on to ask him about the problem of reducing the costs of environmental protection. He wanted to know if the administrator thought the Muskie subcommittee was an obstacle to cost-reducing proposals. Costle replied, "Senator, this committee has in no way been a handicap in this respect . . . I have never felt inhibited in any way about coming up here to discuss these things."[171]

Senator Bentsen submitted a whole set of written questions on the evidentiary basis for the standard. In answering these, EPA said it "did not identify DeLucia and Adams . . . as a pivotal study." The Agency said it relied as well on Kagawa and Toyama, Wayne's study of runners, von Nieding, the Makino and Mizoguchi study of self reported symptoms, and the animal infectivity studies.[172]

In all, congressional review of the ozone decision was neither helpful nor edifying. Senator Muskie gave the administration witnesses a hard time. He was defensive and hostile and had difficulty responding convincingly to Schultze's arguments. The hearings hardly represent democratic deliberation at its best. Costle tried hard to avoid antagonizing the senator and neither his verbal responses nor the agency's written answers were particularly informative or explicit.

OZONE IN COURT

Both environmental and industrial groups took EPA to court. The challenges were consolidated into a single case, *API v. Costle*, which was argued in the D.C. Circuit Court of Appeals on February 26, 1980. Briefs were submitted by the API, the Chemical Manufacturers Association, the states of Virginia and Oklahoma, the city of Houston, the Natural Resources Defense Council (NRDC), the Sierra Club, the American Lung Association, as well as two local environmental groups, among others.[173]

The petitioners presented both procedural and substantive issues. Procedural questions included the role of the risk assessment study, the use of the Science Advisory Board and the Shy Panel, and the propriety of postcomment period comments. Under the Clean Air Act, the court could only invalidate the standard because of a procedural error, if "the errors were so serious and related to matters of such central relevance to the rule that there is a substantial likelihood that the rule would have been significantly changed if such errors had not been made."[174] Obviously, the court pointed out, Congress was concerned that "EPA's rule making not be casually overturned for procedural reasons."[175]

The Court held that although the SAB had played a proper role in the development of the *Criteria Document*, it should, legally, also have had the opportunity to comment on the standard itself.[176] Since the SAB had reviewed the *Criteria Document*, however, the court decided that a SAB review of the standard would probably not have significantly changed the outcome. As for the Shy Panel, the Court pointed out that even if its activities were in violation of the Federal Advisory Committee Act, the administrator did not follow its advice. The Court further determined that there was substantial enough evidence for the 0.12 ppm standard, so that, even without the Shy Panel report, the same result would plausibly have been reached. Thus, all of the procedural objections failed.

The court refused to consider NRDC's challenge of the postcomment period involvement of the White House, on the narrow grounds that NRDC had not raised it in a timely manner to EPA as required by law.

There were several different challenges to the substantive basis of the standard. The API contended that the standard was not rational because no adverse effects had been proven below 0.25 ppm. The city of Houston argued that the standard was arbitrary and capricious because natural levels of ozone and other physical phenomena in the Houston area prevent it from meeting the new standards. The NRDC challenged the adequacy of the established margin of safety, arguing that it did not protect sensitive individuals from easily predicted risks.

In response, the court found that the record was "replete with support" for the final standard, and that there was no reason to hold that the administrator had abused his discretion, Houston's argument could not be upheld because attainability and feasibility were not relevant considerations in setting National Ambient Air Quality Standards. The court faulted NRDC for choosing to rely only on the studies that favored its position, and found that the established margin of safety was adequately supported by the record and rational judgment on the part of the administrator. In the spring of 1982, the U.S. Supreme Court declined to hear an appeal.

REFLECTIONS

Our interest in the ozone decision lies less in what was done than in what was said (or not said). The decision of the EPA was surely defensible. One could argue that the Clean Air Act embodied a conscious congressional decision to place health above all other objectives. DeLucia and Adams did see respiratory function effects in two individuals at 0.15 ppm with strenuous exercise. Von Nieding saw something too, although his measuring techniques may have been somewhat flawed. Although some other studies did not find much impact at 0.20 or 0.25 ppm, they did observe some respiratory and blood biochemistry effects. And perhaps their asthmatics were too healthy to define the subgroup to be protected. Given the high school runners, the Japanese school children, and the radiomimetic and animal infectivity studies (so this argument continues), and considering the "sensitive population" and need for a margin of safety, setting the standard at 0.12 ppm was well within the range of defensible choices.

On the other hand, it is also possible to argue that a 0.12 ppm standard was unwise. The effects shown below 0.20 ppm, from this viewpoint, are too rare in the population and medically insignificant. The fact that some studies found no effects at 0.20 ppm, even on some relatively ill asthmatics, suggests to this camp that a standard of 0.12 ppm is more stringent than needed to protect any significant fraction of the population from meaningful effects. This conclusion is reinforced by noting the tendency of the Shy Panel to err on the side of safety, the misinterpretation of the risk assessment study, the limits of the epidemiologic studies, and the high costs of actually trying to reach 0.12 ppm. This argument would lead one to favor 0.14 ppm or 0.15 ppm—or perhaps the Eizenstat compromise of 0.12 ppm with four exceedences annually. The claim is that if some severe asthmatics, especially those who had newly moved to high altitudes, find strenuous athletics or heavy manual labor uncomfortable four days a year instead of one, this is a price worth paying to save 10 billion dollars.

One could argue, however, that Costle was too lax. From this viewpoint, the statistical insignificance of some of the harmful effects did not matter since false negative risks matter more for standard setting purposes than they do for strictly scientific purposes. Given the difficulties of doing epidemiological research, the lack of more definite findings at lower concentration levels was only to be expected. Furthermore, the Clean Air Act takes a clear stand on the matter of cost considerations, and Costle should have been even more categorical in his rejection of them.

In our view, simply focusing on the one hour standard, with one allowable exceedence, was an oversimplification driven by the need to make the issue so simple that even nonexperts could bargain about it easily. Eizenstat opened up a more general set of issues that deserved more consideration: the relative importance of peak and average concentrations, the duration of peaks, the importance of taking interaction effects among pollutants into account, and just who it is that the standard is trying to protect and against what types of harm.

Whatever one makes of the merits, there are reasons to be disappointed with EPA's internal decision making and even more with EPA's explanations. As the first ambient standard to be reviewed under the 1977 Clean Air Act Amendments, ozone provided an important occasion for policy clarification, both inside the government and beyond. It was time for Congress and the environmental community to confront the conceptual weaknesses in the Clean Air Act, and for the public at large to understand what was entailed in deciding on such a standard. By failing to explain what was at stake, EPA short-changed its educative functions.

Nowhere did the EPA explain how it decided which effects of ozone, and on whom, were to count. It was not even clearly stated that the agency had those decisions to make, beyond some not very illuminating remarks in the *Federal Register* that these choices required "judgment" and "prudence." Technical points, too, were not always well handled. The treatment of the risk assessment exercise and of the Japanese studies in particular left much to be desired. Problems like sample size, significance levels, and study design were not openly addressed.

Most importantly, the important strategic considerations raised by all of these particulars went resolutely unacknowledged. The relevant conceptual, empirical, and ethical difficulties simply were not discussed publicly in ways that improved the citizenry's grasp of these matters.

What lay behind the murkiness of EPA's external communications? Perhaps Costle was afraid that being open about the inadequacies of the studies, and about the unavoidably large role of values in the decision, would undermine EPA's scientific or political credibility. Perhaps he was afraid that candor implied risk for the agency's relationship with Congress, and especially with Senator Muskie. Greater explicitness also seems to have been resisted by the EPA's lawyers. In rulemaking, after all, the language of the public record is controlled by lawyers, who are likely to interpret their responsibility defensively.

For whatever reason, the written pronouncements of the EPA actually became less explicit and informative as it went from *Proposed* to *Final Rulemaking*. Instead of pointing out the difficulties lurking in the words "sensitive group" and "adequate margin of safety," its statements tended to obscure these points. EPA thus repeatedly reinforced the view that safety was both possible and desirable, something that medical experts could define without regard to social and economic concerns.

Beyond all this, any attempt at being more explicit might have exacerbated EPA's internal struggles, which no administrator would have wanted to rehearse in public. Moreover, to have overridden the prudent calculations of rulemaking-as-usual, the administrator would have needed his senior aides to help him see his role in that way. Instead, Costle's appointees, as he had intended they should, helped him strike a balance and play the pluralist game.

Bureaucratic combat, however, proved to be a poor way to illuminate difficult and interlinked technical and policy choices. Too many agency participants seemed to view their responsibilities in terms of overcoming internal opponents rather than learning from them, which surely was not conducive to deliberation as we defined it in the first chapter. As a result the agency did not learn as much as it could have from the rulemaking process. It produced compromise more than understanding; agreement, but not responsible choice. And, when it had to explain these issues to the rest of the government or to the public, EPA did not frame the debate in ways that highlighted the key policy choices, namely: who to protect, from what, and at what cost.

Clarifying these matters would have required the agency to challenge some of the major myths and simplifications that had, and have, dominated environmental politics. What we regret in the ozone rulemaking is not the specific standard chosen, but the failure to make explicit issues that neither Congress nor the public adequately appreciated.

We do not mean that EPA should have thrown up its hands and told Congress that the statute was impossible to implement. The agency had a standard to set, and the country needed EPA to do the best job it could—being as honest as possible about how it had interpreted incoherent statutory language and dealt with technical uncertainties. The burden would then have been on Congress and the courts to object to the agency's decisions. But that is always the case. Furthermore, we think the EPA could have been politically successful if it had chosen to face these difficulties squarely. The country very much needed leadership in this highly complex arena, and only the agency had the expertise and the credibility to exercise that leadership effectively. Such leadership would have made possible a more deliberative and integrative process internally, and in turn a more deliberative democratic politics externally.

In making this suggestion, we are under no illusion that it will be easy to overcome the simplifications the public is accustomed to using. As we described in Chapter 1, because categories organize experience and make it meaningful, people come to see them not merely as convenient filing boxes for sorting phenomena, but as descriptions of the way the world *is*. When experience does not conform to those categories, individuals can easily become frustrated. Consider the dilemma confronting an environmental activist who used Zelac's results to support a "no threshold" damage function, and hence the 0.08 ppm standard, and then had to explain just how low EPA should go in setting the standard. If Senator Muskie could feel trapped and baffled by this complexity, what about ordinary citizens?

Even the most sophisticated quantitative models that scientists construct suffer from imperfection and oversimplification. Like all attempts to make sense of experience, such analytical structures are surpassed by experience itself. Thus, using any set of categories or theories to understand actual phenomena requires some combination of judgment and intuition to appreciate when and where experience departs in important ways from theoretical formulations. The relevant skills are similar in some ways to those exercised by a traditional craftsman, an outstanding quarterback, or a good cook. Experts and scientists rely heavily on such "tacit knowledge" to supplement formal theoretical arguments. The need to make such judgements is what the SAB Subcommittee was struggling with when it tried to decide what physiological impacts were "medically significant" or how to reconcile apparently conflicting studies.

This same problem of translating over-simplified generalizations into practice applies as well to our ethical ideas. Those ideas, at any one moment, are usually both complex and incomplete. They contain combinations of fundamental goals ("promote democratic government"), more detailed commitments ("obey the law"), and specific rules ("stop at red lights"). Since such norms, acquired by experience, are seldom fully articulated, individuals typically subscribe to rules and principles that are inconsistent with each other. As a result, ethical judgments in any given situation tend to be rich in ambiguity and subject to competing and conflicting principles.

Given the complexity and inconsistency of ethical systems, each individual can see the same problem from more than one perspective. Should people view themselves as cooperators or rivals? Do they have obligations to one another as a function of kinship or citizenship, or are they self-sufficient individuals who just happen to live in the same jurisdiction? When considering the ozone standard, should each of us focus on various private concerns? Do I, or people I love, have asthma? Do I work for an automobile company? Alternatively, should we ask about our obligations as citizens to contribute our share to protect the health, and hence the opportunity for self- development, of our fellows?

The late E.E. Schattschneider, a political scientist, used to say that the people speak only when spoken to. Thus we must ask, what questions did EPA ask of citizens and legislators on ozone, and was their ethical and technical sophistication elevated by the resulting debate?

Formulating questions effectively is pivotal because the process of considering them is as important as the answers that are reached. As the range of our experience increases, judgment comes to be better grounded. Ethical and political principles reveal their actual meaning through their application to concrete circumstances. Just how much should society spend to protect those with sensitive respiratory systems and how frequently should they be allowed to suffer limitations on their activities, discomfort, or even more serious symptoms? Is cost per life saved a, or the, relevant measure? Are there rights here, and if so how could we know what they are? Only the experience of debating such decisions gives meaning to broad, but general, principles of reciprocity, solidarity, and obligation and helps us discover what our values actually are.

Unfortunately, in the course of such debates, some technical experts will claim more authority than they are legitimately entitled to. Experts, too, may enjoy exercising power and influence, and as citizens they may (like Carl Shy) feel impelled to participate in the dispute over policy. After all, the line between science and policy is often unclear and, not infrequently, the parties in conflict eagerly make use of one or another expert's credibility. Indeed, decision makers can easily invite experts to "over reach," as this story illustrates.

This risk only reinforces the need for managers to manage the experts in ways that enhance their own political accountability. The less clear the questions that experts are asked, the more they will tend to (indeed have to) rely on their own interpretations and definitions in formulating answers. And those interpretations and definitions are powerful vehicles for injecting personal views and values into an ostensibly technical analysis.

When this happens repeatedly, society runs the risk of discrediting all expertise and reducing all technical analysis to the status of partisan opinion, leaving no common ground for understanding options and consequences. Science has sometimes instilled remarkable deference. But the democratic viewpoint can also yield skepticism about specialized knowledge, making experts vulnerable to envy, anxiety, and resentment that must be constrained if democratic government is to avail itself of whatever technical competence does exist. Devising ways to use expertise to facilitate democratic processes thus should be one of our continuing concerns.

NOTES

1. This chapter makes extensive use of "Implementing the Clean Air Act: The Environmental Protection Agency's Revision of the National Ambient Air Quality Standard for Ozone," Valle Nazar, Masters Thesis, Harvard University School of Public Health, July 1982.

2. Clean Air Act as amended 1977 (84 STAT 1676) 109 (b) (1).

3. Ibid. 1102

4. *A Legislative History of the Clean Air Amendments of 1970*, prepared for the Senate Committee on Public Works, 93rd Congress, 2nd Session, 1974, 410. See also R. Shep Melnick, *Regulation and the Courts* (Washington, DC: Brookings Institution, 1983), 244–245.

5. For these reasons and because it enabled the Agency to dismiss a number of arguments from city governments that their mixture of photochemical oxidants was unique and required a unique standard, it was changed to an 'ozone' standard.

6. C.E. Schoettlin and E. Landau, "Air Pollution and Asthmatic Attacks in the Los Angeles Area." *Public Health Reports* 76 (Jun 1961):545–548.

7. National Academy of Sciences, "Air Quality and Automobile Emission Control," prepared for Senate Committee on Public Works, 93rd Congress, 2nd Session, 1974, Vol 1, 325.

8. *BNA Environmental Reporter*, 16 Jan 1976, 1584.

9. Letter from Frank Ikard, President, American Petroleum Institute, to Russell Train, EPA Administrator, 9 Dec 1976. Since the Clean Air Act Amendments of 1977, these standards must be reviewed every five years. See PL 95–96–106.

10. *Fortune*, May 1977, 295–296.

11. *Fortune*, May 1977, 295–296.

12. Douglas Costle, interview with Marc Roberts, Cambridge MA, 31 July 1981.

13. *National Journal*, 12 Mar. 1977, 382–384.

14. *Washington Post*, 12 Apr. 1977, A4.

15. *Washington Post*, 1 June 1979, A2.

16. On RARG, see Susan J. Tolchin, "Presidential Power and the Politics of RARG," *Regulation*, July/Aug. 1979, 44–49.

17. The Clean Air Act PL 91–904, Section 108(a)2.

18. Lester Grant, Criteria and Special Studies Office, Office of Air Quality, Planning

and Standards, EPA, telephone interview with Valle Nazar, Research Triangle Park, NC, 11 June 1982.

19. See The Environmental Research, Development and Demonstration Authorization Act of 1978 (PL 95–155; 8 Nov. 1977) section 8 (91 Stat. 1260)

20. Members of the Subcommittee, beside Dr. Whittenberger, were Eileen Brennan, Plant Pathologist, Rutgers; Edward F. Ferrand, Assistant Commissioner for Science and Technology, New York City Department of Air Resources; Sheldon K. Friedlander, Professor of Chemical and Environmental Health Engineering at California Institute of Technology: Jimmye S. Hillman, Agricultural Economist, University of Arizona; William K. Kellogg, Scientist, National Center for Atmospheric Research, Boulder, Colorado; James N. Pitts, Chemist, University of California at Riverside. Consultants were Dr. Robert Frank, Professor of Environmental Health, University of Washington; Dr. James McCarroll, Clinician in Preventive and Community Medicine at Electric Power Research Institute, and Dr. Sheldon D. Murphy, Toxicologist, University of Texas Medical School. The plant specialists were to review material for the national secondary ambient ozone standard, which would protect the public welfare. J. Whittenberger, interview with Marc Roberts, Harvard School of Public Health, Boston MA, 15 Sept. 1981.

21. Joseph Padgett, telephone interview with Marc Roberts, 16 Sept. 1981.

22. Robert Rauch, Environmental Defense Fund, staff attorney, interview with Marc Roberts, Washington D.C., 14 Sept. 1981.

23. Padgett interview.

24. Other members of the Shy Panel were Stephen M. Ayres, St. Louis University School of Medicine; David V. Bates, University of British Columbia; T. Timothy Crocker, University of California College of Medicine at Irvine; Bernard D. Goldstein, New York University School of Medicine; and John R. Goldsmith, California State Department of Public Health. Dr. Goldstein had been retained by American Petroleum Institute to assist in the preparation of its December 1976 petition for rule making. All of these individuals were consultants to the preparation of the *Criteria Document*.

25. Transcript, *Hearing on EPA's Proposed Ozone Standard*, Washington D.C., 18 July 1978.

26. *Summary Statement from the EPA Advisory Panel on Health Effects of Photochemical Oxidants,* prepared for the Environmental Protection Agency under the supervision of the Institute for Environmental Studies of the University of North Carolina at Chapel Hill, January 1978, 18. Hereafter referred to as *Summary Statement*.

27. 42 FR 42372 (23 Aug. 1977). Hereafter referred to as the First Draft of the Criteria Document

28. Transcript, *Meeting of USEPA's Science Advisory Board*, 10–11 Nov. 1977, 1–9. (Hereafter referred to as *SAB Meeting*).

29. *SAB Meeting*, 1–6.

30. *SAB Meeting*, 1–100.

31. *SAB Meeting*, 1–106.

32. *SAB Meeting*, 106.

33. *SAB Meeting*, 107.

34. *SAB Meeting*, 131.

35. *SAB Meeting*, 131.

36. 42 FR 65264 (30 Dec 1977), col. 2.

37. Whittenberger, 15 Sept. 1981.

38. Stephen Gage, Assistant Administrator for Research and Development, telephone interview with Valle Nazar, Washington D.C., 3 June 1982.

39. Gage telephone interview, 3 June 1982.

40. Costle interview, 31 July 1981.

41. Costle interview, 31 July 1981.

42. "A Method for Assessing the Health Risks Associated with Alternative Air Quality Standards for Photochemical Oxidants," External Review Draft, Strategies and Air Standards Division, Office of Air Quality Planning and Standards, USEPA, Research Triangle Park, North Carolina, 6 Jan. 1978. The final version was released by EPA in July of 1978.

43. Those interviewed were: Drs. David Bates (Clinician, Shy Panel), Jack Hackney (Clinician), Robert Carroll (Epidemiologist), Robert Chapman (Epidemiologist, EPA), Timothy Crocker (Toxicologist, Shy Panel), Richard Erlich (Toxicologist), Bernard Goldstein (Toxicologist, Shy Panel), Steven Horvath (Clinician), and Carl Shy (Epidemiologist, Shy Panel).

44. "Review of 'A Method of Assessing the Health Risks Associated with Alternative Air Quality Standards for Ozone,' A report of the Subcommittee on Health Risk Assessment," SAB, Sept. 1979.

45. EPA, "Revisions to the National Ambient Air Quality Standards for Photochemical Oxidants: Final Rule Making," (8 Feb. 1979) 44 FR 8210. Hereafter referred to as *Final Rulemaking*.

46. Gordon Hueter, Associate Director, Health Effects Laboratory, EPA, telephone interview with Valle Nazar, Research Triangle Park, NC, 11 June 1982.

47. Whittenberger interview, 15 Sept. 1981.

48. Drs. Frank and Pitts urged another review of the entire document. Dr. Murphy wanted some additional discussion of risk. See EPA, Respondent's Brief, *API v Costle*, 23.

49. *Science* 202 (1 Dec. 1978): 949–950.

50. Letter from J. Whittenberger, to Robert Chapman, Health Effects Laboratory, EPA at Research Triangle Park, NC, 8 Apr. 1978.

51. Whittenberger interview, 15 Sept. 1981.

52. Gage telephone interview, 3 June 1982.

53. See testimony offered by NAS as cited in Report of the House Interstate and Foreign Commerce Committee, May 12, 1977, House Report No. 95–294, May 12 1977(to accompany H.R. 6161), 110.

54. A.J. DeLucia, and W.C. Adams, "Effects of O Inhalation During Exercise on Pulmonary Function and Blood Biochemistry," *Journal of Applied Physiology: Respiratory Environmental Exercise Physiology* 43 (1977): 75–81.

55. Environmental Criteria and Assessment Office, USEPA, *Air Quality for Ozone and Other Photochemical Oxidants,* Apr. 1978, 6. Hereafter referred to as *Criteria Document*.

56. G. Von Nieding, "The Acute Effects of Ozone on the Pulmonary Function of Man," *VDI Berichte* 270 (1977): 123–129.

57. First Draft of *Criteria Document*, 9–11.

58. Statement of Dr. Neil K. Weaver, the Director of Medicine and Biological Science Department of American Petroleum Institute, *SAB Meeting*. EPA Docket File OAQPS 78–8–IIA-F-1–C.

59. *Criteria Document*, 192.

60. *Final Rule Making*, 8207.

61. W. Linn, et al, "Health Effects of Exposure in Asthmatics," *American Review of Respiratory Disease*, 117 (1978): 835–841.

62. Blood chemistry measures included integrity of erythrocytes, hemoglobin and other oxidizable substances, and activity of enzymes necessary for the maintenance of normal intracellular redox potentials, including acetylcholinesterase.

63. Jack D. Hackney, "Experimental Studies on Human Health Effects of Air Pollutants," *Archives of Environmental Health* 30 (Aug. 1975):373–381.

64. *API v. Costle*, U.S. Court of Appeals for the District of Columbia, No. 79–1104, at V.

65. J. Kagawa and T. Toyama, "Photochemical Air Pollution: Its Effects on Respiratory Function of Elementary School Children," *Archives of Environmental Health* 30 (Mar. 1975):305–320.

66. J. Kagawa, et al, "Pulmonary Function Test in Children Exposed to Air Pollution," in A.J. Finkel and W.C. Duel, eds., *Clinical Implications of Air Pollution Research* (Acton, MA: Publishing Sciences Group, Inc., 1976), 305–320.

67. I. Mizoguchi, et al, "On the Relationship of Subjective Symptoms to Photochemical Oxidant," in I.B. Dimitriades, ed., *International Conference on Photochemical Oxidant Pollution and its Control*, Proceedings, Vol. I, EPA 600/3–77–001a, Jan 1977: 477–494.

68. *SAB Meeting*, II-73.

69. *Air Quality for Ozone and Other Photochemical Oxidants,* Environmental Criteria and Assessment Office(MD-52), Research Triangle Park NC 27711, Apr. 1978, 8.

70. David Hawkins, Assistant Administrator for Air and Radiation, joint interview, 31 July 1981.

71. W.S. Wayne, et al, "Oxidant Air Pollution and Athletic Performance," *Journal of the American Medical Association* 199 (20 Mar. 1967):151–154.

72. C.E. Schoettlin and E. Landau, "Air Pollution and Asthmatic Attacks in the Los Angeles Area," *Public Health Reports* 76 (June 1961): 545–548.

73. W.H. Durham, "Air Pollution and Student Health," *Archives of Environmental Health* 28 (May 1974):241.

74. D.L. Coffin, et al, "Influence of Ozone on Pulmonary Cells," *Archives of Environmental Health* 16 (1968):633–636.

75. *Criteria Document*, 10.

76. R. Brinkman, et al, "Radiomimetic Toxicity of Ozonized Air," *The Lancet*, 18 Jan. 1964, 133–136.

77. R.E. Zelac, et al, "Inhaled Ozone as a Mutagen," *Environmental Research* 4 (1971): 262–277.

78. T. Merz, et al, "Observation as of Aberrations in Chromosomes of Lymphocytes from Human Subjects Exposed to Ozone at a Concentration of 0.5 ppm for Six to Ten Hours," *Mutagen Research* 3, 975:299–302.

79. Science Advisory Board Executive Committee Subcommittee on Scientific Criteria for Photochemical Oxidants, 10–11 November 1977, "Summary of Comments on Proposed Revisions of Health Effects Appraisal." EPA docket file OAQPS 78–08–IIA-F-1–d.

80. *Criteria Document*, 11.

81. Memo from M. Jones, Strategies and Air Standards Division, Office of Air Quality, Planning and Standards, EPA to G. Hueter, 13 Dec. 1977. Copies were sent to Robert Chapman and Joseph Padgett, among others.

82. Hawkins interview, 31 July 1981.

83. Walter Barber, Director, Office of Air Quality, Planning and Standards, interview with Valle Nazar, telephone, 11 June 1982.

84. William Drayton, Assistant Administrator for Planning and Management, interview with Marc Roberts, Washington DC, 17 Sept. 1981.

85. Costle interview, 31 July 1981.

86. For an account of the press conference, see *BNA Environmental Reporter*, 13 June 1978, 235–236.

87. "EPA Proposed Rule on Air Quality Standards, Changes in Terminology Forms," (22 June 1978) 43 FR 26962–26985. Hereafter referred to as *Proposed Rulemaking*.

88. *Science* 202 (1 Dec. 1978): 949–950.

89. Ibid.
90. *Proposed Rulemaking*, 26962.
91. Ibid., 26964.
92. Ibid., 26965.
93. Ibid., 26966.
94. Ibid., 26966–67.
95. *Summary Statement*, 18.
96. Statement by Robert Rauch, Staff Attorney, Environmental Defense Fund, from a transcript of USEPA public hearing on Proposed Ozone Rulemaking in Washington D.C., 18 July 1978.
97. Statement by Carl Shy, public hearing on Proposed Ozone Rulemaking.
98. Statement by Axel T. Mattson, public hearing on Proposed Ozone Rulemaking.
99. Malcolm Hawk, API, public hearing on Proposed Ozone Rulemaking.
100. Tolchin, "Presidential Power and the Politics of RARG," 44–49.
101. Robert Litan, former staff member CWPS, interview with Marc Roberts, Washington D.C., 1 Oct. 1981.
102. Tolchin, "Presidential Power and the Politics of RARG," 48. See also Lawrence White, *Reforming Regulation* (Englewood Cliffs, NJ: Prentice Hall, 1981).
103. The official making this statement requested anonymity.
104. For accounts of the cotton dust controversy, see Robert E. Litan and William D. Nordhaus, *Reforming Federal Regulation* (New Haven, CT: Yale University Press, 1981), 59–76; and Christopher DeMuth, "Constraining Regulatory Costs - Part I: The White House Programs," *Regulation*, Jan./Feb. 1980, 13–26.
105. Hawkins interview, 31 July 1981.
106. Litan interview, 1 Oct. 1981.
107. Costle interview, 31 July 1981.
108. "Environmental Protection Agency's Proposed Revisions to the National Ambient Air Quality Standard for Photochemical Oxidants," Report of the Regulatory Analysis Review Group, Submitted to EPA by the Council on Wage and Price Stability, 16 Oct. 1978. Hereafter referred to as *RARG Report.*
109. *RARG Report*, 4.
110. *RARG Report*, 6–12.
111. *RARG Report*, 16–23.
112. The RARG Report estimates that the marginal cost of moving from a 0.12 ppm standard to a 0.10 ppm standard is $1.9 to $3.8 billion per year.
113. *RARG Report*, 23–31.
114. "October 10th Regulatory Analysis Review Group Meeting on the Ozone Air Quality Standard," memo for the files by Mike Jones, Pollution Strategies Branch, USEPA, 26 Oct. 1978. Hereafter referred to as *RARG Oct. 10 Meeting*. EPA Docket File OAQPS 78–8–IV-E-7.
115. "Council on Wage and Price Stability/Regulatory Analysis Review Groups Critique of the Proposed Ozone Ambient Air Quality Standard," memo from Walter Barber, Director, Office of Air Quality Planning and Standards, USEPA, to David Hawkins, Assistant Administrator for Air, Noise and Radiation, USEPA, 8 Nov 1978. EPA Docket File OAQPS 78–8–IV-C-3.
116. Hawkins interview, 31 July 1981.
117. Gilbert Omenn, interview with Marc Roberts, Washington D.C., 15 Sept. 1981.
118. Dr. James McCarroll, consultant to SAB Subcommittee on Photochemical Oxidants, telephone interview with Valle Nazar, 25 June 1982.
119. Omenn interview, 15 Sept. 1981.
120. *Final Rule Making*, 8207–8213.

121. *Final Rule Making*, 8207.

122. *Final Rule Making*, 8206.

123. Drayton, 17 Sept. 1981.

124. Hawkins interview, 31 July 1981.

125. Letter from Douglas M. Costle, Administrator, USEPA, to James McIntyre, Director, Office of Management and Budget, 13 Dec 1978. EPA Docket Number OAQPS 78–8–IV-E-5.

126. Memo, Hawkins to Drayton, 25 Jan. 1979.

127. Costle interview, 31 July 1981.

128. Richard Dowd, former Secretary to the SAB, interview with Marc Roberts, Washington DC, 15 Sept. 1981.

129. Stephen Gage, Assistant Administrator for Research and Development, interview with Marc Roberts, Washington DC, 12 Dec. 1981.

130. Costle interview, 31 July 1981.

131. Gage interview, 12 Dec. 1981.

132. Costle interview, 31 July 1981.

133. Kenneth Lloyd, Strategies and Air Standard Division, Office of Air Quality, Planning and Standards, "Meeting with the Executive Office of the President to Discuss the Ozone Standard," memo of the 21 Dec. 1978 meeting, 3 Jan. 1979. EPA Docket Number OAQPS 78–8–IV-E-2.

134. *RARG Oct. 10 Meeting*.

135. Litan interview, 1 Oct. 1981.

136. Costle interview, 31 July 1981.

137. Costle interview, 31 July 1981.

138. Memo from Costle to Hawkins, 8 Jan. 1979. Dowd talked to Frank and Murphy sometime during the week of January 8. It is unclear whether or not he called them prior to the writing of Costle's memo. However, they were called after Costle apparently made his decision on Friday, January 5, 1979. EPA Docket File OAQPS 78–8–IV-E-4.

139. Costle interview, 31 July 1981.

140. Memo from Blum to Eizenstat, 9 Jan. 1979, Central Docket Section, USEPA.

141. Charles Schultze, interview with Marc Roberts, Washington D.C., 2 Oct. 1981.

142. Gilbert Omenn interview, 15 Sept. 1981.

143. Costle interview, 31 July 1981.

144. fCostle interview, 31 July 1981.

145. "Oral Discussions with Members of the Science Advisory Board and Other Scientists on Ozone," memo to the files from Richard Dowd, received 26 Jan. 1979, EPA Docket File OAQPS 78–8–IV-1.

146. Costle interview, 31 July 1981; Simon Lazarus, former staff member Domestic Policy Council, interview with Marc Roberts, Washington DC, 15 Sept. 1981.

147. "Interpreting Health Effects Data under the Clean Air Act," memo from Costle to Eizenstat, 23 Jan 1979, EPA Docket File OAQPS 78–8–IV-C-4.

148. At a February 27 press conference, President Carter pledged not to undermine environmental laws through his policies to fight inflation and reduce the burden of federal regulation. "I have not interfered in the process" the President said. "I have a statutory responsibility and a right to do so, but I think that it would be a very rare occasion when I would want to do so." Carter said that Costle had the "authority to administer the law in the most effective way." See the *National Journal*, 3 Mar. 1979, 364.

149. *BNA Environmental Reporter*, 2 Feb. 1979, 1813.

150. *The Wall Street Journal*, 24 Jan. 1979.

151. *BNA Environmental Reporter*, 2 Feb. 1979, 1813.

152. *BNA Environmental Reporter*, 2 Feb. 1979, 1813.

153. *BNA Environmental Reporter*, 2 Feb. 1979, 1813.
154. *BNA Environmental Reporter*, 2 Feb. 1979, 1813.
155. *Final Rule Making*, 8203.
156. *Final Rule Making*, 8203.
157. *Final Rule Making*, 8217.
158. *Final Rule Making*, 8217.
159. *Final Rule Making*, 8217.
160. *Final Rule Making*, 8209.
161. *Final Rule Making*, 8209.
162. Litan interview, 1 Oct. 1981.
163. Rauch interview, 14 Sept. 1981.
164. *BNA Environmental Reporter*, 26 Jan. 1979, 1767–1768.
165. *New York Times*, 27 Feb. 1979, 1.
166. Senate Committee on Public Works, Subcommittee on Environmental Pollution, *Executive Branch Review of Environmental Regulations*, 96th Congress, 1st Session, 26–27 Feb. 1979, Committee Serial No. 96–H4, 329–349. Hereafter referred to as *Senate Hearings*.
167. *Senate Hearings*, 343.
168. *Senate Hearings*, 243.
169. *Senate Hearings*, 244.
170. *Senate Hearings*, 250.
171. *Senate Hearings*, 252–253.
172. *Senate Hearings*, 263.
173. *API v Costle*, U.S. District Court of Appeals for the district of Columbia, No. 79–1104, at V.
174. *National Journal* 15 Sept. 1979, 1526–8.
175. API v. Costle, U.S. Court of Appeals for the District of Columbia, No. 79–1104, at V.
176. *Environmental Law Reporter* 9 (1979):20753–54.

4

Writing the Resource Conservation and Recovery Act Regulations

We turn now to the long, tortuous, and at times numbingly repetitious story of the Environmental Protection Agency's efforts to develop regulations under RCRA (pronounced "rick-ra"), the 1976 Resource Conservation and Recovery Act.[1] This tale bears a distant resemblance to those "Pop Art" paintings that used repeated, if slightly varied, images (like Andy Warhol's Campbell soup cans) to make us see familiar objects in a new light. The narrative of RCRA rulemaking, in all its intricate detail, has much to teach us about the current functioning of the American government.

The first part of our story shows how, under the right circumstances, a few individuals can control the legislative process in Congress. When an emerging issue is neither well nor widely understood, the apparent "experts" on a subcommittee and its staff can be left to function in splendid isolation. This was the case in 1976 with hazardous waste .

Those experts however were not very expert. In particular they lacked a clear vision of the "problem" that they wanted EPA to solve. As a result the agency was instructed to establish an exceedingly ambitious program with little guidance about how that was to be done. By also specifying an utterly unrealistic time schedule, Congress put EPA in a classic "no win" situation.

In the previous chapter we saw how difficult it was for Congress to constrain an agency even with regard to a relatively simple decision such as an air quality standard. The irreducible ambiguity of language and the legislature's desire to avoid unpleasant truths lead to a situation in which the executive retained substantial power to make policy through the process of implementation. In the case of RCRA, EPA's discretion was truly enormous. It produced hundreds of pages of regulations, embodying dozens of significant policy choices, all on the basis of the most elliptical statutory language and the sparsest of legislative records.

To do that job effectively, the agency needed a clear formulation of the problems it wanted to solve and an explicit strategy to deal with them. This would have communicated to everyone, inside and outside EPA, how decisions were being and should be made. RCRA was simultaneously about many things. It was about residuals management—that is, dealing with the wastes produced by existing treatment systems.[2] It was about evaluating the hazards produced by different kinds of wastes and determining what constituted "safe" disposal for each of them. It was about creating new disposal sites in the face of community opposition, and about getting waste sources to use them despite their higher costs. And, it was, at least potentially, about finding ways to reduce the volume of hazardous waste in the first place. In the face of such complexity, only an explicit strategy could have produced an acceptable combination of internal coordination and external accountability. Yet the agency failed to produce such a strategy in the four long years it struggled with RCRA.

This failure was a product of both limited knowledge and poor management. No one knew very much, and too few resources of the wrong kind, and with too little supervision, were devoted to the task. Fragmentation within the agency made it difficult to marshall the personnel and insights needed to design an adequate program. Only quite unusual administrative centralization ultimately made it possible for EPA to issue regulations at all, and then only after extensive conflict and turmoil.

As months of delay beyond statutory deadlines turned into years, neither Congress nor the Courts could overcome EPA's own managerial weaknesses. The Congress was more concerned with specific sites and local interests than with the big picture. It looked for scandal, and it gave aid and comfort to some dubious whistle blowers, while avoiding any constructive dialogue with EPA over central policy choices. The courts were not much help either. They could carp and complain, but they could not produce a regulatory program. As part of the process of judicial review, the requirements of the "new administrative law" were a mixed blessing—forcing some useful outside review of agency actions while also diverting scarce resources into socially unproductive, defensive document writing. In the end, these requirements, like the agency's own internal routines, were, perhaps wisely, at least partially subverted to get the job done.

The regulations that emerged were not always well crafted. Those who approved licenses for new facilities were given so much discretion as to undermine political control over those decisions. Siting new facilities remains difficult because of public mistrust that the regulations actually insured safety, while existing sites were allowed to continue to function under relatively lax interim licenses. Insofar as hazardous waste posed health hazards when RCRA was first passed, the first four years of activity under the Act may not have done much to remedy that situation.

The EPA did not fare much better from an educational perspective. The public still does not want to hear that these questions are difficult, that our knowledge is limited, or that the very notion of "safety" is an oversimplification that must be grounded in political, not just technical, judgments.[3] Many citizens view such arguments as industry propaganda or bureaucratic self-justifications for incompetence. No explicit analysis of the risks of hazardous waste, and no coherent account of the agency's strategy for dealing with them, have ever been produced by EPA. The public remains uncomprehending and anxious, never having been clearly told what there is and is not to fear. This was perhaps the most serious shortcoming of all of the RCRA rule writing process.

LEGISLATIVE ORIGINS

The environmental revolution of the 1960s had left solid waste largely untouched. The problem was viewed as a matter of materials handling, routing garbage trucks, and bulldozing landfills. As such, it was viewed as the province of local governments. Those who bothered to consider the associated health hazards generally thought about insects, rodents, and infectious disease. In 1970, however, the army began transporting ammunition and nerve gas from depots in the west to Florida for disposal at sea.[4] Congress, concerned about where to put such materials, gave The Department of Health Education and Welfare (HEW) two years to study the feasibility of creating a national system of sites for the disposal of hazardous substances.[5]

The study, which ultimately took three years to complete, found that safe, if expensive, methods for the disposal of hazardous waste were available. It also concluded that these methods would never be used without the prohibition of less safe and less expensive alternatives. An effective national program to regulate hazardous waste dumps, the report argued, would obviate the need for a national disposal site system. The report recommended establishing such a regulatory program and included as an appendix a draft statute that significantly presages parts of RCRA.[6]

At about the same time, in 1972, others also became interested. Roger Strelow, on the staff of the Council for Environmental Quality, became concerned about where the wastes generated by air and water pollution control facilities would wind up.[7] He drafted a legislative package and, after a good deal of conflict within the administration, a revised form of Strelow's proposal was sent up to the Hill. Jennings Randolph then co-sponsored it as a Senate bill.

On the House side, interest developed in the Transportation Subcommittee of the House Commerce Committee. The chairman, Fred Rooney, was from Bethlehem, Pennsylvania, home of the steel company of the same name. The chief counsel, William Kovacs, recalls that industry began lobbying for a uniform national system to avoid having to deal with fifty different state programs. (California had just adopted the nation's first statewide hazardous waste regulatory program.)[8]

Over the next two and a half years congressional action was sporadic. Committee hearings focused on recycling, using waste for power, or federal "bottle bills." The newly formed EPA, however, continued to worry in a minor way about the health issue.[9] In June 1975 the Office of Solid Waste (OSW) began issuing "Hazardous Waste Site Damage Reports." The first of these described an incident of arsenic poisoning that made eleven people sick, another in which three head of cattle died, and a third that caused burns on the skin of a disposal site bulldozer operator.[10]

Even at this time, there were different views about how to deal with hazardous waste. In March 1974, Russell Train, the EPA administrator, told the House Commerce Committee that states, not the federal government, should do the regulating.[11] In 1975, John Lehman, a division director in OSW, urged a national "cradle to grave" manifest system.[12] Still, interest in the question was limited. Solid waste was not high on the agenda of most environmental groups, apart from the then fashionable "limits to growth" argument and the question of materials recycling.

Toward the end of 1975, Congress considered the Toxic Substances Control Act (TSCA), which regulated the introduction of new hazardous chemicals. As that discussion continued, members began to ask where such materials, once introduced, would eventually wind up. On March 4, in a letter to OMB Director James Lynn, Train again suggested relying on state programs operated in compliance with EPA criteria.[13] But the Senate was not eager to give the states such responsibility. In June, it approved a comprehensive Resource Conservation and Recovery Act, including a hazardous waste management provision.[14] Although the language differed in detail from the version that later became law, the Senate bill provided for an equivalent national regulatory system.

Meanwhile, on the House side, subcommittee hearings in April and June seemed to be producing a consensus. But in July a dispute over funding appeared to have killed the bill.[15] . Once the Senate had acted, however, Rooney's subcommittee compromised its differences. The subcommittee's proposal on hazardous waste was approved by the whole committee on September 9, 1976, which joined it with a House Science Committee bill that dealt with resource recovery and recycling. This was one day before the end-of-session deadline for scheduling bills for debate.[16]

The Rules Committee scheduled RCRA for floor action on September 27. Unfortunately, this left no time before adjournment, on October 1, for a conference to reconcile any interchamber differences. Kovacs and his counterpart on the Senate Public Works Committee, Phil Cummings, decided they had to reach a compromise *before* any House action. They met over the weekend of September 25 and 26 at Cummings' house and did a "cut and paste" job, combining the bill the Senate had already passed and the bill due to come up in the House the following Monday. The version they developed late Sunday evening used the House language on hazardous waste almost in its entirety. The one exception involved omitting the word "reasonably" in those places where the Rooney draft had called for standards "as may be necessary to reasonably protect human health and the environment."[17]

The new bill, in the form of a typescript with handwritten corrections, was presented to the House the next day as a substitute.[18] It seems unlikely that many members of the House read it before the vote. Nonetheless, it was approved by the House on September 27 and by the Senate on September 30. The Congress adjourned the next day and President Ford signed the bill into law on October 21, 1976.[19]

Senior EPA officials later characterized RCRA as having "popped up" at the last minute and surprised the agency.[20] That may well have been true of top management, which was preoccupied with other legislative priorities. But at the working level, EPA people were involved. Kovacs remembers that throughout 1976 he had conversations with people in OSW, including William Sanjor and Hugh Kaufman who will be referred to again later.[21] Various managers in OSW also talked to staff people on the Hill.[22] Kovacs, however, apparently had "great difficulties" getting to anyone at the top.[23] The EPA never formally commented on the various drafts that were sent to them. Elsewhere in the administration only the Department of Defense (DOD) noticed the law. The DOD strongly advocated a presidential veto, arguing that for the first time it authorized state control of federal facilities.

The obvious pressure groups were also relatively uninvolved. Kovacs remembers that there was very little interest, with the exception of some industry people.[24] The environmental community was much more interested in TSCA. On the other hand, until the weekend meeting at Phil Cummings' house, there was little reason for anyone to believe that legislation would emerge so late in the session.

THE RESOURCE CONSERVATION AND RECOVERY ACT OF 1976

Having passed with so little attention, the scope of the program created by RCRA was not widely appreciated at the time, either inside or outside of Congress. Almost no one saw the hazardous waste provisions in subtitle C as the most significant portion of the legislation. Love Canal was two years in the future. Of the sixteen findings of fact that preface the 1976 Act, only one, the ninth, deals with hazardous waste:

> Hazardous waste presents, in addition to the problem associated with non-hazardous solid waste, special dangers to health and requires a greater degree of regulation than does non-hazardous solid waste.[25]

The scheme mandated by Subtitle C was deceptively simple. EPA was to decide what was a hazardous waste.[26] Then, it was to establish standards to be met by all treatment, storage, and disposal facilities (TSDFs) that handled such waste.[27] These were to be enforced by issuing permits only to conforming facilities.[28] Finally, all generators and transporters of hazardous waste had to comply with certain handling requirements as well as fill out manifest forms for all transactions involving hazardous material.[29] This was to insure that such materials were only sent to facilities permitted to receive them. Thus the Act created the kind of system Jack Lehman had urged in 1975. The Act also allowed the administrator to delegate authority to operate the program to any state that created an "equivalent" effort.[30] As we will see the actual problem was far more complex and subtle than the language of the Act suggested.

Many other sections of RCRA also related to hazardous waste in various ways. Some of these included:

> An "imminent hazard" authority which allowed EPA to intervene immediately to abate active hazards which presented "an imminent and substantial endangerment to health and the environment."[31]

> The requirement that state or regional solid waste plans be developed in accord with federal guidelines, with technical assistance and grant support from the Office of

Solid Waste. These provisions, under Subtitle D, included a requirement that the state make an inventory of all open dumps and upgrade them to sanitary landfills.[32]

The creation of a national resource recovery and conservation panel to provide technical assistance to state and local governments.[33]

Application of federal, state and local solid and hazardous waste disposal requirements to all federal facilities.[34]

The provision of grants to fund the purchase of tire shredders to ameliorate the discarded tire problem (inserted at the request of the House Subcommittee Minority Counsel).[35]

The operational content of this regulatory program turned on two major decisions. First, which substances were to be covered? Deciding this involved developing "criteria" as required by Section 3001 and then applying them to the universe of potentially discarded materials to "identify" and "list" those to be treated as hazardous waste. Second, EPA had to write the rules governing TSDFs under Section 3004 to make sure that they could accept such wastes while still protecting "human health and the environment." This determined what actually happened to those wastes that were brought into the system.

As EPA began to grapple with the task of writing the rules, it discovered serious oversimplifications in the fundamental ideas on which the whole scheme was based. The problem lay with the apparently plausible notion that it was possible classify wastes as either hazardous or not hazardous.

Hazardousness is not a dichotomous characteristic. Nor does the hazard posed by a specific waste depend in any simple way on its chemical and physical properties. Instead, there is a multiplicity of possible effects that vary with the dose an organism receives. As toxicologists put it, "The dose makes the poison." Many toxic materials are not injurious in small quantities.[36] Conversely, even relatively benign substances like seawater can be fatal in large doses. Thus, the harm caused by any hazardous waste depends on the circumstances of its disposal. These, in turn, will determine the concentrations in the environment, the doses humans are exposed to, and the resulting effects. Under RCRA this feature of the world is ignored. Decisions on what to regulate are to be made on the basis of the waste itself, without considering its eventual context.

Determining hazardousness is made even more difficult by the enormous complexity and variety of wastes. Many are industrial process wastes that contain different substances in various forms, mixtures, and combinations. The nerve gas that prompted the initial congressionally mandated study was easy to classify. What about mining wastes with trace amounts of heavy metals or large volumes of industrial process wastes that are slightly acidic, or used crankcase oil? How much hazardous material does a waste stream have to contain before the whole mass of material is declared hazardous? If having only low concentrations of hazardous materials in a mixture makes the mixture exempt, sources might be encouraged to dilute their wastes. Without such an exemption, large volumes of wastes containing only trace amounts of hazardous substances could be subject to the rigorous and expensive requirements of the Act.

The definition of a hazardous waste in the Act provides little guidance:

> a solid waste or combination of solid wastes which because of its quantity, concentration, or physical, chemical, or infectious characteristics, may a) cause or significantly contribute to an increase in mortality or an increase in serious, irreversible or incapacitating, reversible illness, or b) poses substantial present or potential hazard to human health or the environment when improperly treated, stored, transported, or disposed of or otherwise managed.[37]

The distinction this language suggests between intrinsically hazardous substances under clause (a) and those that are only hazardous when "improperly" treated under (b) is itself confusing, as the Act presumes that all materials can be made "safe" if dealt with "properly."

Nor is the language in Section 3001, describing what wastes are to be listed, any more helpful. There the administrator is told to define hazardousness by ". . . taking into account toxicity, persistence and degradability in nature, potential for accumulation in tissue and other such factors as flammability, corrosivity and other hazardous characteristics."[38] But how this "taking into account" is to occur is not explained. Yet, it is the key problem.

The original Senate bill only marginally recognized the contextual nature of hazardousness. Under it, the Administrator was instructed to produce:

> . . . guidelines for defining those quantities of a hazardous waste the disposal of which, in consideration of particular locations, circumstances and conditions, are likely to be harmful to the public health or the environment.[39]

Yet, in the end, that insight was ignored and the Senate bill also required production of a simple list of intrinsicly hazardous substances.[40]

The practical difficulties of deciding what was hazardous were as great as the conceptual problems. How, for example, was one to decide on a test for toxicity without some notion of the relevant dose? Furthermore, animal tests both for acute toxicity and carcinogenicity were very expensive. Various less expensive tests under development at the time were based on the effects of chemicals on cells grown in laboratory conditions, but they were (and still are) imperfect indicators of human health risks.[41]

Designing a good test was, therefore, an astonishingly complex and difficult task. The "tighter" the test, that is the more frequently it classified materials as hazardous, the lower the probability that it would let through a substance that in fact posed some danger (i.e., there would be less likelihood of a false negative result). But this made it more likely that the test would mistakenly categorize a harmless material as hazardous. Such false-positive results would then lead to needless treatment costs, false negative results to additional health risks.

Thus, choosing a test would seem to require forecasting the rates at which each alternative would produce both kinds of errors, and the social costs of those errors.[42] But, do we really know enough to predict the doses to humans, and hence the health risks, that undetected hazardous substances are likely to produce if we use various tests? Can we really also predict the volume of, and hence the treatment costs incurred by, needlessly treated material under each option? Given that more expensive tests may well reduce the probability of both kinds of errors, how can we best decide how much to spend on testing in light of our high levels of ignorance about actual consequences? In sum, the technical and policy problems raised by the detailed specification of a toxicity test are truly alarming.

To complicate matters, some of the different characteristics of a waste may interact with each other or with other materials. This would seem to imply that hazardousness can only be determined, if at all, by taking into account the whole configuration of a waste's properties. For example, the significance of a given compound's toxicity will also depend on whether it degrades in nature or conversely is subject to "bioaccumulation."[43]

In contrast to RCRA, both the Clean Air Act and the Clean Water Act created ambient standards that determine the control requirements placed on individual sources. As a result, the treatment required for any given air or water pollutant depends on the state of the receiving ecosystem. To be sure, there are some nationally uniform emissions rules under both acts (e.g., the New Source Performance Standards under the Air Act). But these are justified on the grounds of fairness or administrative convenience as part of a regulatory regime designed to achieve specified ambient conditions.

Without this conceptual structure in RCRA, EPA had difficult decisions to make on, for example, the TSDF rules. What was the ecosystem that was to be protected, and the critical ambient levels to be achieved? The Congress had not only failed to instruct the agency on these points, it had not even formulated the question in this way. Without clearly specified objectives, how could EPA decide if the rules should be uniform or vary with local conditions? Furthermore, what kinds of rules should the agency use to achieve those goals? Should there be performance or design standards? Who would have the burden of proof that any proposed design was adequate? The absence of a clear sense of the harms to be prevented made this enormous tangle of technically arcane issues difficult to resolve.

If RCRA had been a groundwater protection statute, for example, the objective would have been much clearer. The task of writing the rules would still have been technically difficult, since groundwater is hard to monitor and the behavior of contamination within it is not well understood.[44] But at least the agency would have known what it was expected to accomplish.

The issues mentioned thus far hardly exhaust the choices EPA had to make. When was a state program "equivalent" to the federal program? How could one tell if a dump site operator was "financially responsible?" How complex should the manifest forms be, who should file them, and how often? The EPA faced literally dozens of tough decisions, of great substantive and analytical complexity, as it prepared to implement RCRA.

REGULATORY EFFORTS UP TO THE STATUTORY DEADLINE

At first, EPA's senior managers took little interest in the effort to write the RCRA regulations. When, after more than a year, they did look into the matter, they found both a product they did not like and an impending legislative deadline. In response, EPA began to play "catch up," a pattern that persisted for the remaining years of the Carter presidency. The haste, even frenzy, that resulted was often counterproductive. The EPA's difficulties were compounded by problems of inadequate technical background, insufficient staff, and poor management. All of these made the effort to recoup lost ground more difficult. As a result, by the end of Carter's presidency and Costle's tenure, some substantial portions of the RCRA regulations still had not been issued in final form. The failure of the agency to do a better job in the first year and a half cursed the entire history of the effort.

The EPA's managerial problem involved both the quality and the quantity of available personnel. Somehow there was never enough time or enough people to resolve difficult yet important technical, conceptual, and strategic issues. The few individuals who were both competent and knew the Act were too busy to recruit, socialize, and train others in order to expand the team doing the work.

The OSW was far from the most important or sophisticated part of EPA. It had come to EPA from HEW. where, as the Bureau of Solid Waste, it provided technical assistance and grant money to state and local governments. The able young people attracted to EPA in the early 1970s wanted to go where there was action and opportunity. This was not OSW, whose staff declined from 225 in 1973 to 174 in 1976. The hazardous waste group was even smaller. When RCRA was first passed in 1976, there were twenty one people assigned to relevant activities.[45]

To make matters worse, OSW had never produced a regulatory program. Staffed mainly by engineers who were not the most able writers, OSW had little experience producing the carefully drafted preambles and supporting documents that were essential to defend a rule against the inevitable court challenge.[46] Even a long time OSW manager like John Lehman, who argued that OSW "had participated in other regulation writing processes" and so "had a sense of what a good regulation looked like," admitted that OSW "totally underestimated the magnitude of the job of documenting the proposed regulations."[47]

Congress greatly complicated the problem by setting totally unrealistic deadlines. An EPA study had found that the average time for preparing a regulation was twenty six months. In this case, for an exceptionally difficult job, and in the midst of a Presidential transition, the agency was given only eighteen months.[48]

On November 30, 1976, just over one month after the passage of RCRA, John Lehman, as asked, submitted a plan to his superiors for finishing the hazardous waste rule making within the statutory deadline. Lehman said that, in order to do so "all other work must be deferred or canceled." Even so, the schedule was very tight. For example, external review of draft regulations normally took fifteen months. If the regulations were to be out in 18 months however, only nine months could be allowed for public comments and any needed revisions. Since more than one month had already passed, Lehman noted, the proposed rules had to be published in less than eight months.[49] (Not everyone in the EPA at the time understood the situation. At a December 16th public meeting, Deputy Administrator John Quarles commented that, in contrast to the unrealistic deadlines Congress had set elsewhere, the RCRA deadlines were more realistic.)[50]

The second set of problems that plagued the agency during this period were substantive. Unlike other problem areas, there was little state or federal experience to guide the regulation writers. With the exception of the California program, then still quite new, no state in the winter of 1976 was regulating hazardous waste. Furthermore, the underlying knowledge of the problem was appallingly limited. No one, inside or outside EPA, even knew what wastes were being generated or by whom, let alone how they were being disposed of or what hazards they posed.

Consider, for example, the problem of deciding which wastes were hazardous. The EPA might have tried to specify a set of tests, classify any waste that failed them as hazardous, and put the burden of testing on waste sources. Alternatively, EPA could have made the decisions itself and simply issued its own list. Intermediate schemes were also possible. Differences among these options, in terms of private sector cost and program scope, were likely to be enormous.

It is not clear that anyone in OSW in the winter of 1976–1977 appreciated just how difficult these issues were. Even if all other work was "deferred or cancelled," the background understanding of problems and pathways, of risks and remedies, simply was not available. In retrospect, OSW's ignorance of the depths of its own ignorance was a source of many of EPA's early difficulties.

For example, in a 1976 talk at a public meeting held by OSW, John Lehman suggested:

> The identification method in our view would be standard sampling methods and testing methods which would say whether or not the waste meets these criteria.[51]

The problem was however that such "standard sampling methods" simply did not exist.

The OSW pushed ahead despite the leadership problems produced by the transition. (Douglas Costle did not have his confirmation hearing until March 2.) To obtain input from the senior levels of the Agency, OSW formed two groups known as the "red team" and the "blue team" at the division director and the deputy assistant administrator level, respectively.[52]

According to Lehman, these meetings were "crucial." He said, "We needed answers and decisions. We put a lot of staff time into this." Lehman was disappointed with the results. The meetings degenerated into philosophical bull sessions. "We never got any resolution. Everything was amorphous, and I was frustrated as hell." Various outsiders would suggest alternatives, but because they were not worked out in any detail, OSW found them of little use as it struggled to prepare the regulations.[53]

Chris Beck, later the assistant administrator in charge of RCRA, was at the meetings and had a different, though not incompatible, view:

> . . . there was just a complete language breakdown . . . We tried to force them to think strategically about timing, set out clear objectives, to think about resources. I remember a tremendous frustration on the part of everybody, I mean screaming, I mean they were very vituperative meetings.[54]

Gary Dietrich, later a key player in writing the regulations, put it this way. "I don't think the Office of Solid Waste . . . [had] the experience to appreciate the advice that they were getting and did not come to accept it."[55] a

This experience reflected the general problem at EPA of getting different units to work together. Each tended to develop its own concerns and viewpoint and to be suspicious of advice from "outsiders," who were easily seen as rivals and competitors.

When the consultative process broke down, the only solution was for some superior to intervene. But Costle was still assembling his team, and focusing his attention elsewhere. In a March 26 speech he talked about RCRA primarily in terms of materials recycling.[56]

At about this time the OSW split the rulemaking into several separate packages, one for each section of the statute. Each piece was to be developed by a different set of people. No one was responsible for the overall shape, scope, and effect of the program.[57]

By the end of April, 1977, the slowness of EPA's progress on the hazardous waste regulations was attracting outside attention. At a hearing on April 27, Congressman Rooney charged that "at the highest levels of EPA there is a lack of concern about solid waste problems." He asserted that the agency was developing the regulations without a basic strategy.[58] By May 18, in self-defense, Costle was telling Rooney's subcommittee that he had a "special interest" in the solid waste program.[59] That interest, however, was slow to manifest itself in terms of the guidance and resources that the effort desperately needed.

The administrative staff and structure of the agency slowly came into place. Tom Jorling's confirmation hearing as assistant administrator for water took place on June 13. On July 25, 1977, Costle transferred OSW from Air Programs to Jorling's renamed Office of Water and Waste Management.[60] This was done on the rationale that the land disposal of waste most significantly affected water, which was Jorling's responsibility. In October 1977, one year after the Act was passed, Stefan Plehn was named deputy assistant administrator for solid waste. According to Plehn, when he arrived, the OSW was "in a shambles." Plehn's initial impression was that the difficulties of the regulatory effort were not yet fully perceived by the staff of OSW.[61]

In the summer and fall of 1977, however, OSW forged on. Draft regulations were circulated to industry and environmental groups, to elicit comments and views. On November 18, John Lehman wrote a memo to Steff Plehn identifying the major remaining open issues. These included:

> Will there be adequate sites for the disposal of hazardous wastes given public opposition to site development?
>
> What should EPA do about the hazards posed by existing hazardous waste facilities that would be closed because they could not meet the new standards?
>
> Should the initial list limit the definition of waste to a small sector "commensurate with EPA's current resources" or should the Agency develop a broader definition with de facto phasing of coverage based on enforcement actions?

> How should EPA deal with sewage sludge and industrial waste treatment lagoons already regulated by other EPA programs and offices.[62]

Thus, one year after RCRA's passage, EPA still had not resolved many key issues.

The EPA's own lack of clarity about the program hindered its ability to communicate to outsiders. On December 16, 1977, OSW circulated a strategy paper on RCRA followed by a public hearing on January 19, 1978. But the paper was so unclear that outsiders found it unilluminating. As Bill Anderson of the American Consulting Engineer Conference put it:

> I find it most difficult to comment about the strategy, because to me, after reading this document . . . I don't find specifics which will enable us . . . to know exactly what you are going to do. Words like 'considerable attention,' 'more emphasis,' or 'less emphasis' leaves us in a position where I really can't comment, because I don't know what you are thinking.[63]

On February 1, 1978, as required by law, Douglas Costle submitted the agency's first annual report on its activities under RCRA. That report was modestly optimistic, showing that at this stage top management did not yet fully understand the problem. It suggested that "promulgation will be delayed several months past the April deadline—since most proposed rules won't be ready until March or April."[64] (In fact promulgation of some rules lay not two or three months but two or more years in the future.)

The draft regulations that were being circulated hardly met with universal acclaim. On February 7, 1978, the American Paper Institute (API) told EPA that the proposed toxicity test would mean that certain kinds of wood, pulp and paper materials, and by-products would be classified as hazardous, as well as a long list of foods including baby food, canned asparagus, raw carrots, animal crackers, peanut butter, and zwieback.[65]

Like other industry suggestions during this period, API urged that wastes be classified by their "degree of hazard." The proposed classes were:

> Wastes that present a "constant hazard" and that require special care.
>
> Wastes that present a constant hazard but that "can be handled in a normal manner using standard equipment and containers."
>
> Wastes that present no hazard to the health and the environment "when handled in a prescribed controlled manner." This would particularly include certain flammable and reactive wastes, which when isolated from a source of ignition would present no danger.

Notice, however, that in order to develop a classification system not based on dose, industry's proposals focused on potential variations in appropriate handling and disposal practices (as opposed to the degree of acute toxicity).

Some regulations were issued during this period. On February 1, 1978, EPA promulgated "Proposed Guidelines" under Section 3006 that outlined the requirements a state would have to satisfy to be allowed to run its own proposed regulatory program. Perhaps these regulations were issued so promptly because they said so little:

> . . . the Regional Administrator shall evaluate the published regulations of the state's agencies for their equivalency to those of EPA . . . State regulations may not impose any requirement which is less stringent than EPA's . . . a state is required to demonstrate that is enforcement is . . . adequate, and that the state is able to administer and enforce its program successfully. . .[66]

The likelihood that many states would ultimately come to operate their own hazardous waste programs put EPA's regulation writing efforts in perspective. A main function of the Subtitle C regulations was to provide a "model program" to be used by the regions in making state delegation decisions. Hence, there was no guarantee that carefully worked out details would make much difference, since states would have some discretion to make such judgment calls. Furthermore, until the federal regulations were written, the states were so unsure of what would be required of them that they had every reason to delay formulating their own programs.

FROM DEADLINE TO DRAFT

As the statutory deadline for the final regulations approached, the proposed regulations—a necessary preliminary step—had generally not yet been published. At last RCRA began to be noticed by agency managers, if only because they had to sign off on the drafts before publication and were not happy with what they found. On March 7, Tom Jorling testified before Congressman Rooney's subcommittee that the agency would not be able to meet the April 23 statutory deadline for issuing the rules.[67] Jorling knew the delay would occur because it was he who had decided to not let them appear.[68]

This may sound like a perfectly normal decision, but behind it lay managerial developments that were quite unusual in Costle's EPA. The agency had an elaborate, formal, multilevel review process for all regulations—called the *Red Border Review Process*—that culminated with a final sign-off by all the assistant administrators. This system was based on the expectation that no one except Costle himself could be relied on to see the big picture. The staff writing a regulation did not simply report to its own assistant administrator, but rather to a steering committee composed of representatives from across the agency. Jorling short-circuited that process, and began to shape the RCRA rules himself. He listened to suggestions and advice from others, but he had the final say. This constituted a direct challenge to both EPA traditions and his peers in the agency.

This streamlining was not entirely Jorling's idea, although he very much approved of it.[69] Instead, it was Costle who told Jorling, "to put the pedal to the metal," "red line the engine," and take strong personal control.[70] To facilitate the redrafting, Jorling began trying to borrow additional staff from other programs, a step that was generally resisted by those detailed to the hazardous waste effort.[71]

Jorling's main concern in the spring and summer of 1978 was to produce a workable program that would withstand both legal and political challenges. A longtime environmentalist, Jorling wanted to strike a balance between cost saving and adequate safeguards, to have the regulations perceived as setting sensible priorities. To do that, he wanted to avoid regulating materials that were both not particularly hazardous and that would be inordinately expensive to clean up. The program also had to be within EPA's limited administrative capabilities and based on sound scientific principles. Any other approach, he felt, would leave the whole enterprise exposed to potentially devastating counterattacks.[72]

As he read the drafts and sent them back with detailed comments, Jorling found that some of the staff in OSW were less then thrilled with his interest. They had complained of lack of leadership, but as June gave way to July, and then to August, some apparently began to wish that the boss really would use his rubber stamp instead of his red pencil.

Jorling's major concern was over the scope of the program. He wanted waste sources to know up front whether or not they would be required to comply with the regulations under the Act. This meant creating a list of regulated wastes so that firms could easily determine their status. Instead, the draft regulations specified eight tests that sources were obliged to use: ignitability, corrosivity, reactivity, toxicity, radioactivity, infectivity, phytotoxicity, and mutagenicity. Wastes that failed any of these tests were then to be subject to the Act.[73]

Jorling insisted on shifting the focus from testing to listing. He ordered a search made for lists that had already been established under other programs so that their content did not have to be justified from scratch.[74] And although all eight criteria were ultimately referred to in the proposed regulations, only four (ignitability, corrosivity, reactivity, and toxicity) were defined by test protocols.[75] These were the only ones, in Jorling's view, for which sufficiently defensible testing methods existed. The toxicity test was also changed.[76] A simple extraction procedure was used to determine what wastes would "leach" out as water passed through a material. Then, the extract was tested to see if it contained unacceptable levels of any of fourteen chemicals listed in EPA's Interim Primary Drinking Standards. Complicated biological and animal studies (including mutagenicity tests), called for in the earlier drafts, were dropped.[77] Thus, although it was called a "test," the toxicity protocol was really a method for checking against a list established for other purposes.

This massive reorientation was taking place, however, after the statutory deadline. No matter that the deadline itself was unrealistic. The EPA had once said it was fine and outsiders became unhappy at the continued delay. During June there was a critical report from the General Accounting Office,[78] and in July criticism from various Congressional leaders.[79] Also over the summer, various industry and environmental groups filed notices of their intent to sue the Agency over its failure to meet the statutory deadline.[80]

EPA did not respond to this criticism by explaining in some detail how complex and difficult the task was or by pointing out how unrealistic the legislative deadlines were. Instead it temporized and offered excuses. For example, on August 8, 1978, Costle told Congressman Markey at a hearing that his concerns about agency slowness were "largely unjustified."[81] Costle said that the delay was due to public participation requirements and attempts to coordinate the hazardous waste regulations with the 1977 Clean Water Act Amendments. This was only one of many opportunities for congressional and public education that went unused or misused during 1978.

Meanwhile, as the summer of 1978 proceeded, the context in which RCRA was being implemented shifted drastically. On August 1, 1978, *The New York Times* published a major story about a neighborhood in Niagara Falls, New York, called Love Canal.[82] Later in the month *Time* and *Newsweek* also had major articles on the same subject.[83] Under the prodding of intense media coverage, the public began to view hazardous waste as a serious national problem.

Implementing Jorling's changes, and meeting his standards of draftsmanship, put added strain on the relatively small group writing the regulations. Apparently less than fifteen individuals were involved, including a few senior staff from OSW like Gary Dietrich, and a few lawyers from the Office of General Council.[84] Normally such a complex regulation writing process would have involved four to six times as many individuals. But with deadlines looming or already passed, and the subject matter so esoteric, it was hard to take any of the few knowledgeable and capable people "off the line" to educate new staff. The failure to "gear up" more extensively in 1977 thus produced a lasting legacy of staff shortages.

Indeed, at this time there was only one attorney assigned to writing hazardous waste rules, and then only part time. When she left EPA that summer, James Rogers, the assistant general counsel responsible for the area assigned another lawyer, Lisa Friedman, in her place. Even then, Tom Jorling objected since Friedman had been on his "pet project," the control of ocean dumping of sewage sludge.[85] Yet the TSCA effort, which was no more complex or important, was being much more extensively "staffed up" at the same time.

The problem of actually getting the job done was made more difficult because, according to Friedman, the work coming out of OSW at the time was often "unintelligible."[86] Used to relying on experience, the engineers in the office did not appreciate the need to document and justify their decisions in ways that would withstand legal challenges. As a result, Friedman and Rogers wound up doing a great deal of rewriting themselves, often on nights and weekends.[87]

Despite the creation of a special inter-office task force, Jorling insisted on making most of the final decisions himself. A large and forceful man, he seems (intentionally or not) to have intimidated and antagonized William Drayton, who operated the "red border review" process, and was a passionate defender of it. Jorling characterizes his own attitude at the time by noting that he had no patience with unconstructive criticism. "I told everyone I didn't want to hear about problems unless they also had solutions for those problems."[88]

These disagreements and Jorling's assertion of personal control are both reflected in a September 20 memo from Steff Plehn.[89] Plehn was reporting on a meeting with the Office of Planning and Management (OPM) a few days before, in which Drayton's office had made a whole series of suggestions to diminish the economic impact of the regulations. Plehn reported to Jorling that the OPM had been told that their ideas had already been considered and rejected and that their recommendations "constitute major changes at odds with your direction of the program." Plehn went on to say that "OSW acknowledges the OPM concern and recommendations but did not agree to any changes pending your further instructions." OPM had been told that its points would be discussed in the preamble to the regulations to allow for changes in the final version. So much for consensus, consultation and agency-wide sign offs.

Plehn and Jorling also had to deal with some of their own staff who objected vigorously to the new directions. The key players in this drama were William Sanjor, the chief of the Assessment and Technology Branch, and Hugh Kaufman, who managed the Hazardous Waste Assessment Program. They took strenuous, and eventually public, exception to some of the new policies.

Sanjor, who had come to EPA from HEW with the Bureau of Solid Waste, had played a role in drafting both RCRA and the early versions of the regulations. He opposed the simplified testing rules. He also objected to Jorling's proposals on sludge from municipal sewage treatment plants, and on what came to be called "special wastes"—certain high volume by-products of industrial processes.[90] Because these disagreements further illuminate the substantive problems EPA faced in writing the Subtitle C rules, it is useful to take a moment to explain these issues.

In a sense, the underlying issue was one of "residuals management." Waste treatment technologies generally do not eliminate waste. They transform it in some way, leaving behind some physical residual that still has to be put somewhere. Since solid waste controls came last, in time, they had to confront the tendency in previous legislation to encourage treatment technologies that shifted waste from the air and water into solid form.

Under the Clean Water Act, EPA had required towns to construct secondary treatment plants and funded ninety percent of their capital costs. Local industries often tied into such public systems. The by-product of these plants was a sludge which contained the nonorganic components of the wastes they processed. That sludge often contained sufficiently high concentrations of cadmium and other heavy metals to fail even the amended toxicity test. This meant that such sludges would be classified as hazardous and subject to all the requirements under the Act.[91]

RCRA aside, however, cities and towns were already finding it difficult to dispose of the rising volume of sludge. Within EPA itself there was opposition to any and every disposal option. The air pollution people disliked incineration. Jorling himself wanted to stop ocean dumping. The water pollution program had for years sponsored various demonstration projects using the sludge for fertilizer.[92] Yet, under the proposed RCRA regulation, hazardous waste could not be used on land on which crops were grown for human use.

On September 20, 1978, Gary Dietrich, by then Plehn's second in command, told the staff (including Sanjor) that the proposed standards for sewage sludge would have to be modified to correspond to those being developed under the Clean Water Act (which also covered sludge) and the RCRA guidelines on sanitary landfills.[93] Instead of banning fertilizer use, these other regulations specified limits on the rate of application and/or on the rate of uptake of hazardous materials into food crops. Sanjor was strongly opposed to the change on the grounds that the new policy entailed human health risks.[94] On the other hand, neither he nor anyone else in EPA was responsible for finding the least offensive method for dealing with sludge that was also economically feasible. Instead, each unit had its own turf and programs to defend.

Two days later, on September 22, Sanjor again met with Dietrich to discuss "special wastes."[95] They included the sludge and ash from air pollution control equipment, cement kiln dust, phosphate and uranium mining and milling wastes, and the muds and brines used in oil drilling. Often dealt with by their producers on site, it seemed clear that full RCRA controls on these wastes would be extremely—even prohibitively—expensive. That possibility had already led the potentially affected industries to protest to their friends on Capitol Hill. Yet, how could they be arbitrarily excluded if they failed the relevant tests? The regulation writers' solution was to invent a category called "special wastes" that were subject to less rigorous treatment.

The argument between Dietrich and Sanjor focused on whether oil and gas drilling brines and muds and the waste from flue gas desulphurization units on coal-fired boilers would qualify for such favored treatment.[96] Sanjor wanted to drop these from the special waste list. He later claimed to have argued that the former could be handled by using clay-lined ponds, which were already widely employed. Hence, compliance would be both inexpensive and easy. The latter wastes, in Sanjor's view, had come into being as the result of EPA policy, and hence were not entitled to favored treatment. When utilities built scrubbers, admittedly after much protest, it was with the sludge disposal problem clearly in view. Sanjor lost this argument as well.

The issues Sanjor raised demonstrate the complexity and difficulty of the RCRA rule-writing process. Once the agency decided to establish a single class of hazardous wastes, everything from cyanide to sewage sludge to salt water (brine) had to be treated alike. Making exceptions required considering the risks that would be incurred and the costs that would thereby be saved This was no easy task to be sure, and reasonable individuals could surely reach different conclusions. Yet those who advocated stringent rules internally, and lost, decided that public dissent constituted the path of virtue.

Plehn, meanwhile, was reorganizing OSW. According to him, only forty percent of the office was working on hazardous waste, a figure that rose to seventy percent after the reorganization. As part of this reorganization, on September 28 Sanjor was removed from his job as chief of the Assessment and Technology Branch.[97]

Sanjor retaliated. On October 27 the *Environmental Reporter* ran a long story on EPA, the bulk of which was a detailed analysis of the debate over the treatment of sewage sludge under RCRA.[98] It was replete with numerous quotes from internal EPA documents and approving references to Sanjor's position, which was sketched in some detail.

A few days later, Hugh Kaufman testified before the Oversight Subcommittee of the House Commerce Committee.[99] He had been told, he said, not to look for new sites of possible hazardous waste contamination. From the viewpoint of top EPA officials, this was part of an effort to reduce unproductive duplication of effort among Kaufman's group, the Office of Enforcement, and the regional offices.[100] But Kaufman portrayed the new policy as a sellout and an effort to hush up problems.

As we will see, Sanjor and Kaufman became increasingly vociferous in subsequent months. By the following summer they were official "whistle blowers," protected by the law designed to insulate internal dissidents who go public from retaliation by their superiors. To this day, however, those superiors characterize the pair as poorly performing employees who sought protection in whistle blowing to prevent themselves from being demoted or dismissed.[101]

The continued conflicts between Jorling and Drayton came to a head in the form of a decision memo that was submitted to Doug Costle on November 20, 1978, concerning the scope of the program. Since the two principals could not agree, it contained separate pieces written by Steff Plehn from OSW and Roy Gamse from OPM.[102]

Gamse claimed that industry would have to spend $1 billion per year to comply with RCRA and that in some industries the likelihood of plants closing exceeded seventy five percent. He noted that the proposed program would regulate more than 270,000 individual generators, compared with the 50,000 regulated under the water pollution discharge permit program. Hence, it would be extremely cumbersome bureaucratically. Gamse also pointed to the limited capacity of the existing safe disposal sites. To help, he wanted to exempt all sources that generated less than 1000 kilograms a month from the paperwork and reporting requirements of the rules, and delay implementation of the program for sixty industries (including electroplating and metal finishing) where the projected cost of the program exceeded projected annual profits.

In his half of the memo, Plehn argued that Gamse's suggestions "are not environmentally defensible, lack consistency and equity," and might delay the proposed regulations by as much as a year. If implemented, Plehn contended, "the public and Congress could make a legitimate case that EPA has not been responsible to its RCRA mandates to protect public health and the environment from hazardous waste."

In all of our cases it is hard to find a better example of bureaucratic pluralism. Here are different units within EPA, each representing different and relatively narrow, interests and concerns. Equally striking is the fact that there was no general manager in the structure, below the administrator, to reconcile their views. (Costle, in fact, ultimately rejected most of Gamse's suggestions.)

The EPA's continued failure to issue the proposed regulations at last began to produce serious legal challenges. The solid waste trade association argued that the delay was preventing the development of new disposal facilities. Environmentalists argued that it was harming the environment and human health by allowing improper disposal practices to continue. The state of Illinois argued that the absence of a federal program made it difficult for the states to design and adopt "equivalent" schemes.[103]

In response to these claims on November 7, 1978, Judge Gesell gave EPA a month to file a proposed schedule for promulgation of the rules, and set a hearing on that schedule for January 2, 1979.[104] In that schedule EPA said it would issue proposed rules within the next two months, but that it would take another six months to a year for these to become final.[105]

In its submission to the court, the EPA said, "Any tighter schedule would in our opinion, seriously jeopardize our ability to develop a responsible, workable, national, solid waste management program . . . which would withstand judicial review." The agency pointed to the effluent guidelines program where in order to meet "extremely ambitious promulgation schedules imposed by this court," many of the regulations "had to be developed so hastily that they failed to survive judicial review."[106]

This schedule had been prepared by the staff of OSW. According to Lisa Friedman, it sounded reasonable when she discussed it with Steff Plehn and Gary Dietrich. Only later did she learn that it was "pie in the sky."[107] Perhaps one reason for this optimism was that some of the staff of OSW were reluctant to tell Jorling just how long the task of rule writing was actually going to take.

THE FIRST PROPOSED RCRA REGULATIONS

On December 18, 1978, more than two years after the act was passed, EPA finally issued a significant portion of the proposed regulations under Subtitle C of RCRA. These covered Sections 3001 (identification and listing), 3002 (generators), and 3004 (waste management facilities).[108]

Given the limited time, resources, and knowledge EPA had to work with, the resulting regulations were often unavoidably based on debatable premises, guesswork, or arbitrary decisions. Under the circumstances, Doug Costle might have gone back to Congress and asked for more relaxed deadlines, more money, and a whole series of clarifications and amplifications of the law. Instead he decided to publish the draft regulations in the hope that the comment period would educate the agency by bringing forth new data and additional studies. Costle called this strategy, "Cast thy bread upon the waters and see how soggy it gets."[109] As EPA itself wrote in the preamble to the proposed rules, "we especially need not only comments but ideas."[110]

The EPA's limited knowledge is illustrated by the fact that many of the entries on both the first and subsequent lists of hazardous substances were wastes from specific sources or manufacturing processes. This, EPA said, was because "Industrial wastes tend to be complex mixes containing many different components only some of which may exhibit hazardous characteristics. This approach will relieve waste generators of much of the testing burden and uncertainty."[111] But this approach reflected as well the agency's inability to delineate precisely what it was that made a mixture hazardous.

The regulations also provided for the "delisting" of those wastes that could be shown not to be hazardous. The waste had to pass a complex set of tests, including those bioassays unsuccessfully proposed by the OSW as the basis for listing a waste as toxic.[112] Jorling accepted this asymmetry because he believed that any waste that could pass such a complex set of procedures should be exempted.[113] Again we see pragmatism at work in the face of ignorance. No one knew very much about how the probability of inappropriately delisting (or not delisting) some compound varied with the specification of the tests, or how harmful various rates of false-positive and false-negative results would be. Instead, someone had to make a "gut judgment" to make the test "difficult" and leave it at that.

Another example of EPA's difficulties was the controversy over the small generator exemption. This was to allow producers of less than 100 kilograms of hazardous waste in any one month to avoid the paperwork requirements for that month.[114] Yet, in the preamble, EPA noted that "some" (it did not identify OPM as the advocate) had argued for a 1000 kilgoram cutoff.[115] The problem lay with trying to determine the likely effects of different rules. What impact would the 100 kilogram exemption (or the alternatives) have on compliance and enforcement when all sources were still supposed to obey the disposal rules, even if they did not have to fill out the forms? As the agency itself noted:

> EPA has limited data on the number of generators of small quantities of hazardous waste, the amount and type of such wastes generated, the current management of these wastes and their actual and potential impact . . . Consequently the agency is finding it difficult to resolve this issue.[116]

The standards for TSDFs were perhaps the most controversial portion of the proposal. In the preamble, EPA said that it had considered three different approaches; (1) setting ambient standards for air and water quality in the vicinity of facilities, (2) design and operating requirements, and (3) performance standards. The preamble said the agency had decided to combine all three methods while "relying primarily" on design standards.[117]

Comments on previously circulated drafts had argued that design standards were not flexible enough to cope with the many possible combinations of wastes and locations. Therefore, certain of the design and operating standards were followed by "notes" that described the circumstances under which regional administrators could allow deviations from the standards.[118] The regulations also included certain performance requirements that were to be followed in "exceptional cases," where the design standards seemed inadequate.[119]

This hodge-podge clearly reflected the difficulties EPA was encountering. The agency felt it did not know enough about how hazardous waste produced health risks to rely simply on ambient standards. It also could not specify what made for an adequate design in any and every situation. Allowing permit writers substantial discretion—as the note system did—postponed and decentralized the necessary judgments, and left the requirements to be imposed on new disposal facilities quite flexible.

A particular vexing part of the TSDF regulations dealt with surface impoundments that were part of industrial wastewater treatment systems. Such "pits, ponds, and lagoons" were frequent in certain industries (e.g., paper making), and were already regulated under the Clean Water Act.[120] If they discharged into any "navigable water" they had to have a permit under the National Pollution Discharge Elimination System (NPDES). If they were connected to a municipal sewer system, they had to meet the relevant "pretreatment" requirements. Nonetheless, as the proposed regulations accurately note, "the possibility exists for subsurface discharge and/or air emissions which are harmful to human health . . ."[121]

Here again was the residuals management problem. Many of these facilities had been built to comply with the Clean Water Act. They "disposed" of wastes by allowing them to percolate into the ground or disperse into the atmosphere. While industries were being told that this was no longer acceptable, no one at EPA knew much about the environmental or economic consequences of applying the RCRA regulations to such installations.

Furthermore, all of these rules would have taken years to implement. In the meantime, the short-term impact of RCRA depended greatly on the so-called interim status standards, which applied to sites in existence when the Act was passed.[122] Any existing site could apply for an interim permit if it complied with the "minimal" interim status rules and continued to operate under them until EPA could review its permanent application.[123] But how tough should EPA make those interim standards? Stringent requirements that forced facilities to close would leave waste nowhere to go. Lax rules might allow unsafe conditions to persist. The EPA ultimately chose to impose quite modest requirements, but no one really knew what the public health consequences would be.

Moreover, since even licensed disposal facilities might close, the law required EPA to find a way to guarantee continuing managerial and fiscal responsibility. The draft regulations required that for twenty years after closure, the owners had to maintain a facility's security and containment devices and monitor them for possible leaks. The EPA also wanted to insure that there would be someone to sue for any future damages. This meant requiring operators to post some sort of bond or purchase some sort of insurance. The draft regulations made it clear that the agency was not happy with its own proposal in this regard. The resulting massive insurance costs seemed likely to dwarf those of actually handling the waste, especially since no one knew much about what the rates would be.[124] Again, as with many other decisions we have reviewed in this section, EPA confronted a very complex task while possessing very little information.

FROM PROPOSED TO FINAL RULES:

The Role of the Courts

The eighteen months that separated the proposed rules in December 1979 from the publication of the first substantial set of final regulations in May 1980, was a tumultuous time in the history of RCRA rule-writing. The EPA had to deal with the courts, the Congress and other executive branch agencies, as well as with its own internal problems. The agency had to respond to the enormous volume of comments that poured in during the public comment period and do much redrafting as a result. In this and the next three sections we explore the separate strands of the story to highlight the lessons that this part of the history has to teach us.

With regard to the courts, during this year and a half EPA dealt repeatedly with Judge Gesell over its continual delay.[125] While the judge periodically threatened the agency, and once seemed to consider receivership, his power was in fact quite limited. Perhaps if EPA's managerial deficiencies had been more obvious or bad faith more evident, he might have more seriously considered taking over the agency. But given the substantial problems EPA faced and a congressional schedule that was clearly unrealistic, the court was not about to respond unthinkingly to the "black letter law." This part of the story, thus, shows that an agency must be made politically and managerially responsible for its activities to insure effective program implementation. Legal responsibility alone will not get the job done.

In January, after the proposed rules were published, Judge Gesell accepted EPA's promise to issue the rest of the regulations at the end of the year. In his order the judge stated that EPA "is proceeding in complete good faith and conscientiously." The judge also ordered EPA to file quarterly reports indicating "any departure from the detailed implementation schedule . . . and the reason or reasons therefore."[126]

In his oral remarks on January 2, Judge Gesell noted that it was very difficult to prepare a complex regulatory program, and also difficult for Congress to estimate how much time such a task will require. The essential issues were not legal:

> . . . there is not only a shortage of funds, a shortage of personnel, but a lack of necessary technical information; that the issues are extremely complex; and the scope of the regulations is extensive. Well meaning statutes are not self-implementing.[127]

He pointed out that the Court could not itself appropriate money, let out contracts, or do the necessary technical work. Instead he argued that "the political process" was the right place to settle the issues in question.[128] Thus, for the moment at least, the judiciary was prepared to let the executive and the legislature deal with the problems of implementing RCRA.

On April 2, 1979, in his first quarterly report to the judge, Costle said that most major activities were "proceeding according to schedule." He noted, however, that work on the state guidelines and consolidated permit regulations was already running two months late. He promised to institute a variety of special measures to accelerate progress on these particularly troublesome tasks. Costle added that "the initial analysis" of the public comments was scheduled to be completed by June 30, 1979, "at which time we expect to be able to provide the Court with an informed assessment of where we stand."[129]

Three months of struggling with the mountain of comments left EPA overwhelmed. Without waiting for the next quarterly report, due in July, on June 6, Costle filed an affidavit with the court in which he claimed that, despite "unprecedented" measures, the task was so difficult that EPA "may not be able to meet the deadlines" that the agency itself had proposed barely six months previously.[130]

Costle's affidavit stated that the relevant staff had been "unexpectedly diverted" to work on the RCRA reauthorization and on the development of the Superfund legislation. Also, the clean-up of abandoned hazardous waste sites had drained OSW's time and energy. The affidavit did acknowledge, however, that the task of organizing and analyzing the comments "proved to be much larger than we anticipated." One month later, in the second quarterly report to Gesell, EPA said that RCRA "is to be the agency's highest priority for the next six months." Nevertheless, EPA suggested, it might well not make the end-of-year target.[131]

Apparently because a new assistant administrator (Chris Beck) had meanwhile replaced Jorling, EPA delayed submitting its third quarterly report until October 15. In that report the agency said that the "emergency procedures" designed to meet the December 31 deadline, had not succeeded "because they required us to do too much, too quickly."[132] Beck, who had not signed off on the initial schedule proposed to the court, took his staff's advice and proposed delaying final publication of some key rules for two months until February 1980, and others for four months, until April 1980.[133] Even so, not all of the identification and listing standards would necessarily be finalized by that time. The report to Gesell also noted several issues that were likely to require reproposal, implying that the whole business of requesting, receiving, and responding to public comments would have to be repeated. This was sure to create still further delay.

In an accompanying press release, Costle said that he was "extremely frustrated" by the delay, given that "I consider hazardous waste to be the most serious" environmental problem before the nation. Costle went on to say that one of the problems the agency was facing was that the comments were a "mine field" constructed by "sophisticated lawyers" and that if the agency did not respond to each and every comment, the whole package might be overturned in subsequent litigation.[134]

On October 23, the Environment Defense Fund and Environmental Action, two plaintiffs in the suit against the agency, filed with the court a response to EPA's October 15 quarterly report. In their submission, both groups characterized EPA's posture of "open ended indecision" as "totally unaccepatable" and in violation of both the spirit and the substance of the prior court order. They argued that until the final regulations became fully effective, additional "ticking time bombs" of unsafe waste sites would continue to be created. They pointed out that clean up efforts had begun at hazardous waste sites in several states. They asked what would happen to the waste dug up from these sites if there were no safe, permitted landfills to put them in.[135] The American Petroleum Institute also complained that the strategy of issuing the package in parts made it hard to comment intelligently on the regulations.[136]

On December 3, 1979, the Citizens for a Better Environment asked Judge Gesell to issue a show cause order requiring the EPA administrator to explain why he should not be held in contempt for his continued failure to promulgate the regulations. Judge Gesell scheduled a hearing for December 12, 1979. On December 4, he wrote to the agency asking for a wide variety of detailed information on EPA staffing and activities, with an eye toward determining what resources could be freed up to work on the RCRA rules.[137]

This letter caused considerable excitement within the agency. Some wondered if it was the first step toward having the judge take over the rulemaking, or possibly all of EPA. With some trepidation, the information was assembled and submitted to the court.[138]

In this filing, the EPA again claimed that congressional supervision was diverting resources from regulation writing. The agency said that it had been "in virtually daily contact" with the staffs of four committees and subcommittees having jurisdiction over the program and the offices of approximately six congressmen who "are active on hazardous waste issues."

The December 12, 1979 hearing produced a clear victory for EPA. Despite all the delay, the judge denied the request to issue a show cause order. Instead, Gesell adopted the EPA's revised deadlines. He also suspended the requirement of quarterly reports, in response to the agency's claim that these, too, took time away from the rule-making process.[139]

The nominal supervision of the courts continued but clearly the pressure was off the agency after this decision. For example, on March 31, 1980, Costle, without causing any problems, formally notified Judge Gesell that the agency was unable to meet the deadline on the postclosure financial responsibility regulations.[140] Instead, said Costle, he would repropose these in April and hoped to promulgate them in final form by the Fall of 1980.

Thus throughout 1979 and early 1980, the environmental movement tried to use one of its favorite devices—an appeal to the D.C. Circuit Court—to accelerate the RCRA rulemaking process. But the judge was not willing (or able) to take over the job of actually running the government, which is what adhering more closely to the legislative timetable would have required. Yet, the alternative channels for seeing that the RCRA rules got written were not able to accomplish much either.

The Role of Congress

If the courts were unable to enforce the dictates of Congress, Congress itself was not much better off. During this period, various special interest groups, particularly natural resource industries, lobbied hard to weaken the regulations applicable to their particular situations. Public concern over existing dump sites also occasionally stirred congressmen and senators into action. But there was little intelligent discussion of the broad policy issues that were continually arising in the attempts to implement RCRA.

On March 21, 1979, a few days after the close of the comment period, the Oversight Subcommittee of the House Commerce Committee began a series of thirteen hearings on hazardous waste. Individual congressmen asked about sites in their home districts. There was also some discussion of the then-pending Superfund legislation. There were only occasional comments, however, and those were from state personnel about the delay in the RCRA regulations.[141]

Both the Transportation Subcommittee of the House Commerce Committee and the Senate Public Works Committee began work on the reauthorization. At a House hearing, Jorling told Congressman Florio that although the total budget of the OSW had gone down the preceding year, the number of people working on the RCRA program was continually increasing. He declared that in fiscal year 1977 there were twenty-one people working on hazardous waste; 134 in fiscal year 1979; and that the agency had requested 202 slots for the following year.[142]

As Congress considered amendments to RCRA, political pressures continued to mount for special treatment of certain materials, including oil drilling brines and muds, mine overburden, and waste crankcase oil. Would every service station have to file manifests and participate in the RCRA system? The service station operators were strongly opposed to doing so. The "pits, ponds, and lagoons" question also continued to be controversial. Congressman Swift (from a timber and paper area) at first wanted to exempt such facilities from RCRA altogether.[143] Once convinced that doing so would be going too far, he proposed that "where appropriate," standards distinguish between new and existing TSDFs. The language he ultimately proposed, however, was meaningless since it was permissive, and the administrator almost certainly already had the power it conveyed.[144]

Jorling has since argued that the agency was able to negotiate with industry "from a position of strength" because of rising public concern with hazardous waste. Swift was apparently satisfied with the appearance of doing something—however modest—for his constituents.[145] Yet, as we will see, on special wastes several industries won more substantial concessions.

The agency's managerial weaknesses were sometimes exposed by this Congressional scrutiny. About the same time that Costle was telling Judge Gesell that there might be delays, he testified before the Oversight Subcommittee of the House Commerce Committee that the expedited decision process "will permit us to develop final regulations on or very close to the present schedule."[146] When asked about this inconsistency later, Jorling told Congressman Gore that the preparation of the affidavit and the testimony "were on different tracks."[147] This vignette, however, illustrates that the agency tended to see Congress as an enemy to be manipulated, not as a forum for an honest examination of its (and the nation's) problems.

In the growing climate of media-hype and public hysteria over toxic waste (to which EPA contributed, as we will see in a later chapter), Congress could not ignore EPA's slowness in producing the RCRA regulations. The issue entrepreneurs in Congress (unlike Congressman Swift and some of his colleagues from oil drilling states) focused on simple ideas like "toughness" versus "laxity" and "promptness" versus "delay," because they sought to attract the support of ideological rather than economic constituencies.

The most dramatic confrontation began on April 1, 1979, at a hearing before the Oversight Subcommittee of the Senate Governmental Affairs Committee, chaired by Senator Carl Levin. Hugh Kaufman and William Sanjor each testified at length. Earlier in 1979 Sanjor had written a series of letters to John Lehman detailing his disagreements with the proposed regulations. The EPA, at first, had excluded these comments from the public docket on the grounds that as Sanjor was an agency employee, these papers were internal documents. This was reversed under protest from Sanjor.[148] Now Sanjor and Kaufman sharpened their attacks.

The hearing discussed Jorling's decision, the previous summer, to use listing instead of testing. Sanjor and Kaufman both charged that this was in part a response to White House instructions to limit the scope of the regulations for budgetary and inflation-fighting reasons. They claimed that John Lehman told his staff in mid-June 1978 about such a presidential directive. The result, they said, was an inadequate program.[149] In his last congressional appearance before leaving EPA, Jorling repeatedly called these suggestions "a despicable lie."[150]

In an angry exchange, Jorling insisted that although he knew of these charges, he had always been too busy to ask Lehman exactly what the latter had actually said. Jorling also refused to accept Senator Levin's characterization of the changes he made as a "cutback" in the regulations, declaring, "I will not accept that description. I would say we made them more effective." Finally, he asserted that Senator Levin had been badly served by his staff in preparing for the hearings.[151] Several days later, on August 7, Jorling wrote a letter to the subcommittee apologizing for "what may have seemed intemperate comments on my part."[152]

The EPA was sensitive to the charge that significant volumes of hazardous wastes would not be covered by the proposed regulations. In August, EPA proposed adding twenty-eight wastes, sixteen waste streams, and all human carcinogens listed by the International Agency for Research on Cancer to the list of hazardous wastes.[153]

During the summer of 1979 criticism also came from Congressman Eckhardt's Oversight Committee. In August it submitted a report based on hearings the previous spring. In his letter of transmittal, Eckhardt said:

> EPA has failed to meet statutory deadlines for regulations for disposal of hazardous waste; has failed to determine the location of all hazardous waste sites; and has not taken vigorous enforcement action . . . The report reviews EPA's long awaited proposed regulation and finds them lacking . . . While the Congress may have been unrealistic . . . there could be no excuse for EPA's failure to promulgate regulation in the nearly three years since the statute was enacted.[154]

The subcommittee's report also did not accept the agency's oft repeated claim that a major reason for the slowness was the need to deal with Congress. Instead the report traced the personnel shortage to EPA's own management choices. The EPA's requests for funding had been at levels far below the authorized amount and in addition it "spent considerably less than what Congress had appropriated."[155]

The resulting delay, the subcommittee argued, had serious consequences. New Jersey, for example, had adopted stringent hazardous waste regulations. But, because there was no national program, wastes were just being shipped to Pennsylvania where they were disposed of improperly. Similarly, a private waste disposal firm had cancelled a 250,000 dollars hazardous waste treatment facility in North Carolina "upon learning that North Carolina could not issue it an operating permit because North Carolina's regulation was tied to the yet unpromulgated federal regulations."[156]

In discussing the content of the regulations, the subcommittee drew heavily on the example of California, which had instituted a hazardous waste regulatory program in 1972. The subcommittee criticized the agency for not adopting all eight hazardousness characteristics initially considered and for using simplified testing procedures. Many of the tests the EPA said were not reliable, the subcommittee noted, were used for screening purposes under the Toxic Substances

Control Act. The subcommittee was also concerned about the small generator exemption. "Ultimately," wrote the subcommittee, "EPA's regulations will permit future Love Canals."[157]

Having struggled with the legislation through the fall and winter, on February 29, the House of Representatives approved the reauthorization of RCRA by a vote of 386 to 10. In addition to the Swift amendment, the bill exempted oil drilling brines and muds, and phosphorous and uranium mining and milling wastes from any regulation until further studies were done.[158]

In March the Levin Subcommittee issued its report and made many of the same complaints as had Eckhardt's Subcommittee, but to similarly little avail.[159] For, in general, during these eighteen months Congress could do little more than cajole, harass, or find fault with EPA. It tried to sniff out scandal and responded to some special interests (if not others). A number of congressmen became concerned with specific sites in their districts, and the momentum they provided helped ultimately to pass the Superfund law. But nowhere in the extensive hearings, reports, or floor debates is there any strategic consideration of RCRA. The agency did not present any coherent statement of what it was trying to accomplish with the regulations beyond vague claims such as achieving "safety." Hence, there was nothing for Congress to review except details. This led to a discussion of agency errors and efforts to help specific interests. The executive had done nothing to encourage or even allow congressmen to see their role in a different light.

The Role of Other Agencies

The various White House staff groups that played a significant role in the ozone story are conspicuous by their absence from the RCRA chronicles. In ozone, a single number was in dispute and it was relatively easy to have an opinion. In RCRA the issues were so complex that, in the absence of a major stake, other agencies tended to remain on the sidelines. As for the president, without an interagency conflict to force the matter to his attention, he had little reason to find out why the RCRA rules were not being published on time—if he was even aware of the delay. Nor would it have been easy for the president to determine whether the developing regulations were consistent with his own values and political mandate. As long as EPA did not offer either an explicit problem definition or a specific strategy, the president and his staff, like Congress, were left with nothing to review except an enormous and indigestible mass of arcane detail.

The RARG report on the proposed Subtitle C regulations noted the difficulties in developing the regulations and found few faults with the draft.[160] It did suggest that some sort of degree of hazard system be instituted "during the implementation period as part of the permitting process." The idea would be to give "explicit guidance to permit writers as to the degrees of hazard associated with different wastes and their cost effective means of control in different geographical settings." Whether or not EPA knew enough to do that was not explicitly considered.

The RARG also suggested five other points for further study: (1) the toxicity test, (2) the small generator exemption, (3) coordination with Clean Air Act regulations, (4) paper work reduction, and (5) more work on problems of insurance liability.

The RARG commented that depending on the hazardousness of the waste, different small generator cutoffs might be appropriate (a system EPA ultimately adopted). The RARG also suggested that EPA's projections of compliance costs might involve serious underestimates.[161]

None of these issues was new to EPA and none provoked a head-on confrontation. Unlike ozone, there was no clear ideological or professional division. In part this was because there was no traditional economists' "party line" on hazardous waste. In addition, so little was known about the consequences of various policies that it was hard to have an argument. Injunctions for further study were easy to offer and accept.

The one agency that did raise substantial objections to the RCRA rules was the Office of Management and Budget (OMB), which was very concerned about the possible paper work implications of the manifest system. The OMB asked EPA to both simplify the scheme and repropose the regulations so that it could comment on the issue on the record.[162]

Costle met with OMB Director Jim MacIntyre in March of 1980 to try to resolve the issue. The OMB stood firm. As Costle later reported to Judge Gesell, OMB agreed to review EPA's rules on an expedited basis, but EPA had to agree to reopen the public comment period to receive these comments.[163]

The somewhat ironic outcome of this story is indicative of the complexity of the issues RCRA raised and the difficulty of being a dilettante with regard to them. The OMB did finally get EPA to simplify its manifest forms. But many states adopted the more extensive California version. Thus, OMB inadvertently undermined the goal of nationally uniform record keeping.[164]

The only other organization that took a continuing interest in RCRA was the General Accounting Office (GAO). Eager to move beyond a purely auditor role, the GAO in recent years had taken a selective interest in various program areas. The GAO reports and testimony were among the few comments on the rules that tried for a strategic perspective. For example, at the very beginning of 1979, the GAO argued that there were not enough safe disposal sites, and that without such sites there could not be an effective national program.[165] Yet, this was exactly the sort of strategic calculation that no one in EPA was considering.

But the GAO had no power over the agency. With no one in Congress able or willing to undertake a strategic analysis of RCRA, the GAO comments just blended into the white noise of external criticism that the agency sought to muffle or ignore.

Inside the Environmental Protection Agency

As all these organizations circled warily around RCRA, like animals just beyond the fire light of a jungle camp, and 1978 became 1979, EPA struggled on with developing the final rules. Jorling and Costle, with Judge Gesell and the *Washington Post* looking over their shoulders, did not want to miss any more deadlines. On the other hand, they still had relatively few people working on hazardous waste, and they still did not know very much about the problem. This made it easy for internal disagreements to persist, since no one could refute an opponent's claims and arguments. With Jorling scheduled to return to his teaching job at Williams College by fall, everyone became increasingly concerned during the summer of 1979 with "getting out" something that would survive judicial review. There was still no time for an integrated strategy.

By the time the ninety-day comment period ended on March 16, 1979, the agency had received several hundred comments reportedly totaling "seven linear feet" of shelf space. The Administrative Procedures Act, as interpreted by the courts, required EPA to address each of these comments in the final rule-making. Given the limited resources of OSW, the Mitre Corporation was hired to rearrange the comments by topic so that all of those on each separate part of the rules could be addressed together. In practice this exercise was not particularly helpful. The contractors did not understand the issues well enough, and many of the comments related to more than one part of the proposed rules.[166]

The submissions do show that at this stage the environmental community was less concerned with hazardous waste regulation than was industry. Advocacy groups could mobilize local citizens to support cleanup of a nearby site. But the RCRA rule-making was so complex and technical that the environmental community, like EPA itself, suffered from a lack of expertise. There were some exceptions. Citizens for a Better Environment pursued their crusade to have sewage sludge declared a hazardous waste, and Blake Early of the Sierra Club continued to play a role. But most comments were from companies concerned about their own circumstances.[167]

Many argued that relatively unhazardous materials would be dealt with too stringently. Others asserted that the regulations were confusing or inconsistent. For example, the pesticide residue sufficient to have fruit classified as a hazardous waste would not have it classed as unfit to eat under government agricultural regulations. Hence, a piece of fruit that could be eaten could not be disposed of in a municipal landfill. Industrial "pits, ponds and lagoons" were also the subject of many responses.[168]

As the spring of 1979 progressed and the cherry blossoms appeared along the Potomac, matters were most definitely not looking up for EPA on the hazardous waste front. On March 29, the ABC network broadcast a documentary on the problem with the low key title, "That Killing Ground." It contained no praise of the agency and its slowly developing regulatory program.

Given the increasing public and judicial attention, Tom Jorling became dismayed by the prospect of trying to push the huge mass of revised RCRA regulations through the agency's internal review processes.[169] Normally, if any questions were raised, the lead office or division would have to pursue the suggestions and, if necessary, redraft the regulations. Then the whole cycle would begin again.

To avoid such time-consuming iterations, Jorling proposed that a high level policy group meet over the summer of 1979 in a rented room in the Crystal City Mall, across the Potomac from EPA headquarters. Its goal was to resolve all major policy issues before any additional drafting was done. These meetings were actually run by Gary Dietrich, with one of Jorling's staff there to brief him daily on what transpired.[170] This unusual process was apparently approved by Doug Costle over Bill Drayton's strenuous opposition.[171]

As it turned out, the Crystal City meetings were ineffective.[172] Many of the participants knew relatively little about RCRA and offered badly supported proposals. The lack of an overall strategic plan made it difficult to set priorities. In his October 15 report to Judge Gesell, Costle summarized the experience as "largely unworkable" because EPA had tried to make piecemeal decisions before it had analyzed the available data.[173]

At this time, in the spring of 1979, in an effort to speed up the work, Costle transferred sixteen staff people to OSW from other parts of the agency. The individuals concerned were almost uniformly unhappy with the move. Costle had to talk to several of them personally to get them to accept the new assignments,[174] and most made every effort to depart as soon as possible. Within EPA, RCRA regulation writing was not seen as a path to fame and glory.

When he wrote to Gesell in June of 1979 to announce that the December deadline might be missed, Costle also included a list of unresolved issues highlighted by the comment period. This list is worth pondering because it shows that two years and eight months after the Act was passed EPA was still worrying about:[175]

Whether to employ a degree of hazard system.
How to treat waste water impoundments.
Whether to continue to exempt small generators from some paperwork requirements.
The leachate extraction procedure for the toxicity test.
Whether to make Clean Air Act and Clean Water Act performance requirements binding on hazardous waste facilities, and if so how.
How to insure continued financial responsibility after a site closed.
Whether the "note" system in the proposed regulations was reasonable and/or legally permissible.

On each of these the agency had to deal with numerous and often conflicting public comments. Furthermore, it often lacked the in-house technical capacity to do the necessary background work. Instead, the agency often resorted to outside contractors. For example, in December 1979 a major study by TRW of the implications of excusing small generators from the manifest system was finally finished. The next month, Monsanto Research began to look into design questions on hazardous waste incinerators. The quality and timeliness of the resulting work was variable at best, and by continuing to go outside, EPA did not develop its own in-house capabilities.

As August began and Jorling was preparing to leave, he had a lengthy meeting with his replacement, Chris Beck, at the Albany airport. Their schedules were both so tight that this was their only chance to get together. The formal change of command occurred over the weekend of September 7. Jorling left on Friday and Beck arrived the following Monday.[176]

Beck soon made two major managerial decisions. One was to assign the work of drafting parts of the regulations, and especially the background documents, outside OSW. This was a direct response to the difficulty of getting the best people to accept a transfer into OSW. The second was "burping": publishing the regulations for each section as soon as they were ready and not waiting for the whole package.[177] Costle approved this approach reluctantly since it was in direct conflict to his July 6 memo to Judge Gesell. But with mounting public pressure, he felt EPA had to begin to show some progress as soon as possible.[178]

Beck was particularly interested in the rules for TSDFs. He had supervised hazardous waste programs in Connecticut, and believed that sites were so variable that uniform design rules made no sense. He also did not like the idea of pure performance standards (e.g., fifty year containment). How was an engineer to know when that had been achieved, Beck asked? And why was any single goal universally appropriate? Near drinking water supplies it would be too lenient and where the ground water was already polluted it would be too stringent. Beck wanted a more flexible approach.[179]

Meanwhile, another three months had gone by, and as noted previously, EPA had to again report to Judge Gesell—which it did after a slight delay.[180] Many of the issues noted as open in June and July were still open. Again as noted, Beck asked for an extension of the schedule EPA had agreed to the previous December, partly because he wanted more time to influence the rules before they were issued.

Beck, like Jorling, relied on his personal relationship with Costle to sidestep formal agency procedures to resolve outstanding issues.[181] But these same disputes continued to rise Phoenix-like from the ashes of successive efforts at bureaucratic conflict resolution.

In summary, during these eighteen months, the agency's increasingly frantic efforts seemed only to bury it deeper and deeper in the quicksand of RCRA. This history—especially the Crystal City meetings—highlights just how difficult it was to rely on "bureaucratic pluralism," to control any one unit within an "expert" agency. Not only were Congress, the courts and the rest of the executive branch disenfranchised by RCRA's complexity, but so were those parts of EPA that were watching it from the "outside." Only control of the experts by politically responsible managers seemed likely to produce the sort of timely and responsive decision making often absent from this narrative.

THE FIRST PART OF THE FINAL RULES

As mid-May of 1980 approached, the agency was at last ready to publish a large segment of the final regulations (all but the critical TSDF requirements).[182] The site chosen for the press conference to announce the rules was a New Jersey chemical waste dump. Shortly before the press conference, the dump exploded.[183]

The Rules themselves took up 520 pages in the *Federal Register*.[184] In them, EPA tried to respond to the criticisms and comments it had received. In the toxicity test, for example, the lower limit for a "hazardous" concentration in the leachate for any of the fourteen listed chemicals was raised from ten times the national drinking water standard to 100 times that standard. This change exempted most sewage sludge and, it turned out, many paper mill wastes. Similarly, the small generator exemption was broadened to exclude generators of less than 1000 kilograms of hazardous waste in a given month from the manifest system for that month. Wastes listed in the regulations as being acutely hazardous (122 were listed), however, were treated differently. For these, the small generator exemption was set at only 1 kilogram a month. In anticipation of then-pending legislation, the regulations postponed action on sewage sludge, low-level radioactive wastes, drilling brines and muds, mining overburden and slag, and flue gas scrubber wastes.[185]

Unfortunately, without the permanent TSDF rules, no one could begin to review disposal sites for permanent status. Nor could program authority be delegated to the states, many of which were waiting for federal action before designing their own programs. Because the interim status standards were not particularly stringent, and EPA's own enforcement resources were limited, marginal facilities were not necessarily forced to close, but neither could new and better ones be opened. If hazardous waste disposal presented public health problems in 1976, EPA had so far done little to change that.

The EPA estimated the annual cost of complying with the regulations at 510 million dollars, or less than 0.2 percent of the value of sales of the affected industries. Approximately fifty-nine percent of the cost, or 303.3 million dollars, would be spent for the postclosure regulations, including liability insurance costs. Treatment and disposal costs were estimated to be 56.8 million dollars, with monitoring and testing adding 67.5 million dollars.[186]

While publication calmed Congress, not everyone else was satisfied. The agency soon received fifty two notices of intent to file suit from various industry and environmental groups.[187] Public concern did not diminish either. The continuing attention given to Love Canal only reinforced the view that hazardous waste was a serious public health problem.

The regulatory scheme was so complex that few could understand all of its ramifications. On May 30, the agency held the first of a series of briefings and public meetings on the new rules. At this meeting, in San Francisco, according to the *Environmental Reporter*, "It was apparent there is widespread confusion and anxiety."[188]

Protected by the whistle blower law, William Sanjor continued to find fault. In a June 16 memo to John Skinner, the head of EPA's State Program and Resource Recovery Division, he charged that of the nine damage cases studied by the agency in 1975 and 1976, six would not have been affected by the new regulations.[189]

The most important development of the immediate post-publication period came on August 19, 1980, when the agency announced in the *Federal Register* a new strategy for dealing with the "thousands" of requests for information and clarification concerning the new regulations. The notice said that until now such inquiries had been answered orally and that this "is not an effective way of communicating interpretations of the regulations to the public." Therefore, said the agency, it would now begin to issue "Technical Amendments to the Rules" (TARs) and "Regulatory Interpretive Memoranda" (RIMs). Many of these would be published in the *Federal Register* with an explanation but without an opportunity for public comment. Others were to be published either in interim final form or as proposed rules.[190]

The RIMs were designed to explain how the regulations applied to particular situations. Because they were not supposed to involve policy changes, RIMs were to be issued by the deputy assistant administrator for solid wastes, with the concurrence of the deputy assistant administrator for water enforcement and the associate general counsel for water and solid waste.[191]

Somewhat more significant and general issues were to be handled by TARs, which would require the administrator's sign-off. The TARs, too, would not go through all the various review stages within EPA. In some cases, they were walked through all the internal processes and signed in a matter of days. According to associate general counsel, James Rogers, the use of such procedures was unique in EPA's regulatory practice.[192]

The RIMs and TARs were in fact primarily used to settle the lawsuits filed over the final rules. The EPA engaged in extended negotiations with various parties. Some of these meetings involved literally hundreds of lawyers. When agreements were reached, they were characterized as "interpretations" or "technical amendments" and guided ruthlessly through the streamlined processes. One might have thought that EPA would have been too concerned over possible litigation to use such mechanisms. But the agency calculated that if all the potential litigants were satisfied, there would be no one to bring such an action.[193]

Some of the issues treated in this was were arguably not "minor." One *Federal Register* notice, for example, identified six issues for which the agency expected to issue a RIM or a TAR in the next three months, including:[194]

When is hazardous waste first subject to regulation?

Clarifying inconsistencies between RCRA and the Surface Mining Act in the treatment of coal mine wastes.

Application of the regulations to spills of listed chemicals and hazardous wastes.

Clarification of the regulations as they applied to ocean disposal of hazardous wastes.

Also, at this time, the agency began to utilize "program interpretative guidance"(PIGs) documents to assist the EPA regional offices in carrying out the hazardous waste program. Finally, there were also "program operations memoranda,"(POMs) to deal with generally applicable operational issues.[195]

The effect of these new procedural devices was quite clear. Someone not a party to the negotiations could easily be "out of the loop" as far as these various amendments were considered. The elaborate requirements of the Administrative Procedures Act were short-circuited and a much more informal process of interest group negotiation took its place.

Meanwhile, on September 30, a House Senate conference approved a bill that exempted most "special wastes" from regulation, pending the outcome of a series of studies of their characteristics and disposal options. The bill also delayed the cut-off on eligibility for interim status to November 19, 1980. Finally, as discussed previously, the administrator was given the authority to distinguish between new and existing sites.[196] But it seems that no one proposed any basic changes to the RCRA scheme.

BEST ENGINEERING JUDGMENT FOR TREATMENT, STORAGE, AND DISPOSAL FACILITIES: THE FINAL CHAPTER

The last significant decision of the Carter-Costle regime with regard to RCRA involved the TSDF rules.[197] The rules actually appeared several weeks into a new administration with markedly different views on environmental policy. The story of RCRA under the Carter-Costle Administration thus ended as it began, with EPA functioning essentially on its own during a transition.

Admittedly, EPA faced a difficult choice in deciding among design, performance, or ambient standards for TSDFs, as each faced serious problems. The issue was whether the administrative simplicity of nationally uniform technology-based rules outweighed the inefficiency that resulted from failing to tailor remedies to specific circumstances.

In other cases, environmentalists and state governments had pushed for nationally uniform rules. The former wanted as much cleanup as possible, the latter preferred to avoid pressures from industry for lower local standards. Neither group made the same kind of case here, however. Were environmentalists so unfamiliar with the technology that they were unsure of what to ask for? Were industry pressures to avoid "wasteful" uniformity becoming more effective as the political climate changed? For whatever reason, EPA, and especially Chris Beck, seemed very concerned with avoiding restrictive uniformity in the TSDF requirements.[198]

The device Beck chose for achieving this came to be known as *best engineering judgment* or BEJ. It was the direct successor to the notes at the end of the design standards in the proposed rules.[199] The effect of both schemes was to give significant discretion to those who reviewed permit applications.

The BEJ approach was first made public on October 8, 1980, when the agency published a notice of proposed rulemaking. There the agency discussed what sort of standards to write for TSDFs, with specific reference to the problem of ground water contamination. Having found fault with all the obvious options, EPA announced it had decided to combine several approaches. First, there would be design requirements. Second, facilities would be prohibited from degrading surface and ground waters "below certain standards." Finally, the Agency said it would use "nonnumerical health and environmental standards" to protect against any potential adverse effects not otherwise addressed. This part of the proposal amounted to saying that reviewers would use their best engineering judgment to see if the facility was acceptable.[200] The author of the notice, Gary Dietrich, shed further light on the idea in a speech on October 28. There he commented that some degradation of ground water would be permissible under the "best engineering judgment" standard, particularly where the ground water already was significantly impaired and would no longer benefit from protection.[201]

The BEJ approach was promptly and sharply criticized. The National Solid Waste Management Association argued that facilities operators needed specific numerical standards to know what was actually required of them.[202] The General Accounting Office was similarly critical, arguing that more specific regulations were needed to insure the public safety.[203]

Beck persisted. He found performance standards so disagreeable that he was liable to become angry at anyone who supported them.[204] But the EPA's lawyers actively resisted BEJ on the grounds that it was too vague to implement consistently.[205] The internal battle was so severe that the issue had to be brought to Costle. Having worked with Beck in Connecticut, however, Costle accepted Beck's view that state agencies could wisely utilize the discretion BEJ offered. Costle was also skeptical that EPA could anticipate the diversity and complexity of situations they would find in the field. Trying to prescribe a set of solutions before EPA really understood the problem would enormously strain its credibility. In addition, EPA did not have enough manpower or expertise, so the program had to draw the states into the battle aggressively. Costle reasoned that the more people involved in trying to come to grips with the problem the faster everyone would climb the "learning curve."[206]

As 1980 drew to a close, the atmosphere at the EPA, as in much of Washington, changed appreciably. Carter lagged badly in the opinion polls, and then lost the election by a wide margin. Political appointees began to leave and even career people—especially at EPA—viewed the upcoming Reagan presidency with apprehension. Many of them had been through Republican administrations before, but this one promised to be more hostile to EPA's traditional perspective. Everyone knew that decisions reached now might be reversed in a few weeks or months. Yet, the RCRA team deeply wanted to get as much of the program as possible "out the door" while there was still time.

Still, like other changes before it, the move to BEJ caused delays. On December 9, Costle reported to Judge Gesell that although the agency was "making good progress" it would not be able to promulgate the permanent TSDF standards by the court ordered deadline of Fall 1980. (By early December this was presumably obvious to the judge.) Finally he pointed out that if the TSDF regulations had to be reproposed it would be impossible to predict when they would be issued because "there will be a new Administrator by then."[207]

Yet, agency momentum was sufficient to carry along even with a new administration. Many deputy assistant administrators and even some assistant administrators stayed on for a time. On February 5, 1981, EPA finally issued the TSDF regulations, partially as a reproposal of the previously proposed rules and partially as proposed amendments to those rules. The new rules incorporated the BEJ concept.[208] On February 13, the agency also proposed standards under which *new* facilities could apply for and be granted interim permits during the period until the permanent TSDF rules became finalized.[209] Otherwise, there would be no way to even try to open new facilities until the TSDF rules were finalized.

REFLECTIONS

The long and torturous tale of RCRA rulemaking started with the failure of Congress to define either a problem or a solution. This made it difficult, if not impossible, for EPA to write rules to make sure that hazardous wastes, whatever they were, received proper storage and disposal.

No-one had the information, or the analytic or technical backround, to write a proper statute. The requisite understanding and experience did not exist. Not surprisingly, the outcome was legislation that was ambitious, ambiguous, and conceptual flawed. Certainly Congress was in no position to prescribe in great detail how EPA should implement the law.

Agency discretion was inevitable not only for the logical reasons we stressed in earlier chapters, but also because here there was little statutory language to go on. In circumstances like these it is difficult to know how to ensure that agency officials will make responsible decisions and be held accountable. The RCRA story suggests that the classic answer of pluralist politics—the "squeaky wheel"—is inadequate.

Once the agency began to develop rules, the implications of the statute gradually emerged. Industry saw that the law would require significant new controls on existing facilities, including those constructed to comply with prior environmental regulations. In other instances, the Act seemed to require firms to take very expensive precautions to deal with large volumes of material that no one had shown to pose a significant health hazard.

Congress was unable to resist vigorous industry lobbying, having neither a coherent problem definition nor a strategy of its own. There was no way to distinguish small concessions from large ones. Interest group politics proved to be a piecemeal negative force, not the synthetic positive process that was needed.

Nor were the courts able to use the "New Administrative Law" to produce the desirable degree of responsibility and accountability. In some respects those requirements performed as intended, EPA received "seven linear feet" of comments on the first set of proposed rules, and almost everyone connected with the process acknowledges that grappling with the resulting mass of data, criticism, and suggestions was a useful "learning experience." Indeed, the need to write background documents had a salutary impact on the decisions themselves. The lawyers got involved in reviewing drafts and prodded the Office of Solid Waste to provide explanations and justifications.

Yet very substantial resources were consumed meeting the requirements of administrative procedure and anticipating litigation. As James Rogers later put it, the preamble to any set of rules was EPA's "opening brief" for the law suits that would follow.[210] Moreover, as time wore on, the introduction of new ideas was inhibited by the recognition that they might require the reproposal of a rule, a time consuming process.

The procedural requirements were so burdensome that a family of devices—RIMs, TARs, PIGs, and POMs—was invented to get around them and to settle many of the lawsuits that the final rules provoked. Such devices could be viewed as both an unfortunate breakdown of procedural safeguards and an imaginative response to an overly inflexible legal system. But they may have had the effect of restricting participation in the final round of bargaining about the rules to those who were actual litigants.

In fact, the courts played a modest role. Judge Gesell appreciated the limits of his capacity to make "political" or "technical" decisions. He played the role of a stern yet sympathetic schoolmaster. As Gesell himself put it, in circumstances like these, only the agencies themselves, together with the Congress and the president, can provide the will and the resources to deal effectively with national problems.

Yet, at first, the whole enterprise was ignored by senior officials. Not surprisingly, TSCA and Superfund were the exciting new programs that carried top billing. Only slowly did RCRA graduate from being totally ignored to being treated as an aggravating nuisance. The agency was content for years to keep the group of regulation writers small and its resources limited.

Perhaps RCRA deserved to be given short shrift. But if that was EPA's view, it should have gone to the Congress to ask for delays or changes in the law. Of course, as we shall see in the next chapter, Love Canal and its aftermath made EPA's leaders reluctant to make such suggestions.

Instead, EPA plowed on, and on, and on. Driven first by unrealistic legislative deadlines, and then by unrealistic court deadlines (to which it had consented), it was always short of staff, lacked crucial studies, and was unable to distinguish the central from the peripheral. In the absence of a broad strategy, RCRA rule-making was a self-perpetuating crisis.

The agency's internal procedures proved inadequate to the task of creating such a complex program. Jorling, and his successor, Beck, reduced the "red border review" process and the attendant working groups and steering committees to a hollow ritual. If the Office of Sold Waste did not know enough to think clearly about its own problems, the Office of Policy and Management and the Office of General Counsel knew even less. In the absence of an explicit problem definition and an associated strategy, there was only a mass of intricate detail, a series of esoteric choices whose significance and consequences remained obscure.

How good were the RCRA regulations that emerged from this process? Despite later criticism, Jorling's decision to pursue a listing strategy seems essentially sound. Having EPA itself decide what wastes to include in the scheme avoided the unpredictability, cost, and enforcement problems of requiring generators to test their own wastes.

The final TSDF rules seem less satisfactory. The approval of new facilities was turned over to middle level state people who were given, thanks to the "Best Engineering Judgment" decision, very substantial discretion. Ensuring political accountability for the resulting licensing decisions is difficult. Indeed, accountability may really mean veto, as local opposition to every proposed site and facility continues to flourish.

There remains another strategic issue. The requirements on new sites are so stringent that the costs of using them may induce a significant number of waste generators to use illegal but cheaper disposal options. The difficulties of creating new sites may also foster such noncompliance since the actual stock of disposal facilities may not expand. This suggests that EPA should have been more concerned than it was about creating new sites and fostering new disposal technologies like high temperature incineration. Indeed, EPA could have looked for ways to encourage product and process redesign and recycling in order to lower the volume of waste requiring disposal. As public opposition to all disposal options has built, one is struck by how little attention this approach received from EPA during the Carter years.

The EPA's own lack of internal strategic conversation made it impossible to educate anybody else. Even now, most Americans are in the dark about the issues and choices involved in hazardous waste control. The fundamental conundrum, that waste in some form is physically inevitable and that it will just as inevitably have to be put somewhere, is not widely appreciated. Nor does the public understand that the health effects of hazardous waste, whatever they are, come primarily by the groundwater-to-drinking water route.

The environmental community shares responsibility for this state of affairs. Arguing that every risk is too great, they have made demands that are impossible to satisfy. By opposing every site, calling into question every plan, they have avoided offering any realistic suggestions about what should be done. And in doing so, they have discouraged EPA from being honest with the public, since attempts at straightforwardness have often been characterized as proindustry propaganda.

Yet EPA's leaders are hardly beyond reproach. Long before they were in a position to produce a defensible set of rules, the agency's top managers recognized that the scale and meaning of hazardous waste problems had been misunderstood. Organizational and professional interests, however, proved too strong. The RCRA rulemaking represented the triumph of "getting the job done" over thoughtful strategic analysis and a long run view of political responsibility.

Agency officials need to understand both the ethical concerns and technical realities that lie behind the policies they administer. They need to expound this understanding, and its relationship to their decisions, unambiguously and unapologetically in the full light of public scrutiny. Such acts of candor can put officials at risk of their jobs. But sometimes encouraging reconsideration of an issue is more responsible than passive, apparently successful, management. By ensuring that the debate—both in public and within the agency—is thoughtful and considerate, leadership can create deliberative opportunities.

NOTES

1. The Resource Conservation and Recovery Act of 1976, 42 U.S.C. 6901–6987.

2. On the concept of residuals management, see A. Myrick Freeman III, Robert Haveman and Allen V. Kneese, *The Economics of Environmental Policy* (New York: John Wiley, 1973), 171–175.

3. This issue is elaborated in John D. Graham, Laura C. Green and Marc J. Roberts, *In Search of Safety: Chemicals and Cancer Risk* (Cambridge MA: Harvard University Press, 1988).

4. Jack Lehman, Director of Hazardous and Industrial Waste Division, Office of Solid Waste, Environmental Protection Agency, interview with Marc Roberts, Washington D,C., 16 Sept. 1981.

5. Ibid.

6. Ibid.

7. Roger Strelow, former Council of Environmental Quality staff, interview with Marc Roberts, Washington D.C., 15 Sept 1981.

8. Strelow interview, 15 Sep 1981; William Kovacs, interview with Glenn Roberts, Washington D.C., 16 Sept 1981.

9. Kovacs interview, 16 Sept 1981.

10. Lehman interview, 16 Sept 1981.

11. *BNA Environmental Reporter*, 29 Mar. 1974, 1981.

12. Lehman interview, 16 Sept. 1981.

13. *BNA Environmental Reporter*, 9 Apr. 1976, 2073–2074.

14. *Congressional Record*, 94th Congress, 2d Session, 30 June 1976: S11097.

15. See Kovacs and Klucsik, "The New Federal Role in Solid Waste Management: The Resource Recovery Act of 1976," *Columbia Journal of Environmental Legislation* 3 (1977):212.

16. *Congressional Record*, 94th Congress, 2d Session, 27 Sep 1976:H1182.

17. Kovacs interview, 16 Sept. 1981. *Weekly Compilation of Presidential Documents* 12 (22 Oct. 1976):1560.

18. Kovacs interview, 16 Sept. 1981.

19. RCRA 42 U.S.C. 6901–6987; see also *BNA Environmental Reporter*, 29 Oct 1976, 947–948.

20. Douglas M. Costle, joint interview of participants in the RCRA rule making process, Kennedy School of Government, Cambridge, Massachusetts, 15 July 1981. Other participants were: Thomas Jorling, former assistant administrator for water and Solid Waste; James Rogers, deputy general counsel, Water Quality Division, Office of the General Counsel; Eckhart Beck, former assistant administrator, water and solid waste; Gary Dietrich, former director, Office of Solid Waste.

21. Kovacs interview, 16 Sept. 1981.

22. Lehman interview, 16 Sept. 1981.

23. Kovacs interview, 16 Sept. 1981

24. Kovacs interview, 16 Sept. 1981.

25. RCRA 42 U.S.C. 6292.

26. RCRA 42 U.S.C. 6921, Section 3001.

27. RCRA 42 U.S.C. 6924, Section 3004.

28. RCRA 42 U.S.C. 6925, Section 3005.

29. RCRA 42 U.S.C. 6923, Section 3003.

30. RCRA 42 U.S.C. 6926, Section 3006.

31. RCRA 42 U.S.C. 6973, Section 7003.

32. RCRA 42 U.S.C. 6943–6944, Sections 4002–4003.

33. RCRA 42 U.S.C. 6913, Section 2003.

34. RCRA 42 U.S.C. 6961, Section 6001.

35. RCRA 42 U.S.C. 6914, Section 2004.

36. See Louis J. Cassarett and John Doull, eds., *Toxicology: The Basic Science of Poison* (New York: MacMillan, 1975), Chapter 2.

37. RCRA 42 U.S.C. 6903, Section 1004, Paragraph 5A.

38. RCRA 42 U.S.C. 6903, Section 1004, Paragraph 5A; RCRA 42 U.S.C. 6921, Section 3001 a.

39. S.2150, Section 212, *Congressional Record*, 94th Congress, 2d Session, 30 June 1976:S21430.

40. S.2150, Section 212.

41. House Committee on Interstate and Foreign Commerce, Subcommittee on Transportation and Commerce, *Resource Conservation and Recovery Act*, 96th Congress, 1st Session, 27 Mar 1979, Committee Serial No. 96–31, 103.

42. This is a problem in constructing *any* test. Note for example the ongoing discussion of the effect of false negatives and false positives in AIDS testing.

43. When compounds do not degrade in nature and are found in the food of a species, they may build up in the body of that species and produce high dose levels. This is

especially true for organisms at the top of a food chain. The classic example is the effect of DDT on the reproductive system of predatory birds.

44. On the issue of uncertainty and water pollution transport see Bear J. and Verraigt, A, *Modeling Groundwater Flow and Pollution*(Dordrecht Holland: D. Reidel Publishing Company, 1987), and Wang H.F., and Anderson, M.P., *Introduction to Groundwater Modelling:Finite Differences and Finite Element Methods,*(San Franscisco: Freeman and Company, 1982.

45. Thomas Jorling, former assistant administrator for water and solid waste, joint interview, 15 July 1981.

46. This view was provided by an EPA official who prefers to remain anonymous.

47. Lehman interview, 16 Sept. 1981.

48. Memo from Jack Lehman, to Sheldon Myers, Office of Solid Waste, EPA, 30 Nov. 1976.

49. Lehman to Myers memo, 30 Nov. 1976.

50. Transcript, *First Public Meeting on the Resource Conservation and Recovery Act*, Washington D.C., 16 Dec 1976, 21.

51. *First Public Meeting on the Resource Conservation and Recovery Act*, 45.

52. Lehman interview, 16 Sept. 1981.

53. Lehman interview, 16 Sept. 1981.

54. Eckhart Beck, former assistant adminstrator, Water and Solid Waste, joint interview, 15 July 1981.

55. Gary Dietrich, former director, Office of Solid Waste, joint interview, 15 July 1981.

56. *BNA Environmental Reporter*, 1 Apr. 1977, 1839.

57. Lehman interview, 16 Sept. 1981.

58. *BNA Environmental Reporter*, 29 Apr. 1977, 1982.

59. *BNA Environmental Reporter*, 20 May 1977, 88.

60. *BNA Environmental Reporter*, 29 July 1977, 495.

61. Stefan Plehn, former deputy assistant administrator for Water and Waste Management, interview with Marc Roberts, Washington D.C., 15 Sept. 1981.

62. Memo from Jack Lehman to Stefan Plehn, 18 Nov. 1977.

63. Transcript, *Public Meeting: Strategy for the Implementation of RCRA*, Arlington, Virginia, 19 Jan. 1978, 20.

64. *EPA Activities under the Resource Conservation and Recovery Act: Annual Report to the President and Congress Fiscal Year 1977*, USEPA 1 Feb. 1978, 27.

65. American Paper Institute comments on proposed RCRA regulations, submitted to Office of Solid Waste, USEPA, 7 Feb 1978, Docket Number D-357.

66. EPA, *Proposed Guidelines for State Hazardous Waste Programs*, (1 Feb. 1978) 43 FR 4371.

67. *BNA Environmental Reporter*, 10 March 1978, 1738.

68. Senate Committee on Governmental Affairs, Subcommittee on Oversight of Governmental Management, *Oversight of Hazardous Waste Management and the Implementation of the Resource Conservation and Recovery Act*, 96th Congress, 1st Session, 1 Aug 1979, 452. (Hereafter referred to as *Senate Hearings*.)

69. *Senate Hearings*, 440; see also Jorling, joint interview, 15 July 1981.

70. Costle, joint interview, 15 July 1981.

71. Jorling, joint interview, 15 July 1981.

72. Jorling, joint interview, 15 July 1981.

73. Jorling, joint interview, 15 July 1981.

74. See colloquy between Jorling and Sen. Levin (D-MI.), *Senate Hearings*, 452–472.

75. EPA, *Hazardous Waste Proposed Guidelines and Regulations* and *Proposal on*

Identification and Listing, (18 Dec. 1978) 43 FR 58950. (Hereafter referred to as *Proposed Rules*.)

76. *Proposed Rules*, 58952–58953.

77. *Proposed Rules*, 58952–58953.

78. *EPA Slow in Generating RCRA Subtitle D Regulations*, GAO Report 15 June 1978.

79. *BNA Environmental Reporter*, 22 Sept. 1978, 976.

80. *BNA Environmental Reporter*, 11 Aug. 1978, 591.

81. *BNA Environmental Reporter*, 11 Aug. 1978, 591.

82. *New York Times*, 14 Aug. 1978, 1.

83. *Time*, 14 Aug. 1978, 46 *Newsweek*, 21 Aug. 1978, 25–28.

84. Jorling, joint interview, 15 July 1981.

85. Rogers, joint interview, 15 July 1981.

86. Friedman interview, 15 Sept. 1981.

87. Rogers, joint interview, 15 July 1981.

88. Jorling, joint interview, 15 July 1981.

89. Memo from Stefan Plehn, to Tom Jorling, 20 Sept. 1978.

90. Testimony of William Sanjour, *Senate Hearings*, 330–334.

91. In order to ameliorate this problem, EPA was currently working on "pretreatment" requirements for industrial waste sources which were connected to municipal sewage treatment systems. Such rules proved extremely difficult to enforce in view of the large number and very small size of many of these discharges and their episodic nature.

92. Sanjour, *Senate Hearings*, 351.

93. Sanjour, *Senate Hearings*, 375; also Dietrich, joint interview, 15 July 1981.

94. *Senate Hearings*, 351.

95. *Senate Hearings*, 375–376; see also 43 FR 58991–58992 (18 Dec 1978).

96. *Senate Hearings*, 371.

97. Plehn interview, 15 Sept. 1981.

98. *BNA Environmental Reporter*, 27 Oct. 1978, 1212.

99. Testimony of Hugh Kaufman, House Interstate Commerce Committee, Subcommittee on Oversight and Investigations, *Hearing*, 95th Congress, 2nd Session, 30 Oct. 1978, Committee Serial No. 95–183, 305–307.

100. Jorling, joint interview, 15 July 1981.

101. Costle and Jorling, joint interview, 15 July 1981.

102. *BNA Environmental Reporter*, 1 Dec. 1978, 1367–1368.

103. *Illinois* v *Costle* nos. 78–1689, -1715, -1734, -1899. D.D.C. *Environmental Law Reporter* 9:20243.

104. Order was filed 7 Nov. 1978, received by civil docket 13 Nov. 1978, *Illinois* v. *Costle* Docket Number 78–1689.

105. Proposed Schedule for Promulgation of Solid Waste Regulations and Memorandum in Support Thereof, received by civil docket, *Illinois* v. *Costle* Docket Number 78–1689, 4 Dec. 1978.

106. Proposed Schedule for Promulgation of Solid Waste Regulations and Memorandum in Support Thereof, received by civil docket, *Illinois* v. *Costle* Docket Number 78–1689, 4 Dec. 1978.

107. Friedman interview, 15 Sept. 1981.

108. *Proposed Rules*, 58946–59027.

109. Costle, joint interview, 15 July 1981.

110. *Proposed Rules*, 58947

111. *Proposed Rules*, 58953.

112. *Proposed Rules*, 58953, 58959–58967.

113. Jorling, joint interview, 15 July 1981.
114. *Proposed Rules*, 58947.
115. *Proposed Rules*, 58947.
116. *Proposed Rules*, 58970.
117. *Proposed Rules*, 58982.
118. *Proposed Rules*, 58982.
119. *Proposed Rules*, 58983.
120. RCRA 33 U.S.C. 1842, Section 402.
121. *Proposed Rules*, 58993.
122. RCRA 33 USC 1842, Section 305.
123. *Proposed Rules*, 58984.
124. *Proposed Rules*, 58987–58988.
125. *Illinois* v *Costle*, D.D.C., *Environmental Law Reporter* 9:20243.
126. *Environmental Law Reporter* 9:20244.
127. *Environmental Law Reporter* 9:20244.
128. *Environmental Law Reporter* 9:20244.
129. Administrator's First Quarterly Report on the Status and Development of Regulations Under The Resource Conservation and Recovery Act of 1976. Received Civil Docket, *Illinois* v. *Costle* Docket 78–1689, 2 Apr. 1979.
130. *BNA Environmental Reporter*, 15 June 1979, 227.
131. *BNA Environmental Reporter*, 13 July 1979, 653.
132. *BNA Environmental Reporter*, 19 Oct. 1979, 1423.
133. *BNA Environmental Reporter*, 13 July 1979, 653.
134. Costle press release accompanying third quarterly report.
135. *BNA Environmental Reporter*, 26 Oct. 1979, 1449.
136. American Petroleum Institute comments on 19 Sept. 1979 Federal Register Notice, EPA Docket Number D-1996.
137. Memorandum filed with Judge Gerhard Gesell, received by civil docket 5 Nov. 1979, *Illinois* v. *Costle* Docket Number 78–1689.
138. Rogers, joint interview, 15 July 1981.
139. *BNA Environmental Reporter*, 21 Dec. 1979, 1673.
140. Notification by Administrator of reproprosal of closure and post-closure financial responsibility requirements, received by civil docket Apr. 1980, *Illinois* v. *Costle* Docket Number 78–1689.
141. *BNA Environmental Reporter*, 30 Mar. 1979, 2245.
142. House Committee on Interstate and Foreign Commerce, Subcommittee on Transportation and Commerce, *Resource Conservation and Recovery Act*, 96th Congress, 1st Session, 27 Mar. 1979, Committee Serial No. 96–31, 103.
143. Jorling, joint interview, 15 July 1981.
144. House Committee on Interstate and Foreign Commerce, *Report to Accompany HR 3994, RCRA Amendments of 1979*, 96th Congress, 1st Session, 15 May 1979, H. Rpt. 96–191, 5.
145. Jorling, joint interview, 15 July 1981.
146. Testimony of Costle, House Committee on Interstate and Foreign Commerce, Subcommittee on Oversight and Investigations, *Hazardous Waste Disposal: Part 2*, 96th Congress, 1st Session, Hearings 21,22 Mar.; 5,10 Apr.; 16,23,30 May; 1,4,5,15,18 and 19 June 1979, Committee Serial No. 96–49, 1305.
147. Testimony of Jorling, *Hazardous Waste Disposal: Part 2*, 1673.
148. *Senate Hearings*, 335–401.
149. *Senate Hearings*, 330–334, 385–387.
150. *Senate Hearings*, 455–456.

151. *Senate Hearings*, 455–465. See also *BNA Environmental Reporter*, 10 Aug. 1979, 993–994.

152. Letter from Tom Jorling to Sen. Carl Levin (D-MI.), chairman, Subcommittee on Oversight, Senate Committee on Government Affairs, 7 Aug. 1979.

153. *Hazardous Waste Guidelines and Regulations, Supplemental Proposed Rule*, (22 Aug. 1979) 44 FR 49402–49204.

154. House Committee on Interstate and Foreign Commerce, Subcommittee on Oversight and Investigations, *Hazardous Waste Disposal*, 96th Congress, 1st Session, 27 Sept. 1979, Committee Print 96–IFC 31, III.

155. *Hazardous Waste Disposal*, 35.

156. *Hazardous Waste Disposal*, 46.

157. *Hazardous Waste Disposal*, 39–43.

158. *BNA Environmental Reporter*, 29 Feb. 1980, 2057.

159. *Hazardous Waste Disposal*, 43.

160. *BNA Environmental Reporter*, 23 Mar. 1979, 2135–2136.

161. *BNA Environmental Reporter*, 23 Mar. 1979, 2136.

162. Costle, joint interview, July 15

163. Costle, joint interview, 15 July 1981.

164. Costle, joint interview, 15 July 1981.

165. *BNA Environmental Reporter*, 29 Dec. 1978, 1613.

166. Rogers, joint interview, 15 July 1981.

167. Comments on *Proposed Rules*, EPA Docket Number D630–1698.

168. Comments on *Proposed Rules*, EPA Docket Number D630–1698.

169. Jorling, joint interview, 15 July 1981.

170. Dietrich, joint interview, 15 July 1981.

171. Costle, joint interview, 15 July 1981.

172. Dietrich, joint interview, 15 July 1981.

173. Third Quarterly Report of Administrator on the Status and Development of Regulations Under the Resource Conservation and Recovery Act of 1976, received by civil docket 15 Oct. 1979, *Illinois* v. *Costle* Docket Number 78–1689.

174. Costle, joint interview, 15 July 1981.

175. Affidavit of Administrator Douglas M. Costle Notifying Court that Promulgation of Certain RCRA Regulations According to This Court's Schedule Appears Unlikely, received by civil docket 6 June 1979, *Illinois* v. *Costle* Docket Number 78–1689.

176. Jorling, joint interview, 15 July 1981.

177. Beck, joint interview, 15 July 1981.

178. Costle, joint interview, 15 July 1981.

179. Beck, joint interview, 15 July 1981.

180. *BNA Environmental Reporter*, 19 Oct. 1979, 1423.

181. Costle, joint interview, 15 July 1981.

182. *BNA Environmental Reporter*, 9 May 1980, 35.

183. *The New York Times*, 30 Apr 1980, II:6.

184. *BNA Environmental Reporter*, 9 May 1980, 35.

185. *BNA Environmental Reporter*, 9 May 1980, 35.

186. *BNA Environmental Reporter*, 9 May 1980, 35.

187. *BNA Environmental Reporter*, 3 Oct. 1980, 764; *BNA Environmental Reporter*, 17 Oct. 1980, 813–814.

188. *BNA Environmental Reporter*, 13 June 1980, 237.

189. House Committee on Interstate and Foreign Commerce, Subcommittee on Oversight and Investigation, *Hazardous Waste Matters*, 96th Congress, 2nd Session, 2 July 1980, Committee Serial No. 96–200, 66–67.

190. 45 FR 55386 (19 Aug. 1980).

191. Rogers, joint interview, 15 July 1981.

192. Rogers, joint interview, 15 July 1981.

193. Costle, joint interview, 15 July 1981.

194. 45 FR 44387 (1 July 1980).

195. *BNA Environmental Reporter*, 21 Sept. 1980, 684–685.

196. 40 CFR Parts 122, 260, 264, published 5 Feb. 1981: 11126–11177. (Hereafter referred to as *TSDF Rules*.)

197. Three alternatives are discussed in a document issued by EPA as a supplement to the Notice of Proposed Rulemaking. 45 FR 66817–66819 (8 Oct. 1980).

198. Beck, joint interview, 15 July 1981.

199. 45 FR 66819–66822 (8 Oct. 1980).

200. 45 FR 66819–66822 (8 Oct. 1980).

201. *BNA Environmental Reporter*, 7 Nov. 1980, 989–990.

202. *BNA Environmental Reporter*, 7 Nov. 1980, 990.

203. *BNA Environmental Reporter*, 11 July 1980, 383–384.

204. Rogers, joint interview, 15 July 1981.

205. Rogers, joint interview, 15 July 1981.

206. Costle, joint interview, 15 July 1981.

207. *BNA Environmental Reporter*, 12 Dec. 1980, 1226–1227.

208. 46FR11127, 5 Feb. 1981, reprinted in *BNA Environmental Reporter* 13 Feb. 1981, 1964.

209. 40 CFR Parts 122, 267, published 13 Feb. 1981: 12414–12433.

210. Rogers, joint interview, 15 July 1981

5
Passing Superfund

In November 1980, the U.S. Congress gave final approval to legislation providing for the cleanup of abandoned hazardous waste dumps and spills of toxic chemicals. This law, nicknamed Superfund, represented the major legislative effort of the Environmental Protection Agency during the Ninety-Sixth Congress.[1] At first glance, Superfund seems easy to explain. It was passed slightly more than two years after one of the most widely publicized environmental disasters in history, Love Canal, which first gained widespread media attention in August of 1978. Two months later, an EPA working group completed a draft that was similar to the eventual legislation. The bill was passed in 1980 by the Congress that was elected the same fall that Love Canal came into prominence. This seemingly direct relationship between disaster and remedy, however, does not survive a close scrutiny of what actually took place at Love Canal, or in Washington.

When Love Canal came to public attention, no epidemic had occurred among those living close to the abandoned canal. Then, as now, no reliable epidemiological studies showed that area residents were subject to greater health risks than the population at large.[2] The few studies that claimed to reveal serious problems were immediately and severely criticized within the scientific community and have been largely discredited.[3] Even an administration less concerned with stemming inflation and reducing the federal budget might have waited for more substantial evidence that Love Canal was indeed a catastrophe, and that other similar catastrophes might occur across the country, before creating a major new federal regulatory program.

If Love Canal was not a "ticking time bomb," as it came to be called by journalists and public officials alike, then Superfund requires further explanation. This chapter examines why so many people became convinced that abandoned hazardous waste sites merited such aggressive federal action. It also looks at the roles Congress and the executive branch played in both defining the problem and devising the remedy. Additionally, it tries to assess what this experience has to teach about the government's capacity to deal with matters of great scientific, legal, and economic complexity.

Our evaluation relies heavily on the criteria developed in the first chapter. A prolonged public discussion of an issue like abandoned waste sites provided a fine opportunity to inform all concerned about the nature of the problem and the merits of alternate responses. We want to understand what the Superfund debate taught its participants and the public at large and how well it illuminated the environmental, economic, and constitutional issues at stake.

In addition, we want to know whether the definition of the problem and the policy design that was arrived at permitted and encouraged politically responsible action. Did the law embody a strategy for addressing the problem that was sufficiently coherent to enable Congress and the president, after the fact, to effectively evaluate agency performance? Did it provide the appropriate managerial tools to enable the agency to successfully implement that strategy?

DEFINING THE PROBLEM

August is a slow season in the news business. The visual appeal of the Love Canal incident—purple lawns, multicolored basement walls—certainly increased its allure to the media. But, the phrase "ticking time bomb" was not a journalistic invention. It appeared initially as the subtitle of the first important government report on the situation.[4] During the whole controversy, that phrase came trippingly off the tongues of government officials, newsmen and neighborhood activists alike. Only the chemical industry and a few Republican congressmen, most notably David Stockman, took a different view. The government's adoption of the "time bomb" metaphor, and the problem definition it embodied, was due both to the perceived political risks of trying to refute it, and the opportunities it presented to the State of New York for deflecting responsibility to the federal government.

The sense of crisis at Love Canal was greatly increased by a government report issued in August 1978 and further heightened by a second government study released eighteen months later.[5] The dire conclusions, implicit or explicit, in both these documents were later either modified or repudiated. Yet, the original statements, not the refutations, dominated public discussion. A scientific Gresham's Law was at work. Poorly constructed, poorly presented studies drove out more careful and scrupulous research.

These reports were not intended to meet the test of scientific judiciousness but rather to satisfy specific legal and administrative requirements. The first made use of the phrase "public health emergency" to ensure that the Love Canal area would qualify for federal emergency relief funds.[6] The second was intended to provide government lawyers with just enough evidence to satisfy the legal requirements for proceeding against hazardous waste dumpers.[7] Not surprisingly, these nuances were lost on the local residents, as well as the press, who reported the findings as conclusive proof of serious health damage.

Located in Niagara Falls, New York, Love Canal was an abandoned canal leading into the Niagara River. The Hooker Chemical Company had disposed of more than 21,000 tons of chemical waste into it between 1942 and 1952. In 1953, after the canal became full, it was covered with earth and clay and deeded over to the Niagara Falls School Board for one dollar. The old canal became a school and a playground. The vacant land along the canal was developed into modest two- and three-bedroom homes, with backyards bordering what had been the canal.[8]

Although there were occasional local complaints by residents, the Love Canal situation remained quiescent until 1976. In that year, a joint U.S.-Canada commission responsible for monitoring the condition of the Great Lakes detected the insecticide Mirex in fish from Lake Ontario. The New York Department of Environmental Conservation conducted a study to find the sources and Love Canal was identified as a major contributor.[9]

This finding created much local concern. In October 1976, the local newspaper published an extensive front page story on Love Canal. During the fall and winter the newspaper published several additional articles about the potential public health threat posed by the canal and aired the complaints of area residents concerning damage to basements and backyards. In April 1977, the city of Niagara Falls hired Calspan Inc. to institute a groundwater pollution reduction program.[10]

By this time, the situation had come to the attention of the local congressman, John La Falce. He toured the area in early September 1977, and returned later that month with EPA regional officials. By the spring of 1978, both state and federal environmental agencies had become sufficiently concerned to initiate groundwater and epidemiological studies.[11]

Love Canal was transformed from a local problem into a worldwide news event as a result of an order issued by New York State Health Commissioner Robert Whelan on August 2, 1978, declaring a public health emergency. He ordered the city of Niagara Falls to conduct further health studies and to stop the migration of toxic substances from the canal. He also urged area residents to stay out of their basements and avoid eating anything from their gardens. Pregnant women living on the two streets adjacent to the canal were to temporarily move away, as were families with children under two.[12]

In September, Whelan issued a report providing a chronology of state actions and findings through the declaration of an emergency. This report, "Love Canal: Public Health Time Bomb," gave rise to the time bomb metaphor. It described the situation at Love Canal as "profound and devastating, [a] modern day disaster" constituting a condition of "great and imminent peril."[13]

The alarmist tone and the declaration of an emergency were not based on any startling new discoveries. For the health commissioner to obtain the funds for serious studies, state law required a finding of "great and imminent peril." Likewise, the federal government would only designate Love Canal a disaster area if the commissioner declared an emergency. Apparently no one anticipated the impact that this language would have on the press, the public, and particularly on the residents of the Love Canal neighborhood. This reaction was typified by one resident, Lois Gibbs, who stood up at a public meeting convened by Whelan and exclaimed "You're murdering us."[14]

The anger and fear of community members was quickly channeled into an extremely effective organization. Mustering the tools of dramatization and protest perfected by an earlier generation of peace and civil rights activists, the residents focused worldwide attention on what they perceived to be their plight.[15]

To complicate matters, 1978 was an election year in New York. Governor Hugh Carey was being challenged in the Democratic primary by Lieutenant

Governor Mary Ann Krupsak. Early polls indicated that Carey could be beaten. Coming from New York City, he was considered vulnerable in upstate areas like Buffalo and Niagara Falls, particularly since Krupsak was an upstater.[16]

Once his health commissioner had declared Love Canal "an imminent peril," Carey had to act. Krupsak was already accusing him of ignoring the matter. The primary was a mere six weeks away. Carey created a multiagency Love Canal task force and made several trips to the area. On August 8, he pledged to provide state aid. On August 15, he rebuked the mayor for allowing construction work to proceed at the site before safety issues had been resolved. On August 24, he reassured the residents that any person with health problems would be taken care of. He also agreed that all 239 families living on the two streets immediately adjacent to the canal would be relocated at state expense. Love Canal now had the gubernatorial seal of approval as a major cataclysm.[17]

The federal government also quickly responded to the furor caused by the Whelan Report. On August 5, the administrator of the Federal Disaster Assistance Administration (FDAA), William Wilcox, toured the area accompanied by Governor Carey, Congressman La Falce, and representatives of the home owners association. Two days later, the U.S. Senate approved a resolution endorsing aid to Love Canal residents by voice vote. That evening, President Carter declared an emergency in the area. No funds were provided, but the FDAA regional administrator was instructed to coordinate relief efforts.[18]

The residents' impression that they would quickly be taken care of changed as state and federal governments began to consider exactly what should be done and who should pay for it. First, the regional FDAA administrator announced that the federal government would not reimburse the state for the cost of purchasing the homes and property of the relocated families. The FDAA was apparently adopting a narrow definition of "emergency," and federal assistance would only be available for remedial construction measures.[19]

Governor Carey also became more cautious. The realization that federal money would not be readily available, the increasingly tenuous evidence for the existence of health effects at Love Canal, and his primary election victory all served to change his posture. On October 12, the state announced that the relocation request of nineteen families, who felt themselves to be endangered by the remedial construction taking place, had been denied. In early 1979, having been reelected, Carey was quoted as saying that no additional people would be moved and that people could not be protected from all risk. A Carey aide compared Love Canal's health statistics favorably with those in the South Bronx.[20] In August 1979, the state health commissioner announced that not all families contemplating pregnancy would be moved.[21]

The growing moderation on the part of government could not reverse the momentum generated by its earlier pronouncements. Caution appeared as callousness, as the media renewed its attention to Love Canal. In October 1979, public television viewers were treated to a documentary on toxic waste entitled "A Plague on Our Children" that contained a long segment about Love Canal.[22] A few weeks later, Jane Fonda made a well publicized visit to the area. Tears streaming from her eyes, she called for the evacuation of all local residents.[23]

Fear was also increased by studies undertaken by Beverly Paigen, a cancer researcher at the Roswell Institute in Buffalo, who had consulted for the Homeowners Association since early 1978. In March 1979, she testified before a congressional committee, presenting the results of a door-to-door health survey that she had conducted at Love Canal. Her findings led her to suspect increased rates of miscarriages, birth defects, nervous breakdowns, and diseases of the urinary system.[24]

Her findings were later to be discredited. In the words of a panel convened by Governor Carey to review the scientific studies relating to Love Canal:

> . . . her [Paigen's] data cannot be taken as scientific evidence for her conclusions. The study is based on largely anecdotal information provided by questionnaires submitted to a narrowly selected group of residents. There are no adequate control groups, the illnesses cited as caused by chemical pollution were not medically validated . . . The Panel finds the Paigen report literally impossible to interpret. It cannot be taken seriously as a piece of sound epidemiological research.[25]

But, this review did not appear until the fall of 1980. In the spring of 1979 Paigen's work was the only epidemiological evidence available and was widely reported.

Since Paigen worked for the Love Canal residents, some doubted her objectivity. The EPA itself issued a report in the spring of 1980, however, that appeared to present even more dire findings. That report too was discredited and disavowed almost as soon as it was released, but not before it conveyed the impression that the federal government had substantiated the ticking time bomb view.

In late 1979, the Justice Department, on behalf of EPA, initiated a series of lawsuits against the Hooker Chemical Company. To support this litigation, government lawyers asked EPA to provide evidence of health damage. It was not necessary to link exposures at Love Canal directly to disease. All the government had to demonstrate was that such a connection was possible. As a government lawyer later explained, "The laws we're seeking to implement all have to do with the idea of endangerment. We need to show that there is risk." This view relied on a recent federal court decision that did not demand evidence of actual harm before granting relief. It simply required that there be "reasonable medical concern for the public health."[26]

The lawyers had neither the time nor the need for a well-constructed epidemiological investigation. A chromosome damage study would suffice. There were serious scientific problems with this type of study as chromosome damage is not associated with many of the conditions allegedly present at Love Canal. Nonetheless, evidence of excess chromosome damage would meet the court test of "reasonable medical concern." Furthermore, it could be done quickly and inexpensively.

Dante Picciano, whose firm performed the study, described the decision as follows:

> I met with officials of the EPA's Hazardous Waste Enforcement Division and the Health Effects Division on January 16, 1980. I asked the Enforcement officials the following questions. "What is the question you want answered? Do you want to know if the residents of Love Canal have an increased level of chromosomal dam-

> age, or do you want to know if chemical exposures at Love Canal are causing an increase in the level of chromosomal abnormalities in the Love Canal residents." I explained the differences . . . and they told me that the Justice Department was interested in determining if the residents of Love Canal had an increased level of chromosomal damage. They wanted the results as soon as possible.[27]

Picciano's study suffered from an additional difficulty. Since chromosome damage is quite common, it can be caused by x-rays for example, the only way to determine whether the damage level in a particular group is excessive is to compare it with a control group. Ideally, the two groups would be matched in terms of age, sex, medical history, geography, and other relevant variables. Instead, Picciano simply used data from another population he had studied earlier for an unrelated purpose. Picciano claimed that he had wanted to establish simultaneous controls but that EPA had said that such a step was unnecessarily time consuming.

Several other aspects of the study were also severely criticized by two review panels that EPA subsequently established. The EPA ultimately disavowed the report. Stephen Gage, assistant administrator of EPA for Research and Development, described the study as a "fishing expedition." He attributed its flaws to:

> the urgency to prepare for litigation on Love Canal. The government scientists and lawyers involved did not raise questions about the study's design, conduct and quality with individuals at the appropriate management level for resolution.[28]

Picciano found evidence of what he characterized as excess chromosome damage among the thirty-six Love Canal residents he studied. He forwarded his report to EPA on May 15. Even before the report was received, presidential aide Jack Watson's staff had picked up a rumor that its results would be leaked to the press. On May 16, Watson convened a meeting at the White House to discuss the political and policy implications of the report. The meeting was attended by officials of EPA, Justice, the Federal Emergency Management Agency (FEMA, the successor to FDAA), and various health agencies. A decision was made to delay any response for a week until a review panel, headed by David Rall of the National Institute of Environmental Health Sciences, had a chance to examine the study and report back.

That very day the Picciano report was leaked to the press. The next day *The New York Times* carried a front page story concerning the discovery of chromosome damage among Love Canal residents. To the public and to the residents, "chromosome damage" was synonymous with dread disease. Their worst fears concerning cancer and birth defects now appeared justified. The resulting hysteria left EPA with no choice but to release the report immediately, without scientific review.[29]

The subsequent shockwave of publicity was enormous. From the middle of May to the middle of June of 1980, Love Canal was virtually a daily feature of network newscasts. In addition, it was featured on the news programs "Today," "The MacNeil-Lehrer Report," "Sixty Minutes," and "Good Morning America." Phil Donahue devoted a full hour of his talk show to the story, busing forty area residents to Chicago for the taping, including Lois Gibbs and Mayor O'Laughlin of Niagara Falls.[30]

The New York Times published a scathing editorial attacking the Picciano report. It criticized EPA for choosing an obscure firm to conduct the research. It also mentioned that Picciano had formerly been an employee of Dow Chemical but had left after a bitter dispute with the company about the quality of his work.[31] The editorial critique however, could not supercede the influence of the earlier page one story that treated the reports' findings as fact.

Governor Carey immediately attacked both the report and EPA's handling of it. He estimated that a relocation of the additional 710 families whose risks were similar to those in the study would cost 30 to 60 million dollars rather than the 3 to 5 million dollars estimated by EPA Deputy Administrator Barbara Blum. His previous experience had taught him just how costly it was to provoke people's fears. "Anyone can raise a fire alarm, it takes a lot of money to send the fire engines out," he said.[32]

Carey then launched the single most important effort to defuse the Love Canal situation. He established a blue ribbon panel to review the scientific evidence relating to the area's public health risks. The panel was chaired by the noted researcher and author, Lewis Thomas, chancellor of the Memorial Sloan-Kettering Cancer Center. Members included Arthur C. Upton, chairman of the Department of Environmental Medicine at the New York University School of Medicine, and Saul Farber, that school's dean. They examined the reports issued by Whelan, Paigen, Picciano, and others and found them all to be inconclusive. They were particularly scathing about the Picciano report.

> The design, implementation, and release of the EPA chromosome study has not only damaged the credibility of science, but exacerbated any future attempts to determine whether and to what degree the health of the Love Canal area residents has been affected.[33]

Despite the panel's august membership and the strength of its critique, its report had no detectable influence on subsequent events. The banner headlines prompted by the studies that the report had debunked created a set of convictions about Love Canal that could not be altered.

After almost two years of dilatory government action, the Picciano report refocused attention on the situation. On May 19, the Love Canalers dramatized their plight by holding two EPA officials hostage at the headquarters of the Homeowners Association. Although the officials were well treated, and everyone remained in good humor, it kept the matter in the headlines."[34] On May 20, the White House tacitly acknowledged the political seriousness of the situation by assuming direct control of the negotiations.[35] The following day President Carter announced that an emergency existed at Love Canal and that approximately 700 additional families would be temporarily relocated at a cost of between 3 and 5 million dollars.

Although Carter's order was treated as a great victory by the residents and the press, it was criticized by Governor Carey because it did not include provision for purchasing the homes and property of those relocated. By admitting that a peril existed and that it would accept responsibility, however, the federal government had given the State of New York enormous leverage. Carey intended to use that leverage to exact a tribute far greater than the $5 million figure, which he deemed to be grossly inadequate.

Carter's emergency order represented the beginning of the second major intrusion of electoral politics into the Love Canal story. With the presidential election looming that fall, Carter could not afford to appear callous toward the hapless victims of toxic waste, particularly if they lived in New York. Five of the seven largest states were governed by Republicans. California was governed by Carter's primary challenger, Jerry Brown. Only New York offered the possibility of an ally in an important governor's chair.

It was widely predicted that Carey would support Senator Edward Kennedy. Carey had refrained from doing so, in part at least because his neutrality increased his ability to obtain federal assistance for New York State. Even after the demise of the Kennedy candidacy, however, Carey did not offer a wholehearted endorsement of the president. During the summer of 1980 he continued to withhold political support while at the same time refusing to accept increasingly generous offers of federal assistance for Love Canal. Finally, in late August he met with presidential aides Stuart Eizenstat and Eugene Eidenberg and accepted their offer of 15 million dollars in federal grants and loans for purchasing new homes for the relocated Love Canal residents. The deal was interpreted by *The New York Times* as, "an election year concession." Carey remarked, "That's our system."[36]

The public prominence of Love Canal was thus the result of an admixture of faulty science, bureaucratic maneuvering, and electoral exigency. But those unique circumstances do not explain why it served to spark a national program or why the program was designed as it was.

Comparing the chronology of Love Canal with that of the Superfund legislation makes it clear that the former did not cause the latter. The first national news accounts of Love Canal occurred in August 1978, yet by October of that year the EPA had already formulated its draft version of the Superfund bill. The bill was cleared by the White House in December 1978, months before the next wave of Love Canal publicity generated by Paigen's studies and the television documentaries. By the time the tidal wave of publicity crested in June 1980, Superfund had been cleared by several congressional committees and was nearing debate on the House floor. Except for the very first national attention it received in the summer of 1978, each subsequent wave of publicity appeared to serve less as the source of new initiatives than as the occasion for mobilizing support for those already underway. To find the source of Superfund, one must look within EPA itself.

III THE ENTREPRENEURSHIP OF THE ENVIRONMENTAL PROTECTION AGENCY

The furor caused by Love Canal presented a perfect chance for Costle to implement his "public health" strategy. As we have discussed, Costle sought the "political high ground" of health protection to secure EPA's political base. Nothing less than a major new grant of legislative authority would enable the agency to address the problems posed by hazardous waste sites. Furthermore, the right bill could serve to increase EPA's involvement in other forms of toxic substances regulation, specifically those relating to oil and chemical spills, thereby strengthening its role in health protection generally.

Transforming Love Canal's notoriety into a major new grant of legislative authority for EPA required fast action. The legislative process is painfully slow. A bill has a decent chance of passage only if introduced early in a session. The newly elected Congress would convene in the winter of 1978. To pass a bill before the 1980 presidential election, the administration would have to move very quickly.

Costle also needed to ensure that EPA controlled whatever finally emerged. Historically, the Department of Transportation (DOT) had functioned as the lead agency for hazardous waste emergencies. The DOT was already spearheading an interagency legislative drafting effort to deal with Love Canal-type problems. Costle lost no time in asking his assistant administrator for water and hazardous waste, Thomas Jorling, to create a team to produce an EPA draft. Although EPA pledged to cooperate with the interagency task force, its participation was pro forma. The other task force members were not informed that EPA was preparing its own legislation. Meanwhile, Jorling's team worked at breakneck speed to gets its own proposal on the table first.[37] To ensure that, Costle allowed Jorling to ignore the agency's normal consultative procedures. As Charles Warren, assistant administrator for legislation, later put it:

> I think the only way we managed to get this done is that Doug did say, do it, number one, and authorized Tom to start working on it on a very fast track. .. this was something that was not going to go through . . . all the review processes.[38]

The EPA draft was completed on October 8, 1978, two weeks before the interagency task force draft was released, and was sent to the White House. The EPA then had to ensure primacy for its own role in the subsequent interagency deliberations. It also had to convince Congress of the seriousness of the problem. Pursuit of these two goals presented a subtle dilemma. On the one hand, it was necessary to dramatize the hazardous waste site problem, and show its national scope. Congressmen concerned about an abandoned site in their own districts were surely more likely to support Superfund. On the other hand, the worse the problem was made to look, the more questions would be raised concerning EPA's past performance. If EPA had ignored these sites for so long, could it now be entrusted with the responsibility of cleaning them up? Costle decided to make an "aggressive show of impotence," that is, to demonstrate that the agency was vigorously pursuing the problem and that its problems were due to a lack of legal authority and financial resources, not a lack of will or competence.[39]

To do this, Costle appointed a new task force, headed by deputy administrator Barbara Blum. Making use of personnel in all ten EPA regional offices, it was to use the "imminent hazard" provision of Resource Conservation and Recovery Act (RCRA) and Section 143a of the Safe Drinking Water Act to bring as many cases as possible to the Justice Department.[40]

The agency also launched a nationwide effort to discover new sites. Regional officials received a memo from Blum ordering them to produce a list of sites in their regions and to rank the ten worst. This request inspired deep resentment among regional officials. Like policemen assigned quotas of parking tickets, their sense of professionalism was offended by what they perceived to be an order to find a dump in every congressman's backyard. After the headlines the Blum task force was seeking had faded, these regional officials knew they would be faced with the task of calming the fears of angry residents and local officials.[41]

Despite this dissension in the ranks, the effort to make an "aggressive demonstration of impotence" was a success. Throughout the course of the legislative debate, EPA released lists of new sites in many different locales. The agency warned that hundreds of "Love Canals" existed across the country.[42]

In view of what was known at the time, the notion that abandoned hazardous waste sites presented an acute health danger represented serious public miseducation. Although sufficient evidence had been amassed to justify public concern, a dispassionate reading of the record suggests that the agency would have been on much firmer grounds presenting the problem as a chronic irritant rather than a pestilence. This misleading problem definition, however, has dominated all subsequent debate and shaped the design of the eventual policy.

POLICY DESIGN

Jorling assigned Andrew Mank, a member of his staff, to head the five member team drafting the bill. The team included Marc Tipermas and Peter Perez, also of Jorling's staff, and Mike Christenson from the Coast Guard.[43] Ironically, the DOT, of which the Coast Guard is a part, had not been informed of EPA's initiative and was at that moment formulating its own bill.

The Superfund scheme was both innovative and derivative. Its core notion, "shovels first, lawyers later" was borrowed from Section 311, the hazardous spills provision of the Clean Water Act.[44] The government was to respond immediately to emergencies. Cleanup was to be financed by a revolving fund, which was to be replenished with damage awards collected from those responsible. Hazardous waste dangers would thus be dealt with rapidly, without causing a permanent drain on the federal treasury and without depriving the government of a more leisurely opportunity to recover from those at fault.

Superfund, however, was far more than an expansion of Section 311 to include abandoned sites. It extended EPA's existing authority to cover hazardous spills that produced groundwater contamination and it created an off-budget device, a fee on chemical feedstocks, to finance the program. These proved to be two of the most controversial aspects of the bill.

Each of these two provisions enabled additional committees of the House and Senate to claim jurisdiction. Without them, the bill would simply have been referred to the House Commerce Committee and the Senate Public Works Committee, Senator Muskie, and Representative James Florio, chairman of the relevant subcommittees, were both zealous advocates of toxic waste cleanup.

The spills provision meant that the bill would be sent to two additional House committees, Merchant Marine and Public Works, both hostile to EPA. Ray Roberts (D-TX), chairman of the subcommittee of Public Works to which the bill was referred, was a particularly vocal critic of the agency. Convening his subcommittee's hearings on Superfund, he made clear his concerns:

> . . . a lot of people are attempting to use hazardous chemical cleanup procedure to get control of the groundwater, and any legislation we pass out of this subcommittee will have a provision, I hope, prohibiting any control of groundwater by the Federal agencies.[45]

The fee provision enabled the House Ways and Means and Senate Finance Committees to also claim jurisdiction. Both committees had the reputation of being more sympathetic to industry than the Congress at large. In particular, Senator Long of Louisiana, chairman of the Senate Finance Committee, was considered quite sympathetic to the petrochemical industry that dominated his home state.

Jorling and Warren were congressional veterans who recognized the risks they were taking. Although Warren conferred with the parliamentarians of both Houses in an effort to convince them that a fee was different from a tax, he harbored no illusions about his ability to prevent Ways and Means, and Finance from claiming jurisdiction.[46] Increasing the number of committees involved increased the time devoted to hearings, mark-up sessions, and procedural wranglings. The chances that the congressional session would end before Superfund could be completed were thereby greatly increased. These risks were justified in the minds of EPA strategists by the gains an ambitious bill offered. The scope of the legislation would neatly dovetail with the fee and liability provisions to create a major new weapon in EPA's battle to establish its public health responsibilities.

Scope

The EPA's existing authority over hazardous waste sites and spills derived from Section 7003 of RCRA and Section 311 of the Clean Water Act. Section 7003 was a broadly worded provision giving the agency authority to respond to hazardous waste emergencies.[47] The Office of Management and Budget (OMB) feared that such a broad grant of authority might place a severe strain on the federal budget. It refused to give any money to EPA for this activity. The OMB had designated FDAA rather than EPA to handle the emergency response to Love Canal precisely because it wanted to avoid interpreting any of EPA's current statutory authority as permission for the agency to spend money for such purposes.[48]

The EPA's Section 311 authority was limited to chemical spills into interstate waterways. Since groundwater is a major source of drinking water, the ability to respond to toxic spills that effected the latter was central to establishing the agency's public health mission. Furthermore, Section 311 was limited to spills, and did not include sites, a distinction that EPA thought impractical. According to Swep Davis, deputy assistant administrator for Water and Waste Management:

> It is very common to discover a release of oil or hazardous substance to the environment and to begin emergency response before we know whether the source was a spill, the work of a midnight dumper, or the result of a leaking waste disposal site.[49]

The Superfund draft addressed this problem by covering all *releases* including: "any spilling, leaking, pumping, pouring, emitting, emptying, discharging, injecting, escaping, leaching, dumping, or disposing into the environment." Only releases from motor vehicles, those occurring solely in a workplace, and those made in compliance with permits obtained under certain existing federal environmental laws were exempt.

As a lawyer for the Chemical Manufacturers Association (CMA) noted in a memorandum to his client, this was broad authority indeed:

> Generally the bill applies to any "release into the environment" of any "hazardous substance" from any "facility." All three terms are defined in such a way as to suggest no discernable limits on their scope. For example, does the release of hazardous substance from a facility include the emissions from the painting of a building, the irrigation of a field . . . or any of the other myriad human activities which involve the "release" from a structure, installation, or equipment of a substance which under some circumstances may be regarded as harmful to some living organism.[50]

Ultimately, the courts would be left to decide this issue. The definition "release," however, was far broader than was needed to solve the problems Davis mentioned in his congressional testimony. The intent of the drafters was clearly to achieve a significant increase in the scope of EPA's authority.

The Fee System

Wherever possible, the agency hoped to recover the costs of cleanup from those who created the hazard. Many sites, however, had been abandoned and the responsible firms were out of business. In those instances, cleanup could be financed from general revenue. The chemical industry favored this solution, arguing that, since the public at large benefited from inexpensive chemicals, it should pay for cleanup.

The EPA, however, had strong strategic objections to this approach. At a time of severe budgetary stringency, Congress was unlikely to appropriate enough money to operate a large program, especially after the publicity regarding Love Canal had diminished. The cleanup of abandoned sites would be both arduous and expensive. Therefore, EPA wanted a funding mechanism relatively immune from attack once the fickle public and its representatives began to lose interest in the matter.[51]

Taxing hazardous waste directly seemed impractical, as it would involve assessing and collecting fees from more than 250,000 generators of hazardous wastes. Including spills in the legislation made it arguably unfair to place the whole financial burden on waste disposers, as much of the hazardous material that was transported and spilled was not waste.[52]

A tax on waste sources would also cause serious implementation problems for RCRA. Charging generators and disposers on the basis of the volume of waste they reported would significantly increase their incentives to falsify RCRA manifests and dispose of their wastes illegally.[53]

To avoid these problems, EPA took an enormous political risk. It created a scheme calculated to infuriate the chemical industry and to incite opposition from the Treasury Department and the Council of Economic Advisors. It proposed that the program be funded by a fee on chemical feedstocks.[54] This involved fewer than 1,000 collection points. In addition, the firms that made feedstocks were, for the most part, large, stable companies. They would be relatively easy to deal with compared to the small, marginal firms that were often sources of hazardous waste.

The feedstock fee scheme enraged much of the chemical industry and helped provoke its vigorous efforts to defeat Superfund. The real issue, however, was not narrowly economic. Although the CMA claimed that the fee would hurt industry, it was unable to substantiate this claim. The EPA's position was that the fee would be passed on to consumers.[55] The CMA's main concern was symbolic. A fee levied against chemical manufacturers gave the public the impression that the industry was to blame for the problem of hazardous waste sites and would fan the flames of public hostility toward it.[56]

Liability

The liability provision that EPA crafted was designed to minimize the burden on the fee-based fund. Those held responsible for a release would pay the full cost of cleanup. If they refused, the agency could sue them in federal court and obtain treble damages. These suits would be based on the concept of strict, joint and several liability. Strict liability implied that plaintiffs need not prove that defendants were negligent. The defendant could be held liable regardless of the care exercised to prevent the harm. Joint and several liability implied that all defendants could be held liable for the full adverse consequences of the action of any one of them. The plaintiff did not have to demonstrate how much each defendant in particular contributed to the harm.[57]

In the absence of these two liability doctrines, EPA believed it would be difficult to recover damages—and hence replenish the cleanup fund. Negligence is very difficult to prove, especially when it involves actions in the distant past. Likewise, when several firms have used a particular dump site, it is often impossible to prove that the waste from one of them was responsible for a particular harm. The joint and several liability doctrine also enabled plaintiffs to seek the bulk of damages from those most able to afford it. Since many dump sites were used both by marginal operators and financially sound companies, this provision greatly enhanced the likelihood that court awarded damages could actually be recovered.

The chemical industry opposed these provisions with great zeal. It claimed that strict liability would mean that those who had faithfully complied with all laws and regulations then in force would be held responsible for eventualities that they could not possibly have foreseen. Joint and several liability would unfairly impose costs on those whose contribution to a given site was really negligible.[58]

These claims were not without merit. Taken as a whole, however, they enabled EPA to paint the industry as attempting to shirk its responsibilities. On the one hand, the industry opposed the feedstock fee on the grounds that all producers, innocent and guilty alike, were being taxed. On the other hand, it fought for a definition of liability that was so narrow that only the most flagrantly guilty would be caught in its net. The combined effect of such provisions would be to shift most of the cleanup cost to the public sector, as EPA did not hesitate to point out.

POLICY DESIGN: A CRITIQUE

Despite its apparent logic, the EPA draft was flawed in several crucial respects. Since virtually all of its essential features were retained in the statute, it is worth analyzing those flaws to better understand and evaluate the quality and character of the deliberation accorded to the draft by the executive branch and the Congress.

The draft provided no guidance for making the two most difficult decisions that implementation of the program required: which sites to clean up first and how much to cleanup any particular location. It provided no instructions about how to allocate the limited available resources. Without such principles, there was no way for either the White House or Congress to subject EPA's performance to meaningful oversight. No wonder the debate about Superfund implementation later degenerated to a "numbers game," with critics charging that EPA had only "cleaned up" six sites while defenders pointed to some remedial action at 541 sites.[59]

Love Canal exemplifies this critical flaw. For all their impact on lawns and basements, the chemical leachates did not affect the one aspect of the environment with the greatest potential for creating health risks, the water supply. Love Canal residents used water from distant reservoirs. Despite all the attention and resources expended on Love Canal, it is difficult to believe that it either presented or presents a health problem as great as those sites that do endanger water supplies.

As we discussed with regard to RCRA, focusing on drinking water provides a means for establishing priorities. It calls for concentrating cleanup efforts where water supplies are threatened. If this approach appears too single minded, then other objectives, such as preventing airborne contamination or protecting the food chain, could be incorporated. The problem is to find a way to compare the marginal gains of spending additional money at various sites.

Diverse remedial measures can be used at any given site. These range from barriers to keep people away to the total removal of all contaminated soil and all possible sources of additional leakage. The former might cost only a few thousand dollars, the latter several millions. If the goal is to minimize the risk of human exposure, regardless of costs, then very extensive (and expensive) measures might be indicated everywhere. A drinking water objective, in contrast, would require only protecting existing water sources or providing alternate supplies, whichever was cheaper. In the absence of any standard, decisions on how much to spend have necessarily been made on an ad hoc basis.

The EPA should have anticipated that local residents would step into the conceptual vacuum and clamor for the most extensive removal possible, placing local officials and regional EPA offices under enormous pressure. The result has been the expenditure of large sums of money on a few, highly publicized, sites.[60]

The draft delegated the authority for establishing priorities to EPA. It gave the administrator six months to devise a national hazardous substance response plan, which was to include criteria for determining priorities among releases and the appropriate, cost-effective action for different sites. But the draft merely offered an all-inclusive list of concerns including: the population at risk, the hazard potential of the substances involved, the potential for drinking water contamination, the potential for direct human contact, the potential for destruction of sensitive ecosystems, the preparedness of states to assume cost and responsibilities, and "other appropriate actors."[61] It offered no guidance about how to choose among these various concerns.

The EPA proposal was justified on the grounds that there was a nationwide emergency that required immediate response. If so, the very elaborateness of the planning process it created was self-defeating. Implementation was far slower than befit an emergency response scheme, and it took months to put the law into action. Reagan Administration foot dragging played a role, but it is hard to imagine how the complex planning requirements of the act could have been addressed much more quickly.

If neither the agency nor Congress felt competent to establish a system of priorities, they could have simply created a grant program, providing money to each state for waste site cleanup. Then, governors would have faced the sober task of matching cleanup needs with available resources. Rather than lobby the regional offices for the maximum amount of assistance for each site, they would have had to engage in the very process that the Superfund law shirked, setting priorities and determining expenditure levels. States, in turn, would have had strong incentives to pressure localities to do more to solve their own hazardous waste problems. The results would have meant far more political responsibility then the actual Superfund scheme ultimately produced.

The fee scheme also fostered the impression that ordinary citizens would not bear the cost of the program. As discussed previously, the drafters were well aware that this was not the case. In arguing that the fee would not injure the chemical industry, they contended that most of it would simply be passed on to feedstock buyers and ultimately to retail consumers of chemical products. Since such items are purchased by the bulk of the population—phonograph records, toys, plastic bags, and so on—lower income individuals would probably pay more in proportion to their incomes, then would those with higher incomes. In any case, the public was misled about who would pay for the program.

INTERAGENCY REVIEW

The EPA's leaders realized that the agency's draft would undergo an intensive interagency review presided over by the OMB. This review began soon after the draft was completed in October 1978. How successful was this clearance process in calling attention to omissions and oversimplifications of the EPA draft?

Participants in the interagency review included the Departments of Transportation, Justice, Treasury, and Commerce, the Council on Environmental Quality (CEQ), the Council of Economic Advisors (CEA), and the Council on Wage and Price Stability (CWPS). Conscious of how little time remained in the current session of Congress, EPA was frustrated by its slowness. Other agencies appeared to be "nitpicking." As Tom Jorling put it, "They'd get inside each little question, each paragraph, each phrase, and try to write it the way they wanted it written."[62] Often, the participants seemed to be voicing their personal concerns rather than the institutional interests of their respective departments. Jorling saw no reason for tolerating such time consuming "fishing expeditions."

Formal interagency review was often merely the first stage of a bargaining process that would finally be resolved at higher levels. But this time, cabinet officers did not expend their valuable political credit with the president to try to modify the EPA draft. Despite the nitpicking, the review produced few changes in the original document. On most key issues, OMB sided with EPA. In the absence of contrary signals from the White House, the EPA-OMB position prevailed even when a majority of agency representatives were opposed.[63]

The issue of greatest controversy involved the fee system. It was opposed by Treasury, Commerce, CEA, and CWPS. In the words of CEA economist Lawrence White:

> The fees for the Superfund were inefficient and appeared to be levied on the wrong agents. Spills occurred from vehicles, not from refineries: fees should be levied on oil and chemical transportation companies not refineries. Fees on chemicals would lead to distortions and inefficient substitutions of chemical inputs. Also . . . the chemical companies would not be the ultimate payers; rather it was the final consumers of chemical products who would largely bear the cost . . .[64]

The OMB generally opposed dedicated revenue-raising schemes because they limited future budgetary flexibility. Eliot Cutler, assistant director of OMB for Energy and Environment, however, was a former member of Senator Muskie's staff and thus a former colleague of Jorling. He was sympathetic to Superfund. The EPA drafting team had taken great pains to keep him "in the loop" as the draft was being developed. Within OMB, he strenuously argued for retaining the fee as it was the only way to establish Superfund quickly without "busting the budget."[65] In addition, he felt it was equitable and consistent with the user fee principle that the administration was trying to implement in many policy areas.

As the review process drew to a close, Cutler informed Jorling that he had been overridden by his superior, Deputy OMB Director John White. The OMB would now oppose the fee. White was an economist and sympathetic to the antifee arguments of the CEA.

Warren and Jorling wanted administration endorsement of the fee system to secure the support of the Treasury Department. The tax-writing committees of Congress, after all were very deferential toward Treasury's views. Still, they took solace from the fact that this defeat was by no means final. They could try to restore it in Congress, regardless of the White House position. The concept had already been widely disseminated and it was sure to surface in one or another of the Superfund bills emanating from the Hill. The visibility of the abandoned site problem virtually ensured that President Carter would not veto a bill because of its method of funding.[66]

Jorling and Warren were reluctant to confront OMB. Not only did it have a superior claim to expertise in this sphere, but it controlled the paper flow to the president and could, therefore, caricature EPA's position. Furthermore, OMB might threaten to undo all the other painfully achieved interagency agreements on the draft if EPA had the temerity to lock horns over the fee. Jorling and Warren dealt with OMB on a wide variety of matters and did not want to earn OMB's enmity in a fight that OMB was likely to win.

Nonetheless, they were recruited into a scheme to reverse EPA's defeat. The OMB's memo summarizing the interagency deliberations on Superfund had been returned by President Carter with a note to Stuart Eizenstat, indicating that he had questions about the proposal. His concerns were not entirely clear, but Cutler decided to exploit the opportunity. Congressional hearings on Superfund were scheduled to begin in a matter of days. The sole chance to bring the matter before the president was the annual presidential budget review. Normally only Costle would attend from EPA but, knowing that Cutler intended to raise the fee question, Costle brought Jorling with him.

As Cutler began to present the Superfund proposal, President Carter interjected, "Why shouldn't industry pay?" The federal government had been pressured to take this burden off the states. Was it not both proper and politically expedient to shift it, in turn to the chemical industry? Cutler seized the moment to join Costle and Jorling in offering a defense of the fee system. Both White and Budget Director McIntyre were in attendance. They judged from the nature of Carter's initial query and his subsequent responses to Cutler and Jorling that the president favored the fee scheme. Although they held the opposite view, they remained silent. The fee was restored.[67]

In contrast to the attention the fee scheme received in the administration's review of Superfund, the problem of priority setting was not addressed at all. The stamp of "bureaucratic pluralism" is evident throughout, as the positions taken by various agencies reflected their particular organizational or professional concerns. There was no strategic debate, no reconsideration of the problem definition. As a deliberative exercise, the interagency review achieved only modest success.

THE HOUSE

The administration's Superfund proposal was submitted to Congress on June 13, 1979.[68] Only eighteen months remained before the session ended, a brief time span considering the number of committees, some of them unfriendly, who had to consider the bill. Only Florio's subcommittee appeared favorable towards Superfund. Florio and his staff were well informed about hazardous waste and were interested in establishing him as an expert on this issue. Florio was embarking on a campaign for governor and expected to benefit from close identification with efforts to resolve a problem of particular concern to New Jersey.[69]

The two "spills" committees were not as favorable. Merchant Marine had proposed oil spills legislation that had passed the House two sessions in a row. Each time it had been held up in the Senate because Senator Muskie considered it too advantageous to the oil industry and because it did not cover chemical spills. The committee staff suspected that Muskie was partially representing EPA in this, and were not happy over their experience.

As we have seen, the leadership of the other "spills" committee, House Public Works, had a strong antipathy to the agency and a great reluctance to allow it to acquire any new authority. Chairman Roberts was deeply upset by EPA's actions under Section 404 of the Clean Water Act, which gave the agency authority over the dredging and filling of inland waterways. The EPA regulations had hindered dredging operations in his district. He also resented the Senate's imperiousness with regard to environmental matters and was sensitive to the fact that both Jorling and Warren had worked in the Senate.

Unlike the other House committees, EPA did not have regular contact with Ways and Means. No one, however, expected it to favor any new tax, particularly a dedicated one.

The situation appeared to be much brighter in the Senate, where Jorling retained excellent contacts with his former colleagues on Muskie's staff. Moreover, both Jorling and Warren, as former Senate staffers, felt secure in their understanding of Senate mores.

As Jorling and Warren perceived the situation, their great enemy was time, and their great resource the Muskie subcommittee. The key was to minimize the amount of time the bill spent in the House. Muskie had historically dominated House-Senate conferences. It was far more important, therefore, that a bill pass the House than that it be a perfect bill. Jorling and Warren adopted a "guerilla" approach. To steer through the reefs and shoals of the House committees, EPA would support any bill that appeared capable of success regardless of how far it diverged from the administration position. Defects could always be remedied in conference, if only a conference had time to take place.[70]

This approach would prove hard to explain to EPA allies who were not as optimistic about the agency's ability to get its way in conference. It might also strain relations with the White House, if it were discovered that EPA was violating the terms of the interagency agreement. Given the difficulties of steering a bill through two houses of Congress (including four committees of the House) in less than eighteen months, this strategy had the virtue of recognizing how little control the agency had over the legislative process.

Thus, when James Florio chose to introduce his own Superfund bill, EPA acquiesced.[71] Because it covered only sites, it did not need to be referred to Public Works and Merchant Marine. Although EPA desperately wanted Superfund to cover spills, it chose to accommodate a valuable ally. If necessary, spill coverage could be added in the Senate.

Moreover, Florio's action created important incentives for the other committees to pass some sort of spills bill.[72] If they did not, and the House did pass Florio's bill, they would be excluded from the House-Senate conference.

Merchant Marine did not wait for the administration proposal. In May 1979, it passed a new version of the oil spills legislation that the House had passed in the previous session. The bill, HR 85, made no mention of hazardous chemicals. Chairman Biaggi acknowledged his support in principle for chemical spills legislation. But he insisted that the Congress first complete the work it had started on oil spills.[73]

Biaggi's bill went to Public Works, which was receptive. To demonstrate good faith toward Biaggi, EPA's Swep Davis testified in its favor, requesting that the bill be widened to include chemical spills.[74]

The EPA did have a problem justifying this expansion of its authority. Section 311 of the Clean Water Act granted it the power to respond to hazardous spills in navigable waters. The EPA had taken seven years to issue regulations implementing Section 311. This delay was due to the difficulty of defining exactly what constituted a "harmful quantity" of substances designated as hazardous.[75] The agency sought to resolve this impasse through the use of a method so arbitrary that a federal court enjoined it from implementing the regulations.[76]

It was only through the cooperation of Public Works that EPA had been able to amend Section 311 in the final hours of the previous Congress to remove the requirement for a precise definition of harmful quantity. The new language specified that EPA could intervene wherever a discharge "may be harmful."[77] The committee was understandably reluctant to grant the agency a whole new type of authority to clean up chemical spills before it demonstrated that it could use its recently expanded authority effectively.

The EPA's hopes of overcoming this reluctance rested with John Breaux (D,LA). A conservative southern Democrat, Breaux's district included one of the largest concentrations of chemical manufacturing facilities in the nation and he was proud of being a friend of the chemical industry. Since Congressman Breaux's constituents inhabited the area surrounding these facilities, however, he had become increasingly concerned about the threat from hazardous waste. As the author of the Section 311 compromise, he was well aware of its limitations. Not only was it limited to navigable waters, but it was to be funded by the normal appropriations process. If his constituents faced grave risks from hazardous materials, it might prove very difficult to obtain enough funds for cleanup and compensation from federal appropriations.[78]

Despite his conservative principles, Breaux considered a cleanup program paid for by a feedstock fee to be an acceptable solution. Since it would be passed on to consumers, it represented a subsidy to chemical manufacturing regions—like his district—from nonmanufacturing regions.[79]

Breaux's stance reassured moderate and conservative members that they could vote for the addition of chemical spills to the bill without fearing retribution from the chemical industry. Still more was needed, however, to win over Chairman Roberts. Breaux also proposed an amendment to HR 85, reducing EPA's authority to write dredging and filling regulations under Section 404 of the Clean Water Act. Roberts accepted the Breaux amendment as his price for supporting the inclusion of chemical spills in the bill.[80]

The EPA officials thus found themselves supporting a severe curtailment of agency authority in one area to obtain added authority in another. They did so because they were confident that the Breaux amendment would be deleted at a later stage of the legislative process. Nonetheless, it proved highly embarrassing to have to persuade their friends not to try to defeat this limitation upon agency prerogatives.[81]

On May 7, 1980, nearly a year after receiving it from Merchant Marine, Public Works reported out the revised HR 85 by a voice vote. The bill created two separate funds, a 200 million dollar oil cleanup fund, and a 100 million dollar fund for hazardous chemical spills. Sixty percent of the chemical fund would come from fees on chemical feedstocks; the rest from fees assessed on various hazardous substances. It also provided for strict joint and several liability, but set specific dollar limitations on the liability of certain categories of oil and chemical carriers. It provided for the recovery of damages due to the destruction of natural resources and real or personal property, but not for personal injury or illness. Claims were limited to damage incurred after the effective date of the law.[82]

In the fall of 1979, Florio introduced his own bill that was limited to sites. It called for a 1.3 billion dollar cleanup fund with sevent-five percent of the cost to be financed by a feedstock fee and twenty-five percent by general revenues. It also included strict joint and several liability, and established the right of those who were victims to sue for damages in federal court.[83] Although Florio's subcommittee had a Democratic edge of 5 to 3, he was unable to gain a majority for his bill.[84]

At the start of the new year, Florio decided to break the stalemate by significantly modifying his proposal. His new bill, HR 7020, deleted victims' rights to sue in federal court, and halved the size of the fee-based fund. It also imposed important limitations on strict joint and several liability. These changes were sufficient to win the support of two Republicans, ranking minority member Edward R. Madigan of Illinois, who voted for Florio's version after his own amendment deleting the fee was defeated, and Gary Lee of New York.[85]

The full Commerce Committee approved Florio's bill on May 13, 1980 by a 21 to 3 vote. In lengthy negotiations the preceding night, Florio was forced to accept several additional changes, including a 50–50 split between fee-based revenues and appropriations. This last change was bitterly opposed by environmentalists who felt it would enable the industry to diminish the size of the fund by lobbying OMB or the White House. The key to victory was a coalition between Florio and ranking minority member Madigan. Florio argued against an effort to double the size of the fund while Madigan fought off a move to gut the bill.[86]

The House Ways and Means Committee then chose to exert its jurisdiction on the grounds that the fee was really a tax. The EPA team was dismayed because this guaranteed that the parallel Senate committee, Finance, would likewise claim jurisdiction.[87] House rules set a time limit over how long Ways and Means could hold a bill. The Senate had no such restriction. Although the EPA feared that the chemical industry was responsible for Ways and Means' claim of jurisdiction, this does not appear to have been the case. Industry lobbying efforts before that committee were poorly organized.[88] More likely, Ways and Means simply acted to defend its prerogatives. Once the committee began its deliberations, it did not feel compelled to restrict its attention too narrowly. Rather than the cursory overview that HR 7020's champions had hoped for, the committee met for several sessions and scrutinized the bill exhaustively.[89]

In the course of these deliberations, EPA's guerilla strategy confronted an unexpected challenge. Thomas Downey (D,NY) a staunch EPA friend, sought to restore the size of the fund to 1.2 billion dollars. Larry Snowhite from EPA tried to convince Downey that he risked alienating the Commerce Committee. Downey remained adamant. His district faced a serious abandoned site problem. He had dispatched a staffer to compile a dossier of dangerous sites in each committee member's district and he was confident he could obtain a majority for the expanded fund.[90]

Failing to convince Downey, EPA reversed course and provided him with sufficient data to make a convincing case. An agency intern was assigned full time to uncover sites in each member's district. The tardiness of CMA's response to these findings left members with no clear response to questions from constituents about why they were not striving to get funds to defuse ticking time bombs in their own backyards. The CMA's problems were exacerbated by a "conspiracy of silence" among committee staffers and Chairman Al Ullman as to when the final committee markup would actually take place. The CMA had less than fourty-eight hours before the session to make a last minute appeal. Many members did not even receive the CMA mailing that was intended to refute Downey's campaign. On June 18 the Ways and Means Committee approved HR 7020 with the doubled fund by a vote of 30 to 0.[91]

The Ways and Means Committee also chose to exert its jurisdiction over the fund provisions of the spills bill (HR 85). Fearing that large damage awards would result, it removed the compensation provision of the bill that authorized the fund to pay for personal damages suffered by victims. This provision had been sought by the maritime industry as a means of limiting its own insurance expenses and liabilities. It was one of the major objectives of Merchant Marine in passing the bill in the first place. Chairman Biaggi and his staff were irate, as were his allies on the Public Works. Against the advice of his staff, Biaggi agreed to support the bill on the grounds that, at this late stage, it would only pass if the Ways and Means version was accepted. Because of EPA's low repute among Public Works leaders, Breaux and other moderate and conservative members were charged with convincing them to accept the Ways and Means Committee changes.[92]

On August 27, 1980, the House Rules Committee cleared both Superfund bills for floor action. The rules allowed Congressman Breaux to offer a substitute bill that deleted all inconsistencies among the various spills bills, including the deletion of victim compensation, and specified that no amendments would be allowed to the financing provisions.[93]

Led by David Stockman, a small minority of the Commerce Committee launched a spirited attack on the bill. Stockman offered a substitute that provided a subsidy of 500 million dollars to the states, allocated on the basis of population and industrial output. Stockman argued that, unlike air and water, sites did not move and thus did not cross state boundaries. Therefore, the problem was primarily one for the states to handle.

Stockman also questioned the failure to provide more explicit guidance about what sort of cleanup and how much cleanup to do at various sites:

> Nowhere in this bill of 74 pages will you find any specific criteria whatsoever to determine what are the hazardous threats posed by inactive or abandoned sites and what would be the defined and limited scope of EPA activities.[94]

Florio's response stressed the incapacity of many states to deal with hazardous waste problems and the fact that the National Governor's Association and the National Council of State Legislators had endorsed the measure. In responding to Congressman Gore of Tennessee, who made the same argument, Stockman replied: ". . . but can the gentleman recall any problem where the State governments have not come in and said that they could use some Federal money to help them solve the problem?"[95]

Breaux and the EPA lobbyists succeeded in convincing the various committee notables to stand by their earlier commitments to support the legislation. This made passage a foregone conclusion. HR 85 was passed by the House on September 19, 1980, by a vote of 288 to 11. HR 7020 was passed four days later by a vote of 351 to 23.[96]

THE SENATE

The EPA had not intended to apply the guerilla approach in the Senate, where its trusted ally, Senator Muskie, held sway. Events, however, conspired to make Superfund's progress through the Senate even more of a "long march" than in the House. Committee and subcommittee staff play an even larger role in the upper chamber than in the lower one. Also, formal rules are far less important in determining the pace of legislative action. Both these factors created difficulties for Superfund.

The EPA's central problem was that the Muskie subcommittee staff advocated a far more ambitious and controversial program than the one it had developed. Only a series of industry blunders and last minute compromises allowed the bill to pass at all. Moreover, the version that passed was weaker than the House bill and far weaker than the bill that could have been passed if Superfund's friends had come to a more timely agreement.

Unlike congressmen, senators serve on a bewilderingly large number of subcommittees. They are, therefore, highly dependent upon their staffs to determine policy and to draft statutes. A Muskie subcommittee staffer described the process this way:

> The staff served the normal role of a staff which is to drive it (the bill) and yet leave members some room if they have to have it. Sometimes they (staff) drove it pretty far. In fact the members were always able to rein it in.[97]

The Muskie subcommittee had been interested in hazardous waste for several years, and had included hazardous chemical provisions in previous Senate versions of oil spills legislation. It expressed its distinct view of the problem in S1480 introduced by Senator John Culver (D,IA) on July 11, 1979.[98]

Culver's bill provided for compensation for both the property and the medical costs incurred by victims. The definition of a hazardous substance was drawn very broadly, including unregulated substances that "may present a substantial danger to public health or the environment." The definition of a *release* covered virtually any emission of a toxic substance into either a workplace or the environment, and included releases covered by permits issued under other existing laws.[99] The EPA version exempted emissions in compliance with such permits.[100]

The EPA objected to S1480 on both political and administrative grounds. The agency felt that the bill was too ambitious to be passed in the short amount of time remaining. Industry would fight fiercely against both its broad scope and the medical compensation provisions. Although it called for a fee based fund that was several billion dollars larger than that requested by the agency, it would still be insufficient to cover both cleanup and compensation costs.[101]

The Senate Subcommittee staff did not necessarily disagree with this assessment. It was less concerned, however, than EPA with either time constraints or administrative feasibility. On the contrary, the subcommittee staff wanted to use S1480 to create a comprehensive toxic emissions control act that would transcend the limitations of the Toxic Substance Control Act (TOSCA), the Federal Insecticide, Fungicide and Rodenticide Act (FIFRA), and RCRA. The federal government was to be given the capacity to respond to the dangers of chemical releases wherever and however they occurred, regardless of whether these risks came from spills, disposal sites, workplaces, air pollution, or contaminated foods.[102] Although rooted in public health, the underlying goal of S1480 was far more ambitious than Costle's strategy for EPA. It was part of an effort to legislatively establish every individual's right to a "safe" environment.

This same commitment was behind the Senate staff's enthusiasm for victim compensation and for allowing injured parties to sue in federal court. In addition S1480 proposed to significantly shift the burden of proof in such actions. Instead of forcing a plaintiff to prove that an illness was caused by a particular exposure, as the current tort law required, plaintiffs would only have to demonstrate that their illness and the exposure were "reasonably related." This liberal standard was to apply both to claims for compensation from the fund and to suits brought by alleged victims in federal court.[103]

The subcommittee staff was aware that the feedstock fee was too limited a revenue source to finance both cleanup and compensation. Nonetheless, the principle of compensation for environmentally induced illness and property loss was too important to sacrifice simply to accelerate clean up. No great environmental law had been conceived and passed in a single legislative session. Lacking the agency's sense of urgency, it did not seek to hasten or abridge the consultative process.[104]

To conciliate the chemical industry and produce a committee majority, the staff ultimately proposed significant revisions in S1480. Its choice of areas for compromise reflected the strategic gulf that lay between its conception of Superfund and that of EPA. The staff compromised on the liability rule, one of the two provisions, along with the fee, that the EPA regarded as least open to modification. It retained two provisions that differed most from the agency bill: breadth of scope and medical compensation.[105]

In the revised staff bill, "joint and several liability" was modified to permit a defendant to demonstrate "by a preponderance of evidence" that his contribution could be "distinguished and apportioned," thereby limiting his liability to that portion of the damage to which he actually had contributed.[106] This modification was in response to industry complaints that, otherwise, the owner of a single drum at a dump site could be held responsible for the entire cleanup.[107] The EPA strongly opposed this change as diminished receipts from liable firms would make the agency more dependent on Congress for additional appropriations and, in all likelihood, diminish cleanup efforts.[108]

The revised bill did include some modifications to the medical compensation provision. Payouts from the fund would be limited to out-of-pocket emergency medical expenses incurred within six years of exposure. This effectively excluded payment for the long term care of chronic conditions such as asthma, and for illnesses with long latency periods such as cancer. The staff also accepted the idea that the standard of proof for linking exposure to illness be made more stringent, changing it from "reasonably related" to "substantial likelihood." Payouts for property damage would include compensation for loss of natural resources, and the cost of replacing real property—and include the cost of permanent relocation if necessary. The bill retained the provision that allowed victims to sue in federal court to recover damages not covered by the fund.[109]

With regard to scope, the Senate staff refused to limit itself to sites and spills. It did agree to exempt workplace exposure (presumably covered under laws enforced by the Occupational Safety and Health Administration [OSHA]), voluntary exposure to products during their use, and exposure to diffuse mobile sources. But on the key point of whether releases conducted under federal permits should be exempt, it chose to remain firm.[110]

Despite the substantive and procedural problems that Superfund was encountering in the Senate, EPA was confident that Muskie would, at the appropriate time, exercise his traditional leadership. When he left the Senate in the spring of 1980 to become Secretary of State, this confidence evaporated. John Culver, the new chairman, was preoccupied with reelection and uncertain about how hard he wished to push such controversial legislation. He had a particular problem because of the large role of Monsanto in the agricultural chemical field, and the company's great importance in his home state of Iowa. It was one of only two domestic manufacturers of polychlorinated biphenyls (PCBs), which were among the most widespread serious chemical containments. Under strict joint and several liability, therefore, Monsanto faced the prospect of having to pay cleanup costs in virtually every abandoned site in the country.[111]

In an effort to save time, the ranking Republican on the Environment Subcommittee, Senator Robert Stafford of Vermont, moved that the subcommittee markup be halted and that key issues be settled in full committee. This move would also deprive industry lobbyists of the opportunity to get "two bites at the apple," seeking modifications at both the subcommittee and committee levels. Unfortunately, Culver chose to interpret this action as a power grab by the committee minority and was infuriated. The staff was able to convince Culver that he was mistaken. By the time tempers had cooled, however, and the complex problem of rescheduling the markup had been overcome, six additional weeks had been wasted.[112]

Full committee markup finally began in the beginning of May. The committee as a whole made several important additional changes. Senator Bentsen (D,TX) succeeded in exempting federally permitted releases from the liability provisions of the law, although they remained subject to the cleanup provisions. Senator Domenici (R,NM) got the committee to limit compensation for personal injury or loss of income to losses occurring after January 1, 1977. For losses prior to that date, the victim would have to appeal to traditional tort law. The fund was also reduced from 5 to 4.1 billion dollars, with 3.6 billion dollars of it coming from the feedstock fee and the rest from government appropriations. Joint and several liability was modified to allow a court to apportion liability according to the amount and degree of toxicity of various contributions to a site and in light of the degree of cooperation exhibited by individual contributors in cleaning up. Coverage of oil was omitted.[113] This, however, did not appear to be a significant omission because oil was covered in the House version. It was widely assumed that oil would be added either by amendment on the Senate floor or as a "bone" tossed to the House in conference.

Several changes were made to placate individual senators. To reduce the burden on the copper industry the committee accepted an amendment by Senator Dominici (R,NM) to reduce its contribution to the feedstock fee from 25 million to 1 million dollars. Senator Gravel (D,AL) obtained passage of an amendment requiring Superfund to reimburse losses to fishermen resulting from toxic spills. Senator Culver obtained a similar amendment regarding livestock and agricultural products.[114]

The most acrimonious issue proved to be granting victims the right to sue in federal court. In the view of the committee majority, the importance of this provision had grown as a result of the limitations on federally funded compensation that had been adopted. Without access to federal funds, victims of chronic and long latency diseases needed ready access to federal court to pursue their claims.

A minority composed of Howard Baker (R,TN), Domenici and Lloyd Bentsen objected strenuously. For them, creating a federal toxic tort law represented yet another needless intrusion of the federal judiciary into the states' domain.[115] Since Domenici and Bentsen had already obtained the changes they most wanted, however, their opposition to this provision did not prevent them from voting for the measure. The revised version of S1480 was cleared by the committee on June 27, 1980, and reported to the Senate on July 11.[116]

Despite these modifications, S1480 was still far more ambitious than the House version, encompassing many releases in addition to sites and spills. The principle of compensation, and of allowing victims to sue in federal court, also remained. As it emerged from committee, the draft could still be viewed as an environmental rights rather than a cleanup bill.

Still, the bill could not move directly to the floor. The Senate Finance Committee announced that it would seek jurisdiction. The Senate, unlike the House, had no procedure for limiting the time a committee could spend on a bill. There was no easy way to prevent the Finance Committee from holding the bill until after the presidential election in November, or even until the congressional session ended in December.

The staffs of Public Works and Finance tried to negotiate a time limit on the referral.[117] This proved difficult since Senator Long (D,LA), Chairman of the Finance Committee was a close ally of the chemical industry, and had received the second largest contribution of any senator from that source.[118] Facing a primary challenge in September, he had no reason to alienate an important segment of his constituency before that date. The negotiations stalled. Charles Warren of EPA wanted to bring the matter of a time limit to a vote on the Senate floor, but he was overruled by the Public Works staff, that did not relish a power struggle with a senator as influential as Long.[119]

By the end of September, Long's attitude changed. He had won his primary and had received a mailgram from the attorney general of Louisiana, William Guste, claiming that their state had more need than any other of the assistance that Superfund would provide in "cleaning up hideous and life threatening waste dumps."[120] On October 1, Long agreed to finish by November 21.[121] Public Works accepted this offer, although it meant that Superfund would not reach the Senate floor until after the election and that less than a month would remain for conference consideration and repassage by the two houses.

The House-Senate conference loomed as a major obstacle to passage. The interests of the House conferees were quite varied. It would be difficult to reconcile them to adopt a coherent strategy for bargaining with the Senate. In Muskie's absence, the Senate contingent threatened to be even less cohesive than its House counterpart.

It was thought to be vital that the conference report a bill before election day. Carter's campaign was in trouble, jeopardizing the campaigns of congressional Democrats. Republican gains might be interpreted even by Democratic survivors as a repudiation of federal regulatory efforts like Superfund. The presence of many lameducks, who might be particularly open to the blandishments of industrial lobbyists, would only exacerbate the situation.

The EPA had already begun to prepare for the possibility that Superfund might reach conference so late in the session that the conference participants would not have time to craft a compromise. The agency's legislative team worked to produce its own draft in the hopes it would prove acceptable to both houses. This effort took on added importance when it became clear that the Muskie Subcommittee bill would not pass the Senate. A new draft was needed to serve as the basis for a bill that could pass. The EPA did not want the credit. Instead, it would be introduced by a cooperative senator at the appropriate time.

This compromise effort required secrecy to avoid undercutting the legislation actually before the Congress and infuriating its sponsors. Auspices for this activity were provided by the White House. Industry representatives were invited as well as staff members of the three Public Works committee dissenters—Domenici, Baker, and Bentsen. The goal was to co-opt these three powerful, moderate senators by bringing them directly into the drafting process and making the changes they felt most strongly about.[122] A week before the election, however, industry, sensing a Reagan victory, withdrew from the meetings.[123]

Although no actual draft was produced, the meetings did prove to be extremely useful. Domenici, Baker, and Bentsen became more sympathetic, and EPA learned what language might prove acceptable to various parties.[124]

With Carter's defeat, and the election of a Republican Senate majority, the future of Superfund seemed bleak. It had not yet cleared the Finance Committee and adjournment was only a month away. The Republicans might wish to postpone further action until they were in power.

Yet, paradoxically, the Republican victory actually improved the prospects for Superfund, albeit in a less expansive form. Those Senate committee staffers most opposed to compromise recognized that once the new Congress convened they would no longer be in control of the drafting process. Indeed, they might not even retain their committee posts. They were, therefore, more willing to expedite matters. After meeting several of the newly elected Republican senators, Stafford, the new chairman of Public Works, concluded that their extreme conservatism would limit his ability to improve upon what the House had already passed. He, therefore, chose to place his considerable influence behind the immediate passage of a compromise. Also, moderates like Bentsen, Baker, and Domenici, who had committed themselves to support of a compromise bill, did not renege on that commitment.[125] Finally, the aggressive tactics employed by the chemical industry backfired.

The CMA had undergone significant changes in the period immediately before the legislative struggle over Superfund. Between 1978 and 1980 it changed its name (from the Manufacturing Chemists Association), purged a large percentage of its staff, restructured its internal organization and nearly tripled its budget. Most importantly, it shifted its legislative posture. Rather than working quietly behind the scenes, it emerged as a loud and aggressive champion of the industry and the positive contribution chemicals made to American society. Under the leadership of Robert Roland, it launched frontal attacks against federal regulatory incursions. In the words of a trade journal: "activist advocate is the new buzz phrase around CMA these days . . .[126]

This shift occurred because industry leaders believed that the low key approach had not effectively advanced chemical industry interests in the face of vitriolic attacks from environmental organizations and their allies in Congress and the EPA. The turning point might well have been the passage of TOSCA, a law that industry felt to be unreasonably burdensome and that gentle persuasion had done little to modify. A group of new young chief executives in major chemical companies, the so-called Young Turks, sought to make the public aware of the enormous cost of environmental regulation, and to use this changed climate of opinion to gain redress in Congress. Superfund represented the first major test of these new tactics.

From the onset, CMA adopted an adversarial posture toward Superfund. It disputed EPA's assessment of the extent of the problem of abandoned sites and of the risks they posed. It claimed that the feedstock fee would cause severe damage to the competitiveness of American chemical products worldwide and worsen inflation. It attacked the concept of strict, joint, and several liability as unfair to those companies whose contribution to a site was small and who had been under the impression that they were operating within the law.[127]

However reasonable some of these claims might have been, they created the impression of massive industry disregard for the problem of abandoned sites. This posture was particularly galling to those legislators who viewed themselves as friends of the industry but who felt under enormous pressure from constituents and their own consciences to take positive action. Congressman Madigan publicly rebuked the CMA and its allies:

> . . . those in the business community who have stonewalled the legislative process, hoping to stop any legislation, have also made it very difficult for those of us who wanted to develop a reasonable bill.[128]

As a result of this stinging criticism, the CMA felt compelled to modify its position and accept the compromise worked out between Florio and the Republican members of his subcommittee. When Ways and Means chose to double the size of the fund, the CMA withdrew its support. On September 11, 1980, however, appearing on the ABC television program "Nightline," Roland appeared to extend CMA's endorsement to include the expanded fund. When asked whether or not CMA would support the version that would shortly be voted upon by the House, he responded:

> That's right. And we think that the 1.2 billion figure is important because the amount that is specified in the Senate bill is excessive for the job that needs to be done. The most important part of the liability provisions that are contained in S1480 are not contained in either version of the House bill. And that's what we have to negotiate.[129]

But, after the defeat of President Carter, the CMA announced that it opposed any Superfund legislation. Florio responded by publicly accusing the CMA of reneging in the hopes of getting a better deal from the incoming administration.[130]

The CMA announcement placed many senators in an awkward position. Although the CMA had never officially endorsed the final House version, Roland's comments left the widespread impression that it had. Those who now voted against Superfund would appear to be tools of the chemical industry. Influential Republican senators like Robert Dole, Peter Domenici, and Howard Baker had no desire to have themselves or their new president seen in such a light.[131]

The EPA urged the White House to make Superfund its top legislative priority. Carter met with Reagan and urged him to endorse Superfund. Reagan was noncommittal, leaving his Republican senatorial allies free to support it.

The key was to quickly fashion a bill that the moderate supporters of Superfund could accept. Under the leadership of Senator Stafford, the Public Works staff produced a compromise that was introduced on November 18 with departing committee chairman Jennings Randolph as co-sponsor.[132]

The Stafford-Randolph plan retained two key ingredients of S1480: coverage of all nonpermitted chemical releases and compensation for out-of-pocket medical expenses. Workplace releases, however, were exempted and a 300 dollar deductible as well as a 30,000 thousand dollar cap were placed on medical compensation. The size of the fund was reduced from 4.1 to 2.7 billion dollars.

The right of victims to sue in federal court, and strict joint and several liability were eliminated. Other damage claims were restricted to natural resource losses and a cap on such claims was set at 500 million dollars.[133]

During the next few days intense negotiations were held. The problem was not obtaining a majority. Few senators would choose to go on record in opposition to the cleanup of ticking time bombs. Rather it was necessary to accommodate senators who might choose to filibuster, as it would prove difficult and time consuming to assemble the sixty votes necessary to invoke closure. To placate Senator Jesse Helms (R,NC) the size of the fund was further diminished to 1.6 billion dollars and all medical compensation was eliminated.[134] These negotiations were conducted in great haste and nerves became extremely frayed. The compromise was so tenuous that the only way to hold the package together was to agree that all floor amendments, no matter how noncontroversial, would be opposed. On November 24, 1980, the compromise version passed the Senate by a vote of 78 to 9.[135]

This ironclad no-amendment rule, however, jeopardized the bill because it provided no avenue for including oil spills. The Public Works version had excluded oil with the understanding that Senator Warren Magnuson (D,WA) would offer a floor amendment to include it. Now this was impossible. Helms offered to loosen his insistence on the no-amendment rule in exchange for further reductions in the size of the fund, but this was rejected by Stafford.[136] Exclusion of oil caused a furor in the House. Three days remained until adjournment. There was no longer time to convene a House-Senate conference. The only way to pass the bill was for the House to accept the Senate version intact.

BACK TO THE HOUSE

On December 2, 1980, Randolph and Stafford sent a letter to Florio urging him to convince his House colleagues not to amend the bill because the fragile Senate majority was already beginning to disintegrate.[137] House leaders were now doubly offended. Oil was the one aspect of Superfund that some of them really cared about and the upper chamber had excluded it. Now they were being deprived of their right to help fashion a compromise, suggesting that once again the Senate was not treating the House as its constitutional equal.[138]

The only way the Senate bill could be brought to the House floor quickly enough was for Speaker O'Neill to grant a suspension of the rules. When he agreed to do so, several prominent representatives announced their intention to oppose the Senate version, jeopardizing the two-thirds majority required to pass a bill under suspension.[139] The EPA team, and the White House, mounted one last intensive lobbying effort. The National Solid Waste Management Association provided crucial support. Through its state affiliates, it contacted more than 300 congressmen in the four hours before the House vote.[140]

The bill was brought to the House floor at 3 PM on December 3, 1980, two days before adjournment. The antagonism toward the Senate was great. For example, Madigan, who had played a crucial role in fashioning the compromise that had enabled the bill to clear the Commerce Committee, now opposed it. He said, "We have been left with a take it or leave it situation, and I would like to recommend that we leave it."[141]

Of all the speeches of support perhaps the most influential came from Biaggi. As the leading House proponent of an oil spills bill, he was the member most humiliated by the Senate action. Nonetheless, he endorsed the bill, arguing that the hazardous waste spills and sites problem was too important to postpone despite the procedural wrong that had been committed.[142] In the end, most members appeared to agree. The bill passed by a vote of 274 to 94, a majority considerably greater than that required by the suspension procedure. On December 31, 1980, President Carter signed the Comprehensive Environmental Response Compensation and Liability Act into law.[143]

REFLECTIONS

Despite Superfund's torturous route to passage, the basic elements of the original EPA draft remained intact. The "shovels first" concept, the feedstock fee, the liability scheme, and the definition of release contained in the ultimate law were essentially what EPA had recommended. This does not necessarily imply that Congress failed to adequately consider the key questions. Such a judgment depends not on the outcome but on the quality of the deliberation in which Congress engaged.

From that perspective, the results were mixed. The ticking time bomb problem definition—that waste sites presented a widespread serious, acute health hazard—was challenged both by the chemical industry and by a small group of conservative Republicans. Neither, however, was able to provoke serious discussion of this most fundamental question. Industry's challenge was discredited because it was too obviously self-serving. Stockman's assertion that the problem was exaggerated was but one small part of a general claim that federal intervention in this area would lead to a new form of bureaucratic dictatorship. This broader charge was so obviously overstated that it undermined those aspects of Stockman's case that deserved serious attention.[144]

The House did debate the scope of the EPA proposal but the discussion was dominated by "turf" questions. Merchant Marine pushed for coverage of oil spills. Public Works traded its approval of broad scope for changes in a totally unrelated program. No one asked how a system of priorities would be constructed to allocate resources among the broadly defined class of releases over which the agency was requesting jurisdiction.

The fee system and the liability rules did receive sustained attention in the House. Here, Congress focused on questions of distributional equity. The fee was approved only after evidence was presented demonstrating that no substantial harm would befall the chemical industry. The joint and several liability provision was modified to provide certain defenses consistent with common sense notions of fairness. And the provision was adopted because congressmen were convinced that there was no other way to recover money from those who

Table 5-1 The Various Versions of Superfund

		Key Provisions			
Version	Scope	Funding	Liability	Victim Compensation	Federal Cause of Action
Administration Proposal	sites & spills	1.2 billion feedstock fee	strict joint & several	No	No
House Commerce Committee	sites only	600 million ½ feedstocks ½ appropriations	apportioned	No	No
House Public Works Committee	oil & chemical spills	300 million 40% feedstocks 60% appropriations	strict joint & several	No	Yes
Ways and Means Compromise (House Version)	sites & spills	1.2 billion 75% feedstocks 25% appropriations	strict joint & several	No	No
Senate Public Works	all releases	4.1 billion feedstocks	strict joint & several	Yes	Yes
Senate Compromise (final version)	all releases workplace & consumer products exempted	1.6 billion 87.5% feedstock 12.5% appropriations	strict joint & several	No	No

had created the problem. The Senate, in addition, extensively discussed medical compensation and enabling victims to sue in federal court for damages from dumpers and/or site owners.

The House and Senate barely touched on three other crucial issues: priority setting, acceptable risk, and state and local responsibility. The Congress did not question simply delegating to EPA the power to decide which kinds of problems to address first and how to cleanup a given site. The question of state and local responsibility was narrowly confined to the determination of how much states would contribute to be eligible for federal assistance. The broader question of how to encourage states and localities to identify problems responsibily and establish priorities in light of available resources, was ignored. Instead, the funding mechanism encouraged states to seek the maximum amount of federal assistance in all cases and to look upon the money received as a free good.

Senate Public Works was particularly irresponsible about priority setting. It favored victim compensation but never specified how the cruel choice between helping victims of past wrongs and protecting people from current and future risk would be made. Indeed, the staff seemed determined to ignore this problem in the hopes that its ambitious attempt to expand environmental rights would succeed without the full fiscal implications of that step being recognized.

The issues about which Congress deliberated well, and those that it did not, differ in kind. The fee, liability, and compensation questions were all distributive, concerned with who pays and who benefits. The geographical basis of congressional representation habituates congressmen to think in terms of whether their constituents will be advantaged or disadvantaged by specific policy proposals. Furthermore, many congressmen are lawyers, trained to consider the impact on their clients of alternative legal formulations. Whatever other difficulties members may have had in comprehending Superfund, they had no trouble in coping with the wording of various liability provisions or in understanding how such language would affect defendants and plaintiffs.

Congress, however, failed to deliberate about the basic strategic choices regarding program design and resource allocation. This failure too was attributable to the structure, membership, and resources of the body. In the House, each committee viewed the bill through the narrow lens of its own particular mandate. None tried to critically examine the structure of the program as a whole. In addition, the House appeared to have neither the intellectual nor the technical resources to critically examine the overall shape of the program. Congress could not research the health risks associated with various kinds of sites and spills. Instead, it had to rely on EPA for such information. Yet, it was the problem definition provided by the executive branch that needed challenging. Even with the aid of the General Accounting Office and the Office of Technology Assessment, it remained largely at the mercy of the agency and the chemical industry for the data on which to base its discussion.

Only the CMA and Stockman's small band of radical conservatives offered a fundamentally different view of the problem. The industry, however, had little data to offer, nor did it mount a carefully reasoned challenge to EPA's account of the facts. In addition, its hardball tactics helped undermine the credibility of whatever information it did provide.

Unlike the House, only one Senate committee, Public Works, had initial jurisdiction over Superfund. Turf problems did not prevent a comprehensive discussion of scope and strategy. Indeed the Public Works Committee drafted legislation that was quite different and much more ambitious than the EPA draft.

The great obstacle to Senate deliberation was the control over the terms of the debate exercised by a few key staff members. These individuals were committed to redistributing resources from polluters to the victims of environmental harm. Even more than the EPA, they viewed the abandoned sites problem as an opportunity to be exploited for wider purposes. Intensely suspicious of the chemical industry, they assumed that the dangers associated with hazardous waste were every bit as pressing as the most extreme environmental advocates had claimed. They looked upon this specific debate as yet another skirmish in the ongoing struggle to fix the blame on the business community for the environmental damage caused by resource extraction and manufacturing. All other issues were trivial by comparison.

To hardened veterans of that campaign, the concern for public education and political responsibility stressed in this book seem both naive and unimportant. After all, there was a battle to be fought and weaknesses in opponents' defenses to be exploited. Compared with establishing new basic environmental rights, what was a little tactical discretion on program costs?

The failure of Congress to consider transferring more responsibility to states and localities was likewise rooted in the outlook and incentives of members and staffers. Contemptuous of the capacity of state and local governments, they suffer from what one former state official has labeled the "Mississippi Syndrome," fixing upon the worst examples of state government behavior and treating those as the norm.[145] Living and working in the highly artificial environment of official Washington does not encourage an appreciation of national diversity, nor of the deprivation that citizens suffer when power and responsibility are removed from local hands.

The very acts of centralization that decrease state and local political responsibility increase the power of congressmen. They can gain prestige and visibility by being identified with a particular policy, as Florio did with Superfund. To sustain such attention, the program has to be designed so that members can continue to exert influence over it. The fifty state houses are not subject to congressional oversight, a federal agency is. Policy entrepreneurs, therefore, have every reason to insure a large continuing federal role in implementation.

The centralization of authority represented by Superfund is characteristic of many other recent policy initiatives. Certain advocates have perceived themselves to be stronger nationally than locally. Also as with Superfund, federal tax sources have tempted fiscally pressed local governments, despite the recognition that federal dollars entail expanded federal control. The geographic scope of problems, from interstate railroads to interstate air pollution, also leads to federalization, as do the outlooks and incentives of members of the Washington community discussed previously. Finally, professional and technical thinking in government may itself promote common standards and unitary control.

With so many pressures toward centralization, decentralization needs to be vigorously defended. The states, as we have argued repeatedly, are not merely administrative subdivisions of the nation, but independent centers of loyalty and identity to be fostered and supported as the political and social resource that they are. This requires more variety in rules and regulations than either environmentalists or national corporations would prefer. To be sure, there are decisions in which national interests are so strong that they supercede local objectives. Local ranchers ought not to be free to operate Yellowstone Park as they wish. But often, local perspectives deserve a major voice. The environmental policy of a nation so diverse in attitude and geography ought to respond creatively to that diversity, while simultaneously reflecting legitimate common concerns.

Decentralization offers several advantages for preserving responsibility and fostering civic education. The national government is remote, both spatially and psychologically. Political processes in Washington are complex, cumbersome, and difficult to influence compared with smaller units of government. While voting rates are lower in local than in national elections, voting is not the only form of participation. As the story of Love Canal illustrates, communities are good at mobilizing to protect their families and their property from apparent threats. There is nothing like debating the local school budget or neighborhood land use issues to teach citizens about both their differences and about the need to combine their various private objectives with legitimate common concerns to formulate good public policy. Community pride and a sense of mutual respect and interdependence tend to flourish most vigorously when the scale of interaction is small.

Similarly, the responsibility of elected and appointed officials to their constituents is more easily maintained locally, at least in principle. Both the costs and the fruits of government action are more visible. Complaints are more easily voiced. To be sure, a citizenry is likely to get only the degree of competence and integrity it wants and works for. But that democratic failing cannot be cured by diminishing democracy or distancing it from citizens.

Limiting federal involvement also discourages the naive notion that those who are not at fault have no responsibility for solving a problem. It helps citizens to recognize that, to an important degree, hazardous waste belongs to that category of nuisances—like crime, and natural disasters—that make demands on the entire community. In particular, a decentralized approach would impress upon individual communities that they have to find suitable disposal sites within their own borders and cannot expect to find other communities willing to solve their problem for them.

The Superfund story suggests that the Executive Branch, not the Congress, will often have to play the central role in framing legislative questions. Therefore, agencies must be prepared to produce data that facilitate a sound formulation of the problem and to present and analyze policy alternatives in light of their impact on a wide variety of criteria. They must combine both expertise and political responsibility in the synthesis and explanation of programs and strategies.

The EPA's capacity to perform this role was undermined by its pursuit of bureaucratic advantage. The opportunity presented by Love Canal seemed too good to be missed. A more judicious account of the risks and a less ambitious policy design might not have allowed EPA the same chance to strengthen its "antitoxic hazard" orientation.

In acting as it did, EPA was neither venal nor hypocritical. It had no difficulty reconciling its treatment of the sites and spills issue with its conception of the public interest. It viewed the effort to heighten public concern about hazardous waste as a necessary antidote to the political clout available to a chemical industry that could make large congressional campaign contributions and threaten to close plants and cut jobs in the districts of key congressmen. A similar justification was offered for the failure to present a cogent scheme for priority setting. The EPA believed these questions to be so complex and subtle that opening them up for congressional consideration would provide the allies of industry with myriad opportunities for obfuscation and delay. Influenced by the Mississippi Syndrome, it viewed program centralization as a necessary protection against the venality and inefficiency of state government.

According to the mores of their world, the behavior of EPA senior managers was impeccable. They operated with flair and effectiveness, seizing upon an issue of great public concern and converting it into new responsibilities and resources for their agency. Therefore, the key problem was not the quality of the agency's leadership, but in how those leaders understood their responsibilities and in the structure and functioning of the executive branch as a whole. For EPA to have conceived of its duty in a manner more conducive to the enhancement of public education and political responsibility, it needed to be part of a larger political system in which statesmanship, not salesmanship, was the mark of a good agency executive.

NOTES

1. The formal name of the law was Comprehensive Environmental Responses, Compensation, and Liability Act of 1980, (PL 96–510): 10 Dec. 1980.

2. See "Love Canal: False Alarm Caused by Botched Study," *Science* 208 (13 June 1980):1239–1242; "Cancer Incidence at Love Canal," *Science* 212 (19 June 1981)1404–1407: "Biological Markers for Chemical Exposure," *Science* 215 (5 Feb. 1982) 643–647: *Report of the Governor's Panel to Review Scientific Studies and the Development of Public Policy on Problems Resulting From Hazardous Waste*, Lewis Thomas, chairman, 8 Oct. 1980 (hereafter referred to as the *Thomas Report*).

3. *Thomas Report*, cover letter, 2; report, 13–23.

4. *Love Canal: Public Health Time Bomb*, A Special Report to the Governor and Legislature prepared by the Office of Public Health, Department of Health, State of New York, Robert P. Whelan, M.D., Commissioner, Sept. 1978 (hereafter referred to as the *Whelan Report*).

5. The first was the *Whelan Report*. The second was a pilot study by Dante Picciano of the Biogenics Corporation prepared for the Office of Research and Development of the EPA, May 1980.

6. *Thomas Report*, 6.

7. See the letter of Stephen Gage, Office of Research and Development, EPA, printed in *Science* 209 (15 Aug. 1980):754 (hereafter referred to as the *Gage Letter*).

8. The full background to the Love Canal controversy is contained in Adeline Gordon Levine, *Love Canal: Science, Politics, and People* (Lexington, MA: Lexington Books, 1982).

9. Levine, *Love Canal*, 7–15.

10. Levine, *Love Canal*, 17–18.
11. Levine, *Love Canal*, 18–19.
12. Levine, *Love Canal*, 27–29.
13. *Whelan Report*, 1,30.
14. Levine, *Love Canal: Science, Politics, and People*, 29; see also Lois Gibbs, *Love Canal: My Story* (New York: Grove Press, 1982), 30.
15. Levine, *Love Canal*, 175–212
16. Levine, *Love Canal*, 42–43.
17. Levine, *Love Canal*, 54–57.
18. Levine, *Love Canal*, 44.
19. Levine, *Love Canal*, 64.
20. Levine, *Love Canal*, 104.
21. Levine, *Love Canal*, 105.
22. *New York Times*, 2 Oct. 1979, III 23:1.
23. Levine, *Love Canal*, 137.
24. House Committee on Interstate and Foreign Commerce, Subcommittee on Oversight and Investigations, *Hazardous Waste Disposal: Part I*, 96th Congress, 1st session, 21 Mar. 1979, Committee Serial No. 96–48:60–86; also, Levine, *Love Canal*, 126.
25. *Thomas Report*, 20.
26. *Science* 209 (29 Aug. 1980):1003.
27. *Gage Letter*, 754.
28. "Love Canal: False Alarm Caused by Botched Study," *Science*, 208 (3 June 1980):1239–1242.
29. Levine, *Love Canal*, 140–147.
30. Levine, *Love Canal*, 171.
31. *New York Times*, 22 May 1980, A-14.
32. *New York Times*, 19 May 1980, B3.
33. Transmittal Letter of *Thomas Report*, Lewis Thomas to Governor Hugh L. Carey, 8 Oct. 1980, 2.
34. *New York Times*, 20 May 1980, A-1.
35. *New York Times*, 21 May 1980, A-1.
36. Levine, *Love Canal*, 207.
37. Thomas Jorling, former assistant administrator for Water and Hazardous Waste, EPA, joint interview held in Cambridge MA, 13 Jul 1981. Other participants were: Douglas M. Costle, former administrator, EPA; Swep Davis, former deputy assistant administrator for Water and Hazardous Waste, EPA; Charles Warren, former assistant administrator for Legislative Affairs, EPA; Eckhart C. Beck, former assistant administrator for Water and Hazardous Waste, EPA; and Andrew Mank, former member of the taskforce which drafted the EPA Superfund proposal.
38. Warren, joint interview, 13 July 1981.
39. Davis, joint interview, 13 July 1981.
40. *BNA Environmental Reporter*, 4 May 1979, 3 and 13 Jul 1979, 652.
41. Davis and Warren, joint interview, 13 July 1981.
42. Richard Hill, EPA Region V quoted in *BNA Environmental Reporter*, 19 May 1979, 1275.
43. Jorling, joint interview, 13 July 1981.
44. Steven Cohen, "Defusing the Toxic Time Bomb: Federal Hazardous Waste Programs," in Norman J. Vig and Michael Kraft, *Environmental Policy in the 1980's: Reagan's New Agenda* (Washington D.C.: CQ Press, 1984), 283
45. House Committee on Public Works and Transportation, Subcommittee on Water

Resources, *Hazardous Chemicals under the Federal Water Pollution Control Act*, 96th Congress, 2nd session, 15 Apr 1980, Committee Serial No. 96–56, 1.

46. Charles Warren, interview with Marc Landy, New York, 22 Jun 1981.

47. RCRA 42 U.S.C. 6901 et. seq.(1983). Clean Water Act 33 U.S.C. 1251 et aeq(1977)

48. Warren interview, 22 June 1981.

49. Swep T. Davis, Associate Assistant Administrator for Water and Waste Management, House Committee on Public Works and Transportation, Subcommittee on Water Resources, *Comprehensive Oil Pollution Liability and Compensation Act - Hearings on H.R. 85*, 96th Congress, 1st session, 26 Sept. 1979, Committee Serial No. 96–26, 197. (Hereafter referred to as *Hearings on H.R. 85*)

50. Letter Edward Dunkelberger, Covington and Burling to Edmund Frost, Vice President and General Counsel, Chemical Manufacturers Association, 25 July 1980.

51. Jorling, joint interview, 13 July 1981.

52. "Generator-based Fee System," EPA, undated.

53. "Superfund Economics—Rationale and Impacts," EPA document from the files of Thomas Jorling, Assistant Administrator for Water and Hazardous Waste, undated, 4.

54. House Committee on Interstate and Foreign Commerce, Subcommittee on Transportation and Commerce, *Hazardous Waste Disposal: Part 2*, 96th Congress, 1st session, 19 Jun 1979, Committee Serial No. 96–49, 133–140.

55. EPA produced an economic impact analysis based upon a study conducted for it by Data Resources, Inc., Lexington, MA. The study claimed that the fee would have a negligible impact on prices. See "Economic Impact Analysis, Proposed Superfund Fee System," an addendum to "Superfund Economics—Rationale and Impacts," undated.

56. *Chemical Marketing Reporter*, 28 July 1980, 8–14.

57. See R.D. Hinds, "Liability Under Federal Law for Hazardous Waste Injuries," *Harvard Environmental Law Review* 6 (1982):1–33.

58. *Boston Globe*, 13 Sept. 1983, 57,68; Dunkleberger letter, 6–7.

59. *New York Times*, 24 June 1984, F-2.

60. A notable example of the power of publicity to elicit EPA expenditures was the case of Times Beach, Missouri, where the government chose to buy an entire town that had been contaminated by dioxin. See "Special Dioxin Report," *St. Louis Post Dispatch*, 19 Nov. 1983.

61. CERCLA, op. cit., Sec. 105, 8a.d (Sec. 105 8a).

62. Jorling, joint interview, 13 July 1981.

63. Warren, joint interview, 13 July 1981.

64. Lawrence White, *Reforming Regulation* (Englewood Cliffs, NJ: Prentice Hall, 1981), 149.

65. Elliot Cutler, former associate director of OMB Division of Natural Resources, interview with Marc Landy, Washington, D.C., 20 Nov. 1982.

66. Warren, joint interview, 13 Jul 1981.

67. This account is based upon Costle joint interview, 13 July 1981. It is supported by the Elliot Cutler interview, 20 Nov. 1982.

68. Jorling, joint interview, 13 July 1981.

69. Sheila Brown, counsel to the Subcommittee on Transportation and Commerce, Committee on Interstate and Foreign Commerce, House of Representatives, interview with Marc Landy, Washington D.C., 22 July 1981.

70. These characterizations of the congressional climate are derived from Jorling and Warren, joint interview, 13 July 1981.

71. *BNA Environmental Reporter*, 2 Nov. 1979, 148.

72. Jorling, joint interview, 13 July 1981.
73. Warren, joint interview, 13 July 1981.
74. *Hearing on H.R. 85*, 195–200.
75. *Hearing on H.R. 85*, 210.
76. *BNA Environmental Reporter*, 16 June 1978, 249.
77. *BNA Environmental Reporter*, 20 Oct. 1978, 210.
78. Jorling, joint interview, 13 July 1981.
79. Rep. John Breaux (D,LA), interview with Marc Landy, Washington D.C., 20 Nov. 1981.
80. Breaux interview, 20 Nov. 1981; also Davis, joint interview, 13 July 1981.
81. Warren, joint interview, 13 July 1981.
82. *BNA Environmental Reporter*, 9 May 1980, 36.
83. *BNA Environmental Reporter*, 2 Nov. 1979, 1477–8.
84. Warren, joint interview, 13 July 1981.
85. *BNA Environmental Reporter*, 2 May 1980, 4–5.
86. *BNA Environmental Reporter*, 16 May 1980, 59–60.
87. Warren, joint interview, 13 July 1981.
88. Toby Meisel, Legislative Assistant, Rep. Thomas Downey (D,NY), interview with Marc Landy, Washington D.C., 20 Nov. 1981.
89. Davis, joint interview, 13 July 1981.
90. Meisel interview, 20 Nov. 1981.
91. *BNA Environmental Reporter*, 20 June 1980. 271–2.
92. Davis, joint interview, 13 July 1981.
93. *BNA Environmental Reporter*, 5 Sept. 1980, 666.
94. *Congressional Record*, 96th Congress, 2d Session, 23 Sept. 1980:H9439.
95. *Congressional Record*, 96th Congress, 2d Session, 23 Sept. 1980:H9445.
96. *BNA Environmental Reporter*, 26 Sept. 1980, 729–30.
97. Andrew Mank, joint interview, 13 July 1981.
98. *BNA Environmental Reporter*, 20 July 1979, 730.
99. Memorandum to Senator Muskie regarding Hazardous Waste Legislation, from Karl Braithwaite, Sally Walker, Andy Mank, and Phil Cummings, 17 Jan. 1980 (hereafter referred to as *Senate Staff Memo*).
100. H.R. 4566, 601(o), 602(b).
101. "Why doesn't the administration proposal deal with personal injury damage for uncontrolled sites?" Memo from Thomas Jorling's files, undated.
102. Mank, joint interview, 13 July 1981.
103. *Senate Staff Memo*, ?
104. Warren, joint interview.
105. *Senate Staff Memo*, ?
106. *BNA Environmental Reporter*, 8 Feb. 1980, 1976–1977.
107. *BNA Environmental Reporter*, 8 Feb. 1980, 1976–1977.
108. *BNA Environmental Reporter*, 22 Feb. 1980, 2038.
109. *Senate Staff Memo*, ?
110. *Senate Staff Memo*
111. Mank, joint interview, 13 July 1981.
112. Mank, joint interview, 13 July 1981.
113. *Congressional Quarterly Almanac*, 1980, 591.
114. *Congressional Quarterly Almanac*, 1980, 592.
115. Additional views of Domenici, Bentsen and Baker, *Report of the Committee on Environment and Public Works, U.S. Senate*, 96–848, to accompany S.1480, 11 July 1980, 119–122.

116. *BNA Environmental Reporter*, 4 July 1980, 327; *Congressional Quarterly Almanac*, 1980, 591.

117. Warren, joint interview, 13 July 1981.

118. *BNA Environmental Reporter*, 5 Sept. 1980, 668.

119. Warren, joint interview, 13 July 1981.

120. Telegram from William Guste to Senator Russell Long, 19 Sept. 1980 (copy in the files of Andy Mank, Staff, Senate Committee on Environment and Public Works).

121. *BNA Environmental Reporter*, 10 Oct. 1980, 788.

122. Davis, joint interview, 13 July 1981.

123. Mank, joint interview, 13 July 1981.

124. Davis, joint interview, 13 July 1981.

125. Warren, joint interview, date?

126. *Chemical Marketing Reporter*, 28 July 1980, 8.

127. For the CMA position, see "CMA Summary of Superfund Costs," March 1980, published by Chemical Manufacturers Association, Washington D.C.; also, testimony of Robert Roland, president, CMA, at hearings, "Superfund," House Committee on Interstate and Foreign Commerce, Subcommittee on Transportation and Commerce, 96th Congress, 1st session, 19 Jun 1979, Committee Serial No. 96–114, 346–417.

128. *National Journal*, 13 Dec. 1980, 2130.

129. *National Journal*, 13 Dec. 1980, 2130.

130. *National Journal*, 13 Dec. 1980, 2130.

131. Costle, joint interview, 13 July 1981.

132. Mank, joint interview, 13 July 1981.

133. *BNA Environmental Reporter*, 21 Nov. 1980, 1041–2.

134. Warren, joint interview, 13 July 1981.

135. *BNA Environmental Reporter*, 28 Nov. 1980, 2097; on the "no-amendments agreement," see *Congressional Quarterly Almanac*, 1980, 592.

136. Mank, joint interview, 13 July 1981.

137. *BNA Environmental Reporter*, 5 Dec. 1980, 1177.

138. Warren, joint interview, 13 July 1981.

139. *Congressional Quarterly Almanac*, 1980, 593.

140. Beck, joint interview.

141. *Congressional Quarterly Almanac*, 1980, 593.

142. Warren joint interview, 13 July 1981.

143. *Congressional Quarterly Almanac*, 1980, 584.

144. In addition to his comments in the *Congressional Record*, 96th Congress, 2d Session, 23 Sept. 1980:H9439, see also a letter he wrote to his fellow House members entitled, "Superfund: A Regulatory Love Canal," 18 Sept. 1980, and his dissenting views attached to the Report accompanying H.R.7020, House Interstate and Foreign Commerce Committee, 96th Congress, 2nd Session, H. Rpt 96–1016, Part 1, 16 May 1980.

145. Anthony Cortese, former commissioner of the Department of Environmental Quality Engineering, Commonwealth of Massachusetts,interview with Marc Landy, Boston, Massachusetts, 10 Aug. 1982.

6

Forging a Cancer Policy: The Interagency Regulatory Liaison Group

with Margaret Gerteis

At a joint press conference held in late summer 1977, the heads of the Environmental Protection Agency, the Food and Drug Administration (FDA), the Occupational Safety and Health Administration (OSHA), and the Consumer Product Safety Commission (CPSC), announced the formation of the Interagency Regulatory Liaison Group (IRLG). The new group was to coordinate the activities of its members and help them reach their common goals more effectively. As another reform exercise in an administration fond of such devices, it received little attention. But IRLG was unusual in that the initiative for its creation came directly from the senior managers of a set of agencies better known for rivalry than cooperation.[1]

In this chapter we trace the history of IRLG and evaluate its accomplishments. In a world of advocacy agencies, was an interagency alliance either possible or desirable? Were the agencies able to work out a constructive synthesis of their views, or did they merely negotiate an uninformative linguistic compromise?

As an example of "horizontal coordination," so beloved by some management experts, IRLG should tell us something about whether such devices facilitate or retard political accountability for key policy decisions. More importantly, what can be learned from this experience about how to encourage the deliberative resolution of problems within the bureaucracy?

We have already encountered President Carter's ambiguous attitude toward environmental regulation. Elected with environmentalist support, and with strong personal environmental commitments, he made his early appointments accordingly. Yet persistent economic problems, and the influence of his close economic advisors Charles Schultze and Alfred Kahn, led Carter in the direction of trying to make government regulation more efficient and less costly. The anti-Washington, antibureaucracy orientation of Carter's campaign, combined with his long-standing interests in managerial reform, and his engineer's concern with efficiency, all reinforced this second thrust. The agencies involved in IRLG, and their leaders, had to respond to this dualism.

Of the four agency heads involved in IRLG, Douglas Costle was the only one with direct access to the president. His agency also had the most complex set of issues to deal with. As we saw in the ozone case, Costle tried to walk a fine line, being "reasonable" toward industry while maintaining his credibility with environmentalists, the technical community, and the Congress.

Eula Bingham, the OSHA administrator, was an environmental scientist from the University of Cincinnati and a newcomer to Washington. She had served as a toxicology expert on several advisory committees, but she had no experience with public office. "I was picked for the OSHA job," she later recalled, "because I was on two of their lists. I was on their 'labor' list and I was on their 'women' list."[2]

As Bingham took over, OSHA was the target of much industry criticism, particularly for its detailed and (in industry's view) often ineffectual safety regulations. At the same time, organized labor was complaining that OSHA was not doing enough to protect workers. As a strategic matter, Bingham was determined to redirect OSHA's efforts from safety (i.e., accidents) toward health—her own area of expertise. By doing so she hoped to show that OSHA could be useful without imposing detailed (and easily caricatured) safety requirements. As the assistant secretary of labor for occupational safety and health she reported to the secretary of labor, Raymond Marshall, a generally supportive ally. But the agency she found when she arrived in Washington lacked the technical capacity to follow her new strategy. Many of her staff had safety, rather than health, backgrounds. Moreover, OSHA's early efforts to develop industry-wide health standards met with as much legal challenge as had the prior safety regulations.

Donald Kennedy, the new head of FDA, had also spent the bulk of his career in academia, mainly at Stanford University, where he had established a formidable reputation as a biologist. During the Ford administration he served briefly as a senior consultant to the White House's Office of Science and Technology Policy. As FDA commissioner he had two advantages, his academic reputation and his lack of prior connection with the pharmaceutical industry.

The FDA, with a history dating back some seventy years, is the oldest of the health regulatory agencies. It was also the one least threatened by any proposed reorganization scheme. Although FDA was less controversial than, say, EPA, as the Carter administration began, it was being criticized both for the hurdles it put in the way of new drugs, and in regard to the "Delaney Amendment" that banned carcinogenic additives in foods. In the latter case, critics charged, the law did not take account of benefits as well as costs, nor did it distinguish between strong and weak carcinogens. As with EPA, the scientific basis of many of the FDA's actions had been publicly and repeatedly criticized, and public confidence in FDA had declined. The news media often referred derisively to FDA's "cancer of the week."

John Byington, alone among the four, did not serve at the pleasure of the president. A lawyer by training and a Republican, he had been appointed chairman of the CPSC by President Ford, with a fixed term that expired in October, 1978.[3] When Byington retired in June 1978, he was replaced as chairman by Democrat and Carter-appointee Susan King, a political scientist and former executive director of the Center for Public Financing of Elections.

The CPSC was the newest and bureaucratically least secure of the four IRLG participants. Even its continued existence was uncertain in early 1977. Both a presidential reorganization plan and proposed congressional action contemplated abolishing it altogether, absorbing its functions into either a reorganized EPA or a Cabinet-level consumer affairs department. This challenge was especially serious since the strong leadership required to defend the commission's role was not easy for the chairman to exert. The commission was a collegial group and Byington was regularly criticized by his fellow commissioners for attempting to engage in "one-man rule." In addition, the CPSC had an even smaller staff, less technical expertise, and less clout than the other agencies.

Several months before the formal interagency agreement was announced, the regulators had begun to meet among themselves. Their contacts at first were casual. Eula Bingham sought out fellow-scientist Donald Kennedy, whom she knew from earlier committee work. She hoped that her own understaffed agency might be able to draw, from time to time, on FDA resources.[4] Costle, too, had begun to meet with Kennedy, recognizing the value of cooperating with someone who was both a respected scientist and the head of an agency with substantial relevant technical expertise. The idea for IRLG began to take shape in May of 1977, when Costle, Kennedy, and Byington found themselves at a joint press conference announcing a coordinated approach to the regulation of chlorofluorocarbons. This joint effort had actually been agreed on during the previous administration. But to Costle it seemed just the sort of "good government" that regulators might initiate to their mutual advantage. Shortly thereafter, Costle suggested that the four agency heads get together regularly but informally. Without fanfare, Costle, Kennedy, Bingham, and Byington began having breakfast together from time to time.[5]

All of them were impressed by the easy camaraderie that emerged in these meetings. Personally, they liked each other's company, and politically, they were sympathetic to one another (the more so when King replaced Byington in June 1978). These informal get-togethers, with no prescribed agenda, produced a spontaneity and flexibility that they valued highly. Word of their meetings soon began to circulate within their agencies, however, and senior staff began to bring to their bosses' attention various unresolved interagency matters. "We came to realize," Costle recalled, "that there was a whole agenda of unresolved issues among our agencies." Informal breakfast meetings, however attractive, could not deal with everything that could profit from their joint attention.[6]

Each of the IRLG principals believed that they stood to gain from selectively combining forces. Their agencies had similar, even overlapping, missions and common enemies—both inside and outside the administration. Both Costle and King believed that the reputations of Kennedy and Bingham would enhance the credibility of any joint pronouncements. Bingham and King both saw the advantage of technical assistance from the larger and more experienced staffs of EPA and FDA. Costle's access to the White House gave him a political strength that none of the others enjoyed. "We saw the wisdom," Eula Bingham recalled, "of circling the wagons."[7]

THE CREATION AND STRUCTURE OF THE INTERAGENCY REGULATORY LIAISON GROUP

The structure and operations of IRLG reflected attitudes developed in the first few months of its existence. The four agency heads wanted the flexibility to deal directly with issues that might otherwise have languished at lower levels. But to get that job done, they needed staff support. How could they establish some sort of ongoing liaison arrangement without creating a new bureaucracy?[8]

What ultimately developed was a three-tiered arrangement. At the top were the "principals" who continued to meet alone, at regular intervals. They, in turn, selected "surrogates," senior staff members who devoted full time to IRLG. Finally, there were "work groups," staffed on a part time basis by technical personnel from each agency, to do the actual joint projects.

The IRLG had no staff of its own. Work groups were formed around particular issues. Each group's efforts were supervised by one of the principals and that principal's surrogate. All staff time was contributed by the constituent agencies. Later, when it was decided that one full time executive assistant was needed to keep track of IRLG business, that person too was "borrowed" from one of the agencies. The IRLG's $1 million budget—which covered certain contracted services, special publications, workshops, and so forth—was contributed by the participating agencies in proportion to their respective levels of funding. In spite of their varying financial contributions, the four participating agencies functioned as equals. The leadership of both the principals' and the surrogates' groups rotated. Decisions were made by consensus.

The IRLG's organizational characteristics gave it an elusive quality. The very name "Interagency Regulatory Liaison Group" referred to several sets of individuals and activities. To the principals, the essence of IRLG was always their private breakfast meetings—the information they shared, their camaraderie, and their mutual support. The term *surrogate* (suggested by Kennedy), implied that while others might act at the principals' behest, authority was never really delegated. Nevertheless, for most formal purposes—in public documents, for example—it was the surrogates (the only participants in the endeavor who were both full-time and more or less permanent) who were listed as the "members" of the IRLG. To the extent that others in government were aware of the IRLG's activities, on the other hand, it was most often through the efforts of the work groups.

The IRLG was as difficult to locate as it was to identify. The office of the executive assistant rotated every six months along with the chair. As a result, given printing lags, throughout its brief existence the address and telephone number listed under "IRLG" in the federal telephone directory was almost always incorrect.

Although the organizational details were being worked out in the summer of 1977, the surrogates' group began to meet. The enthusiasm for collective action that had infected the principals did not at first rub off on the surrogates, Eula Bingham later recalled. But once they got going, they developed some esprit de corps of their own.[9]

The surrogates' first major assignment was to identify areas for cooperation that were both feasible and on the agencies' common agendas. Given the nature of their agencies' responsibilities, toxic substances dominated the surrogates' thinking. Working with the principals, they decided to establish various work groups. Throughout the fall of 1977 they worked to identify people from the agencies to staff the groups they had established. The work groups were not to be "pie-in-the-sky" efforts, but to take on specific tasks that could be completed within six months to one year. The charge given to each group was deliberately broad, leaving each to define its own agenda.[10] In February 1978, the eight groups' work plans were published in the *Federal Register*. They were as follows:

1. *Testing Standards and Guidelines* was to develop uniform methods and protocols for conducting various common toxicological tests, so that agency requirements would be similar at least at that technical level.
2. *Research Planning* was to find out who was doing what toxicological research, both within and outside of IRLG agencies. A long run hope was to coordinate this research, and eventually cooperate in research budget submissions.
3. *Risk Assessment* proposed to develop common criteria and approaches to the scientific aspects of risk assessment techniques, but without "suggesting regulatory responses."
4. *Regulatory Development* had already made the IRLG agencies sensitive to those cases where more than one agency proposed regulatory action on the same chemical, in an effort to insure consistency and coherence. It also intended to continue to survey the horizon and facilitate joint agenda setting.
5. *Epidemiology* proposed to begin with an inventory of epidemiological personnel and research activities. It hoped then to go on to develop study guidelines, investigate legal problems of gathering epidemiologic data, and look at the value of existing data bases.
6. *Information Exchange* was designed to find ways to facilitate this process among IRLG agencies by developing common identification codes for keeping track of existing data and, if possible, coordinating reporting requirements to enhance the usability of new data.
7. *Communication and Education* was in part IRLG's own publicity apparatus. It was to issue press releases, pamphlets, and so on. It also hoped to develop public school curriculum material on public health issues.
8. *Compliance and Enforcement* was intended to facilitate coordination of field efforts. Immediate tasks were developing interagency notification procedures for use in emergencies and mechanisms for sharing laboratory facilities and other equipment and personnel. Longer range efforts involved joint inspector training and hopes of interagency alerts for possible violations, or even "crossover" inspections.[11]

As can be seen from this list, the tasks the work groups proposed varied greatly. Some involved practical, relatively uncontroversial, details—like common computer identification codes—that did not raise issues of high principle. This did not make them unimportant. For OSHA area offices to know who and where to call in EPA when they came across a potential environmental emergency, and to be able to use EPA laboratories when their own were not available, was all to the good.

Other proposed agenda items were quite different. Who would reconcile conflicting intra-agency and inter-agency claims in preparing a budget for new research? Would the White House, including the Office of Management and Budget (OMB), view such collaboration as helpful coordination or as an incipient coup d'etat? Who could resolve policy differences among the agencies (in part due to statutory differences) in considering costs, and in the use of formal mathematical models for risk assessment in regulatory decision making? How would dissenters be "brought along," if at all, and what legal status, if any, would IRLG's pronouncements have? All of these issues were waiting for the principals, as they shared homemade muffins, comradeship, and a view of Washington at sunrise at their periodic breakfast meetings.

IRLG AND RELATIONSHIPS WITHIN THE EXECUTIVE BRANCH

Until the work groups began to complete some of their tasks and circulate their output beyond the agencies, IRLG received little attention. This was due in part to the principals' deliberate attempt to maintain a "low profile." But it also stemmed, no doubt, from the fact that other coordinating committees, clearing houses, liaison groups, and the like were already on the scene. When IRLG was formed, at least twelve other liaison groups were already in existence devoted in one way or another to the task of coordinating policy and research on toxic substances. Indeed, IRLG was part of an extensive competition within the Carter Administration for authority over policy on toxic substances. This struggle, shaped by conflicting bureaucratic interests and fueled by diverse ideological commitments, was in no way resolved by IRLG's development.

Early in his presidency (in May of 1977), just as IRLG was beginning to coalesce, Carter sent an Environmental Message to Congress that announced the formation of a Toxic Substances Strategy Committee (TSSC) under the White House's Council on Environmental Quality (CEQ). The idea for the TSSC had originated with the CEQ, which had helped to draft the president's message. The head of the TSSC was to be the new CEQ chairman, a lawyer named Gustave "Gus" Speth, who previously worked for the activist National Resources Defense Council, which he had helped to found. Speth believed he had presidential backing to use the TSSC to shape policy and coordinate regulatory action on toxic substances.[12]

The TSSC was to include representatives from all of the cabinet departments, regulatory agencies, and research institutions that had anything at all to do with toxic substances (some seventeen in all). Since neither the CEQ nor the TSSC would have much staff to provide support, the members of TSSC were to supply the bulk of the necessary expertise.

When Speth began to discuss his plans, the response of other agencies was lukewarm at best. The idea of a uniform regulatory approach emanating from TSSC was not welcomed by those who (like OSHA) had been working for years on their own particular policies. Speth and his staff began to get the message to "back off" on their grander designs.[13] When news of the IRLG's meetings and the scope of its plans began to circulate, Speth was reported to be "furious" at the regulators' independent initiative, notwithstanding their denials of any hostile intention.

From the point of view of Costle and his regulatory colleagues, however, the IRLG was an expression of their view that clout in government does and should reside with those who make decisions, and not with some "self-appointed" staff group attached to the White House. The latter could not really know what was possible and practical. And since they did not have to live with the consequences of any decisions, such a staff group could not be trusted to balance conflicting claims. Costle always believed that IRLG (not TSSC) had the president's tacit approval and that Carter's support was critical to IRLG's ability to function.[14]

Despite Speth's initial anger, he had few substantive differences with the IRLG principals and few political resources with which to try to displace them. There was, in fact, little he could do but "join 'em." IRLG work groups subsequently opened their meetings to TSSC representatives, and the two groups worked in relative harmony for the balance of IRLG's existence.

The IRLG's efforts to coordinate regulatory activity also brought it into conflict with others in the Executive Branch who had planning and oversight responsibilities, particularly with the OMB. The focus was the proposal of the research planning group to make an inventory of all toxicological research at the regulatory agencies and at nonregulatory government research institutes. It then intended to identify those areas where regulators' needs were not being met or where there was unproductive duplication. The ultimate goal was to coordinate research planning on the basis of this information, and to integrate that planning into the fiscal year 1980 budget process.

The information-gathering phase of this project was reasonably successful. With the cooperation of the various research units, the IRLG work group was able to put together a relatively comprehensive list of research activities. When it came to integrating this information into agency budgets, however, the regulators, the research community, and OMB had notably different agendas.[15]

The principals hoped to use the research planning effort—and the information it generated—to build a case for more resources, and perhaps to lay claim to some of the very substantial funds being spent on toxicology by the National Institutes of Health (NIH). By 1979, however, when the fiscal year 1980 budget was being prepared, OMB was, in Costle's words, "more interested in eliminating waste and redundancy than in filling in gaps," especially as the budget scramble in this austerity year became more intense.[16]

The scientific research community was represented within the White House by Presidential Science Advisor Frank Press and his Office of Science and Technology Policy (OSTP). Apart from their role in substantive problems, Press and his colleagues also saw themselves as defending society's long-term interest in science against shortsighted political or economic considerations. When the regulatory agencies' proposed budgets came in, reflecting the IRLG's targeted research priorities, Press found that from his point of view they placed too much emphasis on "immediate goals." He urged the agency heads to "correct the imbalance" between short-term and long-term research in revised budget submissions.[17]

Meanwhile, the leadership of the Department of Health and Human Services(HHS) was hardly pleased by what it saw as an attempted raid on their budget by the regulatory agencies, and more particularly by OMB. Costle believed that OMB was trying to use IRLG "to get a handle on the National Institutes of Health." This was not successful, however, since OMB backed down at the insistence of HHS's leaders.

After the experience with the fiscal year 1980 budget process, the IRLG abandoned its research planning effort, which instead of promoting collaboration, intensified turf battles and pitted regulators against scientists. When he reflected on the experience some years later, Costle concluded that IRLG had underestimated the strength of entrenched research interests. The regulators' lack of control over research priorities remained, in his view, a weakness of the overall research system. The IRLG proved to be an ineffective device for altering the behavior of noncooperative agencies beyond its own boundaries. In such a context it functioned as a relatively small coalition within the larger game of executive branch pluralist politics.[18]

Still, the purpose that IRLG had pursued was so obviously important that the objective reemerged as HHS's own initiative, the National Toxicology Program (NTP), established in November 1978. This group sought to coordinate all in-house federal toxicological research. The interests of the regulatory agencies in this activity were accommodated by having the heads of the appropriate agencies serve as an executive committee of the NTP.

The role of IRLG as a coalition designed to strengthen the participating agencies against the rest of the government is also clearly illustrated by the group's relationships with various White House economic units. Even in the early days of the Administration, the staffs of the Council on Wage and Price Stability (CWPS) and the Council of Economic Advisors (CEA) often worried more about costs than the regulatory agencies thought appropriate. As the economy failed to improve, the steadily mounting pressures on President Carter to control government spending promoted the decision to create the Regulatory Analysis Review Group (RARG) in March of 1978.

As our earlier discussion of ozone showed, there were well-publicized conflicts between RARG and both OSHA and EPA in 1978 and 1979 over the extent to which the administration should concern itself with the costs of health and safety regulations. In the wake of the president's intervention in OSHA's cotton dust decision, it seemed highly desirable to many in the administration to find a way to shield their leader from the political flak such interventions were sure to generate. In response, Charles Schultze proposed to broaden RARG's regulatory oversight powers.[19]

Given their problems with the existing mechanisms, the IRLG principals were hardly thrilled with the proposal to give RARG still more authority. Senior officials at EPA quickly put together a counterproposal. This became known as the "Regulatory Council," a sort of IRLG writ large. A liaison group that would coordinate the activities of nineteen (instead of only four) regulatory agencies, the council was to promote more cost-effective regulation through mutual cooperation across a wide range of issues. (Skeptics emphasized the irony of asking the regulators to regulate themselves.)

Although the president did not abandon RARG, he agreed to the new idea.[20] The council could, after all, provide a way for the administration to claim that it was trying to control the costs of regulation. If the Council did act, then someone other than the White House would be available to take the heat for decisions aimed at avoiding regulatory excess. The IRLG became, in effect, a health subgroup of this new organization.[21]

The IRLG thus served both implicit and explicit functions. At a practical, working level, it aimed to coordinate regulatory policy and make it more efficient. But it also allowed the principals to present a united front within the administration, and to use this strength in an effort to influence the administration's overall policy on the regulation of toxic substances. Nowhere was this dual purpose more visible—nor were the technical and political complexities that such a strategy raised more evident—than in the IRLG's attempt to draft an interagency "cancer policy."

IRLG AND CANCER POLICY:

Background and Context

Each of the IRLG agencies had some responsibility for controlling chemical substances that might cause cancer. Hence, they all faced the difficult problem of deciding what constituted sufficient evidence to initiate regulatory action. As we will see shortly, this was a problem fraught with great uncertainty and controversy. Both the science and the law involved in controlling chemical carcinogens were filled with pitfalls, ready to trap even the most wary regulator. With all of their decisions subject to increased pressure and attention, the IRLG members searched for a politically tenable and legally defensible basis for their regulatory actions, and they looked to a "scientific consensus" to do that job for them.

The development of a uniform policy among the agencies was complicated by the presence of strong but conflicting attitudes toward the problem.[22] The potential hazards of chemicals in the human environment first became a focus of international public health concern shortly after the end of World War II, when the carcinogenic potential of a few chemicals had been established by studies of specific occupational groups. The increasing use of chemicals, especially as food additives, prompted a series of international conferences in the 1950s and 1960s. The principles, guidelines, and standards that emerged from these conferences and studies, sponsored by the International Union Against Cancer and later the World Health Organization were characterized by caution toward the introduction of new chemicals. These groups advocated laboratory testing, particularly animal toxicology studies, as the principal tool of regulatory decision making. (After all, before the development of animal bioassays, there was no way to test new compounds, except to feed them to humans.)

This European (or international) literature on policy toward chemical carcinogenesis began to influence American policy makers and researchers in the late 1950s and early 1960s. Its most notable early effect was the Delaney provision in the 1958 amendments to the Pure Food and Drug Act that included banning outright the use of any food additives found carcinogenic in laboratory animals.[23]

In post-war America, however, the more general attitude was expressed by DuPont's slogan "Better things for better living through chemistry." Used in pesticides, plastics, synthetic fibers, and food preservatives, chemicals were widely seen as good things, promoting American growth and prosperity. It was not until the 1960s that public opinion began to change, prompted in part by such events as the thalidomide disaster and the publication of Rachel Carson's influential book, *Silent Spring*.[24]

When the IRLG agencies came together to discuss cancer policy, they brought with them a variety of statutory constraints, interest group pressures, and bureaucratic traditions. In this respect, EPA's history was the most complicated. Its jurisdiction over carcinogens derived from several different statutes—including the Clean Air Act, the Clean Water Act, the Safe Drinking Water Act, and the Federal Insecticide, Fungicide, and Rodenticide Act (FIFRA). Each of these had its own definitions and requirements. Each of the various bureaucratic units that were merged to form EPA had its own established jurisdictions, constituencies, and ways of doing business. But many of the agency's central staff were young and idealistic, with a strong commitment to environmental values. In EPA's early days this led to serious conflicts between relocated sub-units and EPA policy-makers, as the latter sought to pursue new directions.

One of the more publicized of these conflicts pitted the Office of General Counsel (OGC) against the Office of Pesticide Programs (OPP) in the implementation of FIFRA.[25] Out of this history came EPA's first efforts to draft some sort of formal cancer policy. Federal power over the licensing of pesticides under FIFRA resided with the OPP, which had long been part of the Department of Agriculture. Oriented more toward agricultural production than health, and with strong ties to agricultural constituencies inside and outside government, OPP had traditionally favored expeditiously licensing new pesticides. In its regulatory deliberations, it had focused on short term toxic effects. As long as the exposure that resulted from a pesticide's use did not exceed safety thresholds as determined by OPP, a substance was deemed safe for agricultural use. Through the 1960s, despite steadily mounting criticism, OPP (as part of the Department of Agriculture) was politically strong enough to continue these practices.

When EPA was formed in 1970, OPP was transferred to it. The office, however, retained its old staff, its old loyalties, and continued its old licensing practices. Rather than challenge OPP directly, EPA leaders sought instead to control pesticides through case-by-case litigation. Since the latter function was controlled by the OGC, the litigation strategy offered the prospect of a more sympathetic locus for action.

The OGC's strategy depended on its ability to make the case that a particular substance was harmful, and that, therefore, its "registration" should be revoked. To accomplish this, OGC decided to focus their attention on cancer, rather than on some less readily demonstrable or less feared harm. Cancer was, after all, the most widely studied health effect of chronic chemical exposures. But even for cancer, epidemiologic studies of the compound in question were often unavailable. The only usable evidence was often from animal tests. To establish the persuasive value of such data, OGC attorneys invoked the principles that had been advanced in the international cancer literature.

After successfully prosecuting several actions, OGC sought to draw together the scientific arguments and the legal precedents they had established in a single set of "cancer principles." These, they hoped, would provide a foundation for future litigation and would obviate the need to start again and again from first principles. In a number of cases, various versions of such principles were introduced, often based on the testimony of, or prepared with the help of, Dr. Umberto Saffiotti, then associate director for carcinogenesis in the Division of Cancer Causation and Prevention at NCI. Saffiotti had also been the delegate from Italy to the meetings of the International Union Against Cancer in the late 1950s.

Even before it began to draft these cancer principles, the actions of the OGC had provoked the OPP and its agricultural constituents into a series of counterattacks. In case after case, scientists from the U.S. Department of Agriculture and various trade groups testified against the OGC position. They were especially vocal in their opposition to Saffiotti's cancer principles. They characterized these as arbitrary and scientifically unfounded. Regulators, they said, would be compelled to make a determination of carcinogenicity on the basis of weak scientific evidence. The policies of the OGC were further criticized for failing to heed the requirement in FIFRA that the EPA consider the benefits as well as the risks of pesticides. The OPP staff members were also unhappy with the situation. They resented their loss of control over policy to a group of lawyers whom they considered both technically incompetent and possessed of an environmentalist bias.

So heated did the controversy become (and so organized the opposition) that by late 1975, EPA Administrator Russell Train appeared to withdraw his support from OGC. He reorganized the efforts aimed at revoking pesticide registrations and put the technical people from OPP back in the lead role in such decision making. Several of the OGC attorneys resigned in protest— among them Anson Keller, former associate general counsel. In the closing months of the Ford Administration Keller joined OSHA, where he focused on drafting a legally defensible generic policy for the regulation of carcinogens in the work place.

Train also signaled a new direction with regard to the substance of EPA's policy. He established a new Carcinogen Assessment Group at EPA, led by Dr. Roy Albert of the Institute of Environmental Medicine at New York University. Albert and his colleagues were to develop a quantitative methodology to assess cancer risk. This method was to provide a basis for weighing both risks and benefits (thus defending the agency against industry critics) and to serve as a foundation for future EPA regulations. This was, in effect, a repudiation of Saffiotti's view that such assessments were not possible.[26]

The EPA's senior science advisor, William Upholt, asked for the National Cancer Institute's view of the previously established principles. In due course, a subcommittee of the National Cancer Advisory Board reviewed them and found them wanting. The subcommittee's main objection appears to have been to the formalistic, rigid nature of the proposed decision rules. They argued that deciding whether a substance is carcinogenic is so complex that the weight to be given to various kinds of evidence could not be specified in advance. Instead, each decision should be made in light of all the details of that particular situation.[27]

This challenge to the premise of the whole activity was not enough to counteract the impetus for devising a generic set of cancer principles at OSHA. That agency's motives were similar to those of the OGC lawyers at EPA. OSHA had hardly begun to draft work place exposure standards for hazardous substances when Eula Bingham arrived, but already the opposition was mounting. The future seemed to hold a repeat of the pesticides litigation, with repeated battles fought over the same fundamental principles. Since much of the legal groundwork had already been laid in the earlier pesticide cases, the lawyers reasoned, OSHA's cancer policy could pick up, in a sense, where EPA's had left off. Moreover, OSHA's statutory authority seemed more clear cut. The onus was clearly on protecting workers, to the extent "feasible." There were no vague imperatives (as there were in FIFRA) to weigh costs and benefits.[28]

In defense of OSHA's emergency benzene standard, in early 1977, OSHA lawyers attacked any quantitative calculation of cancer risks and benefits. They aimed to establish, as a matter of legal precedent, that such efforts lacked scientific validity, would therefore compromise workers' safety, and would thus violate OSHA's legislative mandate.[29] Both professionally and philosophically, Eula Bingham belonged to the same tradition as Umberto Saffiotti. She found a natural ally in Anson Keller and the OSHA attorneys who were busy devising a cancer policy based on Saffiotti's "principles." The position of OSHA against quantitative risk assessment, therefore, contrasted sharply with the new policy direction in favor of such methods at EPA.

In some respects the FDA's experience with drug regulation was similar to that of OPP's with pesticides, although the FDA's jurisdiction was broader and its constituency more varied. The FDA had strong ties with the medical and pharmaceutical communities, and, like OPP, it was under considerable pressure to make new and beneficial compounds available. Like OPP, too, it took a traditional toxicological approach—determining safety levels and dose tolerances for various categories of users. The FDA staff was also accustomed to making at least informal determinations of risks and benefits as they considered whether to approve the introduction of new drugs into the marketplace.[30]

The Delaney Amendment required the FDA to take a different approach in the case of food additives. This statute mandated a zero tolerance level for all compounds found to be animal carcinogens and that no benefits justified their use. Faced with uncongenial concepts and standards, the FDA was not eager to get into this area. Before the saccharin case in the spring of 1977, the Delaney clause had been invoked only infrequently, and some FDA actions under it had been widely perceived as arbitrary and unwarranted. If the FDA could participate in the establishment of a set of guidelines that were both widely accepted and scientifically valid, its future actions would be more credible and perhaps easier to defend.[31]

The CPSC, alone among the IRLG agencies, had no real experience with regulating chemicals. (It had deferred to EPA and FDA in the case of chlorofluorocarbons.) As a result, it had no strong traditions that shaped its approach to the subject, and no particular stake in one outcome over another. Its hand would clearly be strengthened by a common regulatory policy on the subject. But, as one observer later commented, it was for the most part a "free rider" in the IRLG's cancer policy work.[32]

Chemical carcinogenesis was thus on the regulators' agenda before IRLG began to tackle the issue in earnest through its Risk Assessment Work Group in the summer of 1977.

Early Steps and Basic Issues

The initial charge to the Risk Assessment Work Group, drawn up by Richard Bates (then Donald Kennedy's IRLG "surrogate"), had not specified a focus on cancer. Instead it had directed the group to address all of the qualitative as well as quantitative aspects of risk assessment. There were, however, serious problems with such a comprehensive agenda. The work group was large and diverse—including lawyers, economists, engineers, and scientists from each of the agencies. "None of us knew each other," Joe Rodricks, chairman of the group, later recalled, "and it was obvious that we could never produce anything like consensus on such broad policy issues."[33]

The first order of business, then, was to pare down the group's agenda. As Rodricks later explained:

> We decided to focus on those issues where we thought we could reach consensus . . . The state-of-the-art in risk assessment was much more advanced for cancer than in any other area. We thought that if we just addressed ourselves to the scientific issues, leaving the policy issues aside, we could come up with a good consensus document.[34]

The size of the working group was similarly reduced, leaving only those who had direct experience with cancer issues. This core group included Rodricks, who had worked in the area of risk assessment at FDA since the early 1970s, David Gaylor, from FDA's National Center for Toxicological Research, Elizabeth Anderson, executive director of the Cancer Assessment Group at EPA, Frank Kover, also from EPA, and Richard Heller and Joseph McLaughlin from CPSC. Eula Bingham was the IRLG principal supervising the group's work. As Costle explained it, "Eula had the most interest in it because of her own background."[35]

The difficulties the group faced can be understood by asking how they should have gone about answering the question, "Is substance X carcinogenic?" Unfortunately, put that way, the question cannot always be answered since there are genuinely ambiguous cases. The problem of where to draw the boundary is not a purely scientific issue. Instead, it can only be solved by evaluating consequences as well as describing facts. The Working Group's presumption that they were dealing with an essentially technical issue reflected a flawed understanding of the nature of cancer risk assessment.

First of all, many of the individuals exposed to any given carcinogen suffer no ill effects. Hence, all anyone can hope to observe is a compound's role in producing an increase in the rate of cancer incidence. Second, since there are many different types of cancer, a large number of rates must be considered. Third, since cancer primarily occurs in older individuals, changes in the age distribution of the population will affect apparent disease rates. Therefore, the rates to be concerned with are the lifetime age-adjusted rates. Fourth, cancer may have multiple causes. Some compounds function as "initiators," others as "promoters," and risk factors are not additive but interactive. Hence, we cannot ask whether "X" leads to elevated rates without specifying the context. We have to ask, in effect, whether there are any circumstances in which some specified exposure will lead to elevated disease rates. Fifth, populations may vary in susceptibility. This leads us to ask not about rates in general, but about the rates for particular population groups and subgroups.

Finally, and perhaps most critically, the effect of all carcinogens is dose dependent. At small doses the effect on rates can be very small (scientists disagree about whether it can actually be zero for some very low doses). At high doses, however, many of these compounds induce acute (even fatal) toxic effects. Hence, we must ask whether there is any dose, short of the toxic level, at which the elevated rates we are looking for appear. Furthermore, "dose" is not a one parameter phenomenon. Doses vary in intensity, duration, variability, and so forth. Putting all this together, our reformulated question becomes: Is there any dosage (pattern and level) of substance X short of the toxic level that would produce an elevation in the lifetime age adjusted rate of any cancer in any human subgroup under any set of specified conditions with regard to other risk factors?

Note first that it is not at all clear why this formulation as a "yes" or "no" question is the most appropriate one for public policy purposes. Why should it be important to distinguish very small effects from zero effects, and not important to distinguish small effects from large ones? Putting the question this way also pays no attention to either the size of the risk to any given individual, or to the number of individuals at risk. Compounds that might produce small increases in rates under highly unusual circumstances would be characterized in exactly the same way as those that could produce noticeable increases in disease risks for the bulk of the population at currently common exposures. Is it really true that the best way to make public policy is to treat all carcinogens alike, weak or strong, common or rare, and to make great efforts to distinguish all of them from noncarcinogens?

Conceptual difficulties aside, there are also serious empirical problems. Most questions about the carcinogenic effects of chemicals are troubled by poor and ambiguous data. American policy is sufficiently conservative that easily observable risks are clearly unacceptable. Hence, the risks that are typically relevant for policy purposes tend to be small and, therefore, not easy to find. When we operate at the limits of our scientific capacities, we should not be surprised to encounter contradictory studies.

"We need more data" may be a defensible response to ignorance in the laboratory, but not when a policy decision must be made. To decide how to interpret conflicting and ambiguous results, a decision maker needs to compare the relative likelihood and the relative social cost of making various mistakes as a result of answering the question in various ways. The ethical judgments involved in such comparisons are often not explicit. Instead they are contained in words like "conservative" or "prudent" that conceal the central role of ethics and values in the process.

The implication is that there is no way to make a simple separation between the "scientific" and the "policy" aspects of labeling a compound "carcinogenic." Trying to do so can easily compound public confusion and misunderstanding. Yet, deceptive or not, this was exactly the task that the IRLG Working Group decided to undertake.

Specific Issues

The need to consider social values in deciding how to interpret various kinds of evidence appeared repeatedly in the Risk Assessment effort. Again and again, an apparently narrow "scientific" question opened like a magic door into a large and confusing world of epistemological complexity. Yet these were the issues the working group had to resolve to devise a cancer policy.

The Uses of Epidemiology

Good epidemiological studies that confirm an association between human exposure to a chemical and elevated cancer rates have long been considered the best evidence for a chemical's carcinogenicity. Indeed, the earliest identification of such risks came from epidemiological studies of occupational exposures.

But such studies are not easy to do. If the cancer is rare, it will be hard to find enough cases for statistical purposes. If it is common, we may need a large sample to be sure that a small observed increase is not due to random fluctuations. To make matters worse, past exposures to various chemicals are often not well documented. In other cases, history has performed poor "experiments" so that potential subjects have been exposed to multiple substances. In addition, it is difficult to identify control groups that differ from the study population only in their exposure and not in other risk factors. As a consequence, we often have no way to separate out the effects of the substance we are interested in. And following up a population, perhaps decades after exposure, is greatly complicated by mobility and death. In general, it is difficult to isolate the effects on human cancer rates of any given substance, particularly when we would like to know about the whole "dose-response" curve.

These problems give rise to a position the working group explicitly had to deal with, namely, that one should not take seriously negative epidemiological findings. That is, some would be persuaded by epidemiological studies that find a risk, while discounting epidemiological studies that find no risk when there is other evidence for that risk. The argument is that the power of epidemiologic studies is so low that they can easily miss something that in fact is there. Others contend that for all their limitations, adequate epidemiologic studies can set an upper bound to the likely level of risk in humans. Of course, one may have to make assumptions about doses, confounding factors, and many other things. Still, in some cases—this view goes—we can say that a proposed effect cannot be very large because, if it were, we would have seen it in epidemiological studies.

Most of these problems cannot be solved even in the long run by well designed prospective studies. Such studies are often both impractical (because of long latency periods and possible intergenerational effects) and unethical to the extent that they subjected humans to avoidable hazards. And for current decisions of the sort the IRLG agencies faced, better data twenty years from now are not very helpful.

There are also serious controversies in epidemiology over which kinds of statistical analysis to use. Some favor making a "best guess" (i.e., an estimate), while others argue for using statistical "tests" to see if the observed rate is "significantly" different from background. But if one follows the latter approach, just how high a burden of proof do we impose? Just what significance level is required? Similarly, to what extent should we make use of Bayesian methods that utilize information outside of the study itself as a basis for interpreting its results? Again, because different statistical methods could yield different results, "guidelines" must specify what methods and tests are to be employed. (In fact, many of these same issues arise in the interpretation of animal as well as human data.)

The final epidemiological problem involves the use of "mechanistic" information. Do we need to understand the underlying biology before an association between exposure and disease rates can be considered as revealing a causal connection? Insisting on such understanding is a way to guard against spurious or random associations. Alternatively, mechanistic understanding could be treated as desirable but not essential, and its absence as not enough to discard a positive finding.

It is not easy to write general rules of inference that will tell us how to react to any and every possible configuration of study results. Furthermore, uniform rules make it impossible for scientists to use their experience and intuition to make case-by-case judgments. Is it even sensible to try to reduce the problem of weighing ambiguous evidence to a simple set of rules that takes little or no account of the context of each situation? Do we really want to ignore the economic, political, and social setting of a problem in the way that uniform decision rules inevitably do? Unfortunately, these problems in doing risk assessment were (and are) not confined to the interpretation of epidemiology.

The Design of Animal Tests

How else could one assess a chemical's carcinogenetic potential? The answer evolved by the international cancer assessment community, as noted previously, involved animal studies. Animals, especially small rodents, are inexpensive and short lived—although a large sample is still expensive. Since, it is alleged, all human carcinogens are likely to be animal carcinogens, using animal data can be characterized as a "conservative" strategy. Perhaps because toxicologists use animal tests in many other situations as well, this approach has been widely accepted.

The contrary position had—and has—its advocates. This view invokes the general scientific presumption that effects should not be assumed until they are proven. Since animals and humans differ, and the animal test situations are sufficiently unrealistic, automatic generalization from animal studies to humans is not advisable.

There is even more dispute over whether animal tests can be used to estimate dose-response relationships in humans. Many who would use animal tests for assessing carcinogenicity balk at using them to make numerical risk estimates.

Clearly, deciding whether—and how—to make such estimates involve nonscientific policy considerations. Should one make a "best guess" or a "conservative" overestimate? If one chooses the latter, how conservative should one to be? To answer such questions one needs to know what risks one is prepared to run as result of each alternative and how adverse the consequences would be.

For the work group, this question arose in the form of a debate over what criteria, if any, should be imposed on animal tests as a condition for taking them seriously. For example, which species of animals should be tested, and how many positive studies should be required? Some scientists contend that any tests of mammals can reasonably be applied to humans. Others argue against the widespread use of rodents (which are mammals) on the grounds that only larger animals—dogs, cats, or primates—are metabolically similar enough to humans to be appropriate subjects. (The latter proposal has usually been dismissed on practical grounds by those actually doing the testing.) The question, nevertheless, remains as to whether test results from all species are equally significant, and whether results from a single species suffice.

Some have argued that findings from one species should be enough to call something carcinogenic but that findings from two should be required to prove it is not (on the grounds of "public health conservatism"). Such scientists also argue that one negative finding and one positive finding should be treated as a positive result. There was and is, however, considerable disagreement on these points. For many scientists, the persuasiveness of the test results depends on other factors, such as sample size, and the variation in the response rate at different dose levels.

Another major question that confronted the working group related to the use of high test doses in animal testing. The aim of cancer testing has been to maximize the likelihood of discovering carcinogens at the lowest possible cost. As a result, the convention has been to use the highest possible dose levels, even up to the "LD_{50}," the dose level that produces lethal toxic effects in fifty per cent of the experimental animals. Critics have argued that such practices do not nearly approximate the conditions of human exposure. Moreover, unless animals are exposed at various doses, they argue, there is no way to use animal studies to explore the dose-response relationship. Those who argue for high dose testing counter by saying that information on the dose-response curve is irrelevant since all we need, or should need, to know for policy purposes is whether or not a compound is "carcinogenic." In fact, when the working group began its assignment, the use of high dose studies of small rodents was the accepted approach in the cancer testing community.

The Interpretation of Animal Studies

Apart from controversy over the use of animal tests, there also was disagreement about their interpretation, particularly over what constituted a carcinogenic response. Two questions arose repeatedly: (1) were benign tumors to be counted the same way as malignant tumors and (2) did increased tumor incidence always suggest a true carcinogenic process even when there was evidence that the compound might be merely an "enhancing factor?" The conventional answer to both questions—as reflected in international guidelines and in Umberto Saffiotti's cancer principles—was, "Yes." Since many malignant tumors had a benign stage, benign tumors were to be regarded as "precancerous" lesions; and since the underlying mechanisms of cancer were little understood, any significant increase in tumor incidence was considered to be evidence of carcinogenicity.

Both inside and outside of government circles some cancer researchers were suggesting that both of these standard assumptions ought to be qualified. In a set of general guidelines drawn up late in 1976, for example, EPA's Cancer Assessment Group had suggested that some substances produced only benign lesions, and hence need not be considered carcinogenic.[36] There was also the possibility that some positive results would arise for compounds that were only enhancing factors. Such substances produced a high incidence of tumors only in laboratory animals with high background rates of the same tumors. In the absence of confirming data, some researchers argued that a finding of carcinogenicity in these cases was needlessly cautious, as it was possible that the compound would not have carcinogenic effects in more normal circumstances.

As noted earlier, once a count of tumors was made, it was still necessary to decide whether the effect was "real" or due to random factors. The members of the cancer testing community disagreed as to what level of evidence should be regarded as persuasive. This involved both statistical criteria for individual studies and the problem of making sense of the evidence about a given compound when that evidence was conflicting, as was often the case.

When the working group convened, OSHA's cancer policy draft specified that the administrator had the discretion to distinguish between "persuasive" as opposed to "merely suggestive" test data. In contrast to such informal, subjective assessments, several scientists in FDA and EPA were working on more elaborate, formalized methods of categorizing the persuasiveness of evidence. Yet, the question of whether or not that could be done remained open.

The Use of Short-Term Tests

Animal tests for carcinogenicity are expensive and time consuming—especially since investigations of low dose levels require very large samples. As a result, in the early 1970s there was considerable interest in the use of tests based on cultured cells, which could be performed quickly and easily in a laboratory. The first of these tests was developed by Dr. Bruce Ames, based on mutation rates in sensitive strains of bacteria. The theory was that compounds that caused such mutations were likely to initiate carcinogenic processes in humans. The problem these tests raised was, how should they be used, along with animal and epidemiological evidence, to make regulatory decisions.[37]

Early studies by Ames showed that, while compounds that were human or animal carcinogens tended to be positive in his test, the reverse was not necessarily so. His test could easily yield false positive results from the point of view of human risk. Furthermore, there was no real way to link the strength of the reaction observed in such tests to the potency of the possible carcinogen in humans.[38]

Given all that, what role, if any, should be given to such test results in analyzing human risks? Should a positive "Ames test" be enough to prompt regulatory action? Should it be counted instead of (or as equivalent to) an animal test, at least in certain circumstances? Were negative as well as positive results to be considered? And as the number of such short-term tests multiplied (using different test cell cultures and systems), how should we take account of conflicting results in such assays?

Quantifying Risks to Humans

Many of these inference problems came together in the context of arguments over whether it was possible to make quantitative estimates of the risk to humans from various carcinogenic exposures; and if so, how this could best be done.

The view of those involved in the post-war international cancer assessment efforts had been that quantification was impossible. Their view was that the mechanisms of carcinogenesis were too little understood, individual risk factors were too variable, and available data too poor to allow for the meaningful numerical estimation of human risk.

On the other hand, there were several reasons for quantification. First, there was a need to respond to legislative and public criticism. The FDA's actions under the Delaney Amendment had been widely attacked for failing to take into account the magnitude of risks. Several of EPA's statutes required it to take drastic action on all compounds classed as carcinogens, and it was tempting to try to find a way to distinguish "small" from "large" problems. Moreover, the evolving structure of administrative law put pressure on an agency to present a reasoned defense of each of its regulatory actions. That defense could be significantly bolstered by an analysis of the magnitude of the risks that any given decision was designed to deal with. Finally, even though OSHA tended to be strenuously opposed to such quantification, the traditional approach of occupational hygiene had been to find "safe levels" of exposure and use those as a basis for standards. But finding such levels meant, in a sense, quantifying the risk of different exposures.

The extent to which any method for doing numerical risk assessment could be described as scientific remained open to dispute. How could or should one extrapolate from experimental animal data to hypothetical human exposures? How did a given dose in an animal translate to a given dose in a human? Even more controversial, in the absence of data at low doses, what assumption should one make about the existence (or absence) of "threshold" levels in the dose-response curve? To what extent should risk estimators use methods that would tend to overstate, understate, or do neither, with regard to human risk?

One school of thought was that there were no safe levels of exposure to carcinogens. Since a single mutagenic event was sufficient to trigger a carcinogenic process, there was always some (albeit small) level of risk even at very low doses. To those who held this view, it seemed that "conservative public health practice" required the use of linear extrapolations between the lowest recorded dose level down to zero dose. Such an extrapolation reflected the assumption that a positive dose always produced some (even if small) positive harm.

As noted previously, this assumption was contrary to the way that some other regulated pollutants were treated by law and viewed by scientists. Standard toxicologic approaches assumed that most substances were harmless at sufficiently low doses. Those who opposed assuming a linear dose-response curve argued that not all carcinogens acted by causing mutations. Instead some were called *epigenetic* instead of *genotoxic*, and behaved more or less like other toxic chemicals. A parallel argument was that the latency period for tumors varied inversely with the level of exposure to a carcinogen. Hence, at very low levels of exposure, the latency period might be beyond an individual's lifetime. In either case, the extrapolated dose-response curve would be an S-shaped curve, with a threshold. But if there were no experimental data at low doses, how was such a threshold (or its absence) to be determined? And if there were animal studies at low doses, how should animal and human doses be equilibrated?

A second issue in quantitative risk assessment was how to interpret conflicting results from more than one strain or species of animal. The EPA group doing such analyses argued that data from the "most sensitive" animal experiments should be used because doing so would yield a "conservative" estimate of the upper limit of risk to humans. That is, humans were believed to be no more likely to develop cancer at particular exposure levels than the most sensitive animals in any given set of experimental data.

Opponents argued that there was no evidence to warrant this assumption. Given the tremendous variations in species susceptibility, the risk to humans could be either higher or lower, and by several orders of magnitude. Those who believed that the purpose of mathematical estimation should be to approximate the actual risk complained that the "most sensitive animal" approach was likely to yield exaggerated risk estimates, since some inbred rodent species were notoriously sensitive to certain carcinogenic effects. Critics of numerical methods argued that such uncertainty made any specific numbers meaningless or misleading.

IRLG PRODUCES A DOCUMENT

The Risk Assessment Work Group did not contain every shade of opinion on these scientific issues, but it did include a fair sampling of the prevailing attitudes within government. Joining the actual members of the work group at various times were Nathan Karch of the TSSC, Gilbert Omenn of the Office of Science and Technology Policy (OSTP), Roy Albert of the EPA's Cancer Assessment Group (CAG), and, less frequently, scientists from the National Institute of Environmental Health Sciences (NIEHS), the National Cancer Institute (NCI), and the National Institute of Occupational Safety and Health (NIOSH).[39] As Chairman Joseph Rodricks envisioned the task:

> We wanted to make it as complete a discussion as we could of the state-of-the-art in cancer risk assessment. We wanted to include all of the criteria, methodologies, or procedures that could be regarded as scientifically valid.[40]

Trying to do all this took several months of hard work. By early 1978, however, the Risk Assessment Work Group had a draft document ready for agency review. Rodricks was pleased with the outcome. It was, he thought, a rather complete discussion of the scientific issues relating to cancer risk assessment, such that no one could reasonably accuse the group of producing biased or inflexible guidelines. In many areas, Rodricks believed, the group had in fact made a real contribution to the field. He had worked especially hard on a section elaborating criteria for stratifying conflicting experimental data according to levels of evidence. The draft document also contained a more detailed discussion of negative evidence than existed elsewhere in the literature; a discussion of the various factors—including enhancing factors and the appearance of benign lesions—which might lead one to discount apparently positive findings; and a detailed description of the scientific bases for alternative quantitative risk assessment methodologies, both threshold and non-threshold.[41] Although the Risk Assessment Work Group had been working on its draft document, OSHA proceeded to produce its own draft cancer policy. In October 1977, that draft was published in the *Federal Register*, and by early 1978, after comments were received and digested, OSHA's cancer policy was in the final stages of revision. Since joining OSHA in late 1976, Anson Keller had devoted all of his time and energy to this task. Although he was nominally a member of the IRLG work group, he was, initially "pretty quiet," as Rodricks later remarked. Once the pressure of work at OSHA had begun to slacken a bit, Keller thought he could take it easy for awhile. But Bingham's concerns intervened.[42]

Bingham believed that in several key areas the IRLG document conflicted with, and thus threatened to undermine, OSHA's cancer policy. Prepared by lawyers, OSHA's document had quoted primarily from judicial opinions and congressional committee reports rather than from the technical literature. It had tried to portray its principals as emerging from an established body of legal and legislative judgment. The IRLG statement now seemed to reopen many of those "settled" issues. The biggest problem, from OSHA's perspective, was the discussion of quantitative risk assessment. OSHA had eschewed quantification and was in the midst of litigation defending that position. Yet the IRLG had written a document based on the assumption that numerical estimation was a valid aspect of cancer risk assessment.[43]

Bingham recognized the value of the IRLG agencies presenting a united front. But when she read the work group's first draft, she was afraid that it would be impossible to come up with a statement that OSHA could accept. She persuaded Anson Keller to gear up once again to see another draft through; from then on, the two of them would oversee the group's work every inch of the way.

Joe Rodricks was disheartened and dismayed by Bingham's reaction to the first draft. "She wanted the whole thing done over entirely—and this time, she wanted Umberto Saffiotti to do it."[44] There were to be other changes, too. Many of the staff people at OSHA regarded OSTP's Gilbert Omenn—who had been quite vocal in his criticism of OSHA's draft cancer policy—as an unofficial spokesman for industry's point of view. Omenn had kept in close touch with the IRLG work group, but from that point on, he was conspicuously absent. As one group member put it, "Eula put the word out to keep Gil out. We just stopped telling him when and where the meetings were, and he stopped coming." The White House continued to be represented, however, by TSSC's Nathan Karch, who was more sympathetic to the OSHA position and who wanted to see a cancer policy emerge from the IRLG.[45]

There was little enthusiasm within the Risk Assessment Work Group for producing another draft. The position of OSHA was widely perceived as extreme, and in many respects it conflicted with existing practices at EPA and FDA. At the urging of the other IRLG principals, however, who hoped that the Bingham/Keller/Saffiotti leadership would produce a risk assessment policy statement that OSHA could agree to, Rodricks and his colleagues returned to the task.

Saffiotti found scientific reasons to object to many of the points OSHA found unacceptable on legal or strategic grounds. Like Keller, he believed that the document should be a clear statement of the scientific basis for determining carcinogenesis. Although he had no objection to including—as Rodricks had done —discussions of alternative scientific viewpoints, he insisted on unequivocal language requiring a conservative interpretation of the evidence in every instance.[46] "Most of the work on the second draft was spent figuring out the language," Rodricks recalled, "Saffiotti was a stickler on language."[47] Several features of the first draft, however, could not be fixed by tinkering with the language. Saffiotti objected strenuously, for example, to any scheme that characterized different compounds by the strength of the evidence available in each case. Such thinking, he argued, confused persuasiveness of evidence with degree of toxicity, and would wrongly posit the former as a scientific basis for regulation.[48] The lengthy discussion that Rodricks had included in the first draft was thus eliminated. Saffiotti, along with Keller and Bingham, similarly opposed any discussion of quantitative risk assessment methodologies.

Until this point, Roy Albert—the leader of EPA's Cancer Assessment Group— had played only a limited role in IRLG.[49] In the view of some of Albert's fellow risk assessors, he and the CAG generally opposed any effort to make the criteria for cancer assessment explicit, as that would limit their own discretion. Keller characterized the CAG's philosophy (embodied, for example, in their broad 1976 guidelines) as a "Sistine Chapel" approach to cancer regulation:[50] the experts go into the chapel and emerge with divinely inspired judgments. The reputation of CAG within EPA, however, was built around Roy Albert's work in quantitative risk assessment. "We were light years ahead of anyone else," Albert believed. He wanted to make sure that IRLG did nothing to "screw things up."[51]

Saffiotti's objections to quantitative risk assessment were well known. Quantification suggested a degree of certainty that Saffiotti believed simply did not exist. Joe Rodricks, who, as chairman was mainly interested in delivering a document, was willing to acquiesce to almost anything Saffiotti and Bingham decided, even though he strongly supported the discussion of quantitative methods. Albert, however, was not willing to concede the issue.

Finally, Saffiotti agreed to a shortened discussion of quantitative risk assessment, provided the language made the uncertainties explicit and clearly stated the limits of its application. Quantitative methods, the draft ultimately said, could be used to provide a "crude" estimate of the magnitude of cancer risk to furnish a "very rough" idea of the magnitude of the public health problem. If such methods were used, however, then only those models that were "least likely to underestimate risk" should be used—such as the linear nonthreshold dose-response model preferred by CAG.

Saffiotti's concessions, however qualified, were enough to satisfy Roy Albert that CAG's work would not be jeopardized. "We had an eyeball-to-eyeball confrontation," Albert explained, "and they blinked."[52] But Saffiotti was also reasonably assured that the principles of carcinogenesis assessment he believed in had not been compromised. Anson Keller and Eula Bingham were ultimately satisfied that the IRLG statement could be read as consistent with OSHA's cancer policy. Keller had begun as a skeptic. His experiences at EPA during the pesticides battle had left him with little faith in interagency coordination. "But I became a staunch supporter of the IRLG when I saw OSHA's view pretty much prevail. Let's just say that EPA moved a lot more than OSHA did."[53]

To an outsider it is striking that both sides claimed victory. The final document, as we will note in the next section, was a compromise that used ambiguous language to encompass divergent views. The compromise was, to be sure, an achievement. Agencies with distinct traditions operating under distinct statutes could easily have become deadlocked. Instead, the IRLG principals had set in motion a process of horizontal integration to which they were personally committed. That commitment distinguished IRLG from most other attempts at interagency coordination. When stalemate threatened the risk assessment working group, the principals' interest in finding common ground prevailed. But did compromise in these circumstances serve a useful purpose? Was that the right objective? These questions will be discussed at the end of the chapter.

THE DOCUMENT APPEARS

Nine months after work on the second draft had begun, the Risk Assessment Work Group produced its final document, "Scientific Bases for Identification of Potential Carcinogens and Estimation of Risks." For the most part, the work group members were satisfied with their product. The IRLG principals, too, applauded the achievement, although that consensus did not extend down into the agencies they represented. "There was a lot of hostility to the risk assessment document at FDA," Rodricks admitted. That agency had recently adopted formal numerical assessments (a 1 in 1,000,000 lifetime risk) in some of its own rulemaking. "A lot of people thought we had sold out to OSHA," he said.[54] Presumably, the risk assessment document would be circulated now for internal agency review. Rodricks doubted, however, that it would survive such review, either at FDA or EPA.

In the meantime, the principals felt a growing sense of urgency with regard to getting the IRLG document "out the door." After he was excluded from the IRLG's meetings, Gil Omenn at OSTP began work on a separate cancer policy statement. Published as an OSTP staff report on February 1, 1979, it differed in key respects from the IRLG position.[55] Reflecting a laboratory scientist's skepticism, Omenn argued that a determination of carcinogenic potential should be made only when the "preponderance" of scientific evidence so indicated. He further proposed centralizing authority for all risk assessment in the National Toxicology Program at HHS. If OSTP's cancer policy were allowed to stand as the administration statement on the subject, it would undermine not only

OSHA's policy, but also the regulatory philosophy and ongoing work of EPA (especially the CAG) and FDA. "The principals wanted an IRLG policy statement," Rodricks explained, "so they signed off on it." There was no internal review of the risk assessment document at FDA or at EPA.

Nevertheless, the status of the IRLG document was ambiguous. Each agency had its own procedures for internal review, and any publication of regulations had to undergo the "notice and comment" process specified by the Administrative Procedures Act. Without going through such a process the IRLG document had no legal force. At most it could be viewed as a "report" on current practice or as a set of informal guidelines for each agency to consider in making subsequent regulatory decisions.

To try to give the statement as much authority as possible, Costle wanted it published in the *Federal Register*. Such publication, he believed, would help it to be taken seriously as a statement of regulatory collaboration on cancer policy, and as an example of what collaboration could accomplish.[56]

Officials at OSHA, however, objected. While they had been satisfied that the words of the document were consistent with OSHA's policy, the manner of its dissemination was a separate matter. The IRLG guidelines should be circulated to the scientific community, they argued, as a statement of current regulatory practice. But publication in the *Federal Register* would give the incorrect impression that the guidelines represented new agency policy and it would give them the status of a draft, open to public comment and presumably subject to revision.[57] Such publication would risk turning the working group's agreement into something authoritative, which of course was just what Costle wanted to do. In contrast, OSHA already had its own cancer policy and did not want to reopen the whole issue.

Since Eula Bingham was out of the country at the time, the OSHA case was presented to Secretary of Labor Ray Marshall, who agreed with this reasoning and informed Costle that OSHA could not endorse publication in the *Federal Register*. After the work group had spent over a year working out a document that OSHA could agree to, Costle was furious at Marshall's decision and thought it was damaging to OSHA not to proceed. The IRLG guidelines were published in the *Federal Register* in February, 1979 (shortly after the OSTP staff paper appeared), without the participation of OSHA.[58]

Meanwhile, Keller went to work in earnest to get the IRLG guidelines circulated to the scientific community through a peer-reviewed journal. Through contacts at NCI, he helped to accelerate the usually slow editorial processes. The report was accepted for publication on February 15, less than two weeks after it had been submitted. It then appeared in the July 1979 issue of the *Journal of the National Cancer Institute*.

Costle also wanted to give the IRLG document a Regulatory Council imprimatur. From his point of view, this was a logical function for the Regulatory Council to perform, and it would give the Council some visibility. Joe Rodricks was again recruited to supervise the production of a consensus statement on the regulation of chemical carcinogens. Instead of four agencies to contend with, however, there were now many more, and they were able to agree only on the vaguest statements of principle. "Never did a group of people work so hard,"

Rodricks later commented, "to assure that a document said so little." Nevertheless, a "Regulatory Council Statement on the Regulation of Chemical Carcinogens" appeared in late September 1979. The IRLG guidelines were not formally adopted by the Regulatory Council, but they were cited throughout the text of the statement and attached at the end as Annex B.[59]

THE CONTENT OF THE DOCUMENT

The Work Group's report clearly reflects the conflicts and the compromises, not to mention the conceptual limitations, that characterized the entire effort.

First, the bulk of the discussion is focused on the yes-no question, namely whether exposure to a particular substance can lead to any increase in any cancer rate in any human group under any circumstances. Thus, there is much discussion of how one determines whether a compound leads to "carcinogenic risk," "carcinogenic effects," "human cancer risks," and so on. There is the continuing presumption, often implicit and occasionally explicit, that this is the relevant issue for public policy analysis.

Second, the document generally presumes that questions about which tests and methods to use are scientific issues that can be decided on a technical basis. In most cases this view is simply asserted as if it were self evident, without acknowledging that there are other conceptions of how to view the problem.

The authors' concern with not underestimating risks, however, leads them occasionally to statements that reflect some sense of the unavoidable policy-rootedness of such decisions. Some decisions are defended on the grounds that they are prudent or conservative. For example, take the following discussion of the requisite sample size for animal tests:

> Ideally the number of animals required to provide adequate negative evidence would be such that excessive risk would not arise if the test failed to detect carcinogenicity . . . The number of animals tested may need to be increased if the number of humans exposed is large or if a small margin of safety exists between the animal dose and the human exposure.

This can be read as recognizing that "scientific" choices must be made in light of the social gains and losses from making different kinds of errors. But the nature of that decision and how it should be analyzed are never explicitly considered.

In similar fashion the document first recognizes and then glosses over the problem of choosing statistical confidence levels in interpreting laboratory studies. First, the paper notes that picking any one confidence level (e.g., 0.05) is arbitrary and unwarranted. Then it offers no guidance on how the actual level (which experimenters are urged to report) should be chosen. The problem the authors faced is that to have honestly discussed the issue would have led them to consider the risks and benefits of false-positive and false-negative results. That in turn would have highlighted the role of values in making risk assessments, a discussion the document went to some pains to avoid. Indeed, later in the document there is an almost total retreat from the earlier (uncomfortable) insight to the simple view that "the 95% confidence level is widely accepted as a reasonable assurance that the observed effect is real . . . ."

When one views the document's overall thrust, it is clear that the specific decision rules it urges are based on the goal of minimizing the rate of false-negative results, regardless of what that does to the rate of false-positive results. Indeed, unlike a 1977 publication by the EPA's Cancer Assessment Group, the very notion of false-positive results is not really mentioned in the IRLG paper.

This approach has many manifestations, most particularly in the notion that in cases of conflicts between positive and negative evidence, the positive evidence is always to be believed. If animal studies conflict either with each other, or with epidemiology, or if short term tests conflict with animal studies, the presumption urged by the report is always to accept the evidence of carcinogenicity unless the evidence to the contrary is overwhelming. Similarly, in the design of animal tests, all conditions and interpretive rules are oriented toward maximizing the probability of finding an effect:

> . . . bioassays done at doses and under conditions permitting maximum expression of carcinogenicity provide a sound basis for the identification of a carcinogenic hazard.

This is taken to mean lifetime exposures to the highest doses that do not produce competing toxic effects. It also is taken to imply that in the analysis we should count all tumors, even benign or precancerous ones. Any tumor in animals is relevant to potential cancer risks in humans, even if the suspected cancer site in humans is different. The report recognizes that it might be possible to prove that certain animal experiments were misleading, but if that is to be done, the burden of proof is on those who would claim significant interspecies differences.

With regard to estimating the magnitude of human risk and its dose-dependence, the report is not very optimistic about the possibility and value of such an activity:

> . . . current methodologies, . . . permit only crude estimates of human risk . . . Thus risk assessments should be used with caution in the regulatory process.

> . . . the quantitative assessment of cancer risks provides only a rough estimate of the magnitude of those risks . . . a very rough idea of the magnitude of the public health problem.

The document acknowledges, however, that "despite the uncertainties, risk assessments can be and are being made." The view is also expressed that such analyses are useful for "setting priorities for control of carcinogens . . . ." but presumably not in deciding specific policies, such as how much to control. That use is noticeably absent from the discussion.

As for the matter of how to make risk estimates, the whole emphasis in the report is on not underestimating the magnitude of the relevant risks. Again, as in the false positives case, no concern is expressed with the possibility of overestimating such risks.

To use the document's own language:

> Because of the uncertainties . . . and..the serious public health consequences if the estimated risks were understated, it has become common practice to make cautious and prudent assumptions wherever they are needed to conduct a risk assessment.

As an example of this approach, the report advocates using the most sensitive species to extrapolate to humans unless "there are strong reasons to believe that the most sensitive animal model is completely irrelevant to any segment of the exposed human population." Similarly, "an added degree of protection" can be gotten by using the "upper confidence limit" as the basis for estimating the carcinogenic response. (Interestingly enough, the probability to be used in defining that confidence limit is not specified.)

Most controversial perhaps is the persistent advocacy of a linear dose response model, and the assumption that there never is any threshold dose. As the report says, "a prudent approach from a safety standpoint is to assume that any dose may induce or promote carcinogenesis . . . because no threshold level for exposure to a carcinogen can presently be reliably determined for a population, a contributing risk level from any exposure level, however small, must be assumed." Further:

> When it comes to the shape of the curve, the argument is, . . . a linear model will provide an upper bound to curves of this shape and, it is hoped, a conservative estimate of the dose associated with any given level of risk.

Part of the argument on this point is that humans are so varied that even the failure to observe responses below some dose level in an epidemiological study cannot be generally relied upon:

> Variability among individuals makes it very difficult to have confidence that an observed no-effect level of exposure . . . even in a specific human population . . . will be applicable to the total human population at risk.

In summary, the IRLG document posed what was really a set of mixed scientific and policy questions as if it were just a set of scientific issues. The critical political judgments were bundled up in and concealed by words like conservative, prudent, significant, and reliably. No concern with the costs of false-positive results or of overestimating risks was expressed, nor indeed is the reader's attention even called to these issues. What is important to note is that this criticism stands regardless of whether one agrees or disagrees with the IRLG Work Group's specific substantive conclusions. Our point is that a reader does not emerge from these twenty large and closely printed pages with a clarified and critical view of the issues and choices, but rather with some vague thought that all responsible citizens are "conservative" and "prudent," that they do not do what cannot be done "reliably," hence they use upper bounds and linear extrapolations, and always take positive evidence to outweigh negative evidence. Even if one agrees with such an approach, "deliberation" or "education," as we have defined them, are not advanced by the document's failure explicitly to defend and explain its own assumptions.

Moreover, much of what the IRLG group urged is in direct contradiction to the tenets of contemporary decision theory. That whole approach is based on using one's best guess of the magnitude of any risk and on making explicit estimates of one's uncertainty about those guesses. Again, it is striking that nowhere in the document can one learn that there is this alternative viewpoint, or that the IRLG group's work was sure to be questioned by many analysts prepared to penetrate beneath the treaty-like language.

REFLECTIONS

Anson Keller believed that Costle's insistence on publishing the IRLG cancer guidelines in the *Federal Register* ensured their ultimate ineffectiveness.[60] If, instead, they had been quietly promulgated as a set of scientific principles, he argued, they might have stood a better chance of surviving, even into the Reagan administration. In this view, Costle tried to pass the cancer guidelines off as something more than they were. In fact, they were never revised, in spite of the considerable public comment received.

In Joe Rodricks' view the Regulatory Council statement had only political value. "I guess Costle got a press conference out of it," he said. For Rodricks, however, the IRLG guidelines would not have made much difference, regardless of how they had been handled. The "consensus" achieved by the IRLG did not permeate the member agencies. "The principals were all committed to the IRLG cancer policy," Rodricks claimed, "but no one else was—with the possible exception of the work group members." Hence, in spite of the principals' enthusiasm, much interagency inconsistency remained.[61]

Looking back on the experience after leaving government service, however, the principals generally believed that IRLG was a successful experiment in interagency collaboration, and that the cancer policy statement was its largest achievement. "At a time when regulators were being attacked from every quarter," Susan King remarked, "we took preventive cancer regulatory policy further than it had ever gone before."[62] On a more pedestrian level, the IRLG did promote cooperation in the daily work of the agencies in inspection and enforcement, and in sharing information and technical resources.[63] "We had no illusions that we could coordinate from the top down," Costle recalled. "The bureaucracies would have to find it in their own self-interest." Eula Bingham thought that the group's work would survive in this way, even though the IRLG itself lasted only a few months into the new administration:

> I like to think that there are some people out there in the field who got into the habit of working and cooperating with each other; and one day, if someone asks them how they came to do that, they might remember that there was this thing called the IRLG back in the Carter Administration.[64]

It is important to note that in their recollections of the significance of IRLG all of the agency heads referred to more than the risk assessment working group's document.

The price of an interagency generic statement on carcinogen assessment was that it was detached from regulatory decision making. Hard-fought compromise among the members of the working group could neither efface differences among the agencies' legislative mandates nor resolve ambiguities in statutory language. In the end, OSHA did not abandon its own "cancer policy." Indeed, OSHA's benzene brief to the Supreme Court cited the IRLG document as one more authority in building the agency's case against quantitative risk assessment.[65] The risk assessment antagonists, in their determination to find agreeable language, had not moved the agency heads and others to a clearer understanding of, or agreement about, what lay beneath the words.

According to some ex-White House staffers, Costle had sold the Carter Administration on IRLG with the promise that it would help keep Eula Bingham and OSHA "on the reservation." In this respect, IRLG was hardly a success, as it is not clear who was encouraged to stay on, or stray off, which reservation as a result of its efforts. No agency head was ready for IRLG to make his or her decisions. As a cooperative enterprise based on mutual consent, IRLG had no power to coerce its own constituent agencies.

By conceiving of the task before them as they did—to find words that everyone could agree upon—the working group missed an important opportunity. Risk assessment is an enterprise that is neither wholly scientific nor wholly independent of science. How it works, or should work, and what can be expected of it as an aid to decision making was obscure when the working group first began. Unfortunately, answers to these questions were still obscure when it ended. Compromise was not the way to explore and clarify for the public the nature of the disagreements that appeared in the course of the working group's efforts. Where one might have hoped for deliberation, one got confrontational word-smithing.

Stronger bureaucratic and intellectual leadership might have raised the group's sights and lengthened its time horizons. It might have served to produce a document of such obvious clarity and thoughtfulness that the public's view of the capacity of government would have been, justifiably, enhanced. In retrospect at least, we are left wondering what could have been accomplished if attention had been focused on what the nation needed to understand about conflicting views of carcinogenicity assessment, as opposed to what would best suit the four IRLG agencies' bureaucratic purposes.

Such a candid and no doubt complex analysis would have enhanced the president's strategic control of policy, and improved congressional oversight and public understanding (and the understanding of experts themselves). The conundrums of risk assessment presented an opportunity to create new intellectual and political capital in the citizenry and the government. One can easily argue this is too much to have expected of IRLG, given its limited institutional and political resources. But that response only raises a new question. What kind of arrangements, and leadership, would have promoted civic education and enhanced deliberation, which as we argued in the first chapter are central to the health of the republic?

NOTES

1. An official account of the formal beginnings of the IRLG is contained in the IRLG's "Status Report" of April 1981. For a discussion of other coordinating mecha-

nisms, see Dick Kirschten, "The New War on Cancer: Carter Team Seeks Causes, Not Cures," *National Journal*, 6 Aug. 1977.

2. Eula Bingham, interview with Margaret Gerteis and Stephen R. Thomas, Cincinnati, Ohio, 1 Sept. 1982.

3. Linda E. Demkovich, "The Battles of Byington," *National Journal*, 15 Jan. 1977, p. 113.

4. Bingham interview, 1 Sept. 1982.

5. Interview with panel on IRLG at the Harvard School of Public Health, July 9, 1981. The panel consisted of Douglas Costle, Susan King, Joseph Rodricks, Toby Clark, Richard Heller, and John Froines. Faculty participants were John Cairns, David Harrison, Donald Hornig, Marc Landy, Marc Roberts, Stephen Thomas, Milton Weinstein, and James Whittenberger. Also Douglas Costle, interview with Stephen R. Thomas, Boston, Massachusetts, 2 July 1982; Bingham interview, 1 Sept. 1982.

6. Costle interview, 2 July 1982; HSPH joint interview, 9 July 1981.

7. Bingham interview, 1 Sept. 1982.

8. This issue was stressed repeatedly by Douglas Costle in describing IRLG's origins.

9. Bingham interview, 1 Sept. 1982.

10. Emphasized by Richard Heller and Joseph Rodricks, HSPH panel interview, 9 July 1981.

11. *Federal Register*, Friday, Feb. 27, 1978; HSPH joint interview.

12. "Carter Seeks Coordination of Environmental Programs," *National Journal*, 28 May 1977; Robert B. Nicholas, former TSSC staff member, interview with Margaret Gerteis, Washington, D.C., 13 Oct. 1982.

13. Nicholas interview, 13 Oct. 1982; Anson Keller, former OSHA staff member, interview with Margaret Gerteis, Washington, D.C., 13 Oct. 1982.

14. Costle interview, 2 July 1982.

15. Costle interview, 2 July 1982; Hale Champion, former undersecretary of the Department of Health and Human Services, interview with Margaret Gerteis, Cambridge, MA, 26 June 1982.

16. Costle interview, 2 Jul 1982; Champion interview, 26 June 1982.

17. On Frank Press at the Office of Science and Technology Policy, see William J. Lanouette, "Carter's Science Advisor— Doing Part of His Job Well, *National Journal*, 6 Jan. 1979.

18. Costle interview (note 6 above); Champion interview, 26 June 1982.

19. On the origins of RARG, see Susan J. Tolchin, "Presidential Power and the Politics of RARG," *Regulation*, July/Aug. 1979, 44–49; also Christopher DeMuth, "The White House Review Programs," *Regulation*, Jan./Feb. 1980, 13–26.

20. William Drayton, former EPA assistant administrator for planning and management, interview with Margaret Gerteis, Christopher Dunne and Stephen R. Thomas, Boston, MA, 20 July 1981; see also Timothy Clark, "What Will Happen When the Regulators Regulate Themselves?" *National Journal*, 4 Nov. 1978.

21. Costle stressed this point in looking back at what had occurred.

22. This account of the development of interest in chemical carcinogenesis draws on an interview with Umberto Saffiotti of the National Cancer Institute conducted by Margaret Gerteis and Stephen R. Thomas, Washington, D.C., 29 Oct. 1982.

23. The operative language says, "no additive shall be deemed to be safe if it is found to induce cancer when ingested by man or animal." See Peter Barton Hutt, "Use of Quantitative Risk Assessment in Regulatory Decisionmaking under Federal Health and Safety Statutes," in David Hoel, Richard Merrill, and Frederica Perera, eds., *Risk Quantification and Regulatory Policy* (New York: Cold Spring Harbor Laboratory, 1985), pp. 15–29; Linda C. Cummings, "The Political Reality of Artificial Sweeteners," in Harvey M.

Sapolsky, ed., *Consuming Fears: The Politics of Product Risks* (New York: Basic Books, 1986), pp. 116–140.

24. Among these were the "Evaluation of the Carcinogenic Hazards of Food Additives" (World Health Organization, 1961) and "Prevention of Cancer" (World Health Organization, 1964).

25. The following account is drawn from Angus MacIntyre, "Administrative Initiative and Theories of Implementation: Federal Pesticide Policy," in Helen M. Ingram and R. Kenneth Godwin, eds., *Public Policy and the Natural Environment* (Greenwich, CT: JAI Press, 1985), pp. 214–224; and Nathan J. Karch, "Explicit Criteria and Principles for Identifying Carcinogens: A Focus of Controversy at the Environmental Protection Agency," in National Research Council, National Academy of Sciences, *Decisionmaking in the Environmental Protection Agency: Case Studies*, (Washington, D.C.: National Academy Press, 1977), Vol IIa, 119–206.

26. Roy Albert, chairman of the EPA cancer assessment group, interview with Margaret Gerteis and Stephen R. Thomas, Washington, D.C., 29 Oct. 1982; Roy Albert, Russell Train and Elizabeth Anderson, "Rationale Developed by the Environmental Protection Agency for the Assessment of Carcinogenic Risks," *Journal of the National Cancer Institute*, vol 58, no. 5 (May 1977), 1537–1541.

27. The story is told in Nathan J. Karch, "Explicit Criteria for Identifying Carcinogens," pp. 176–182.

28. Bingham interview, 1 Sept. 1982.

29. Bingham interview, 1 Sept. 1982.

30. Saffiotti interview, 29 Oct. 1982; Joseph Rodricks, interview with Margaret Gerteis, Washington, 12 Oct 1982.

31. Linda E. Demkovich, "Saccharin's Dead, Dieters are Blue: What is Congress Going to Do?" *National Journal*, 4 June 1977.

32. HSPH joint interview, 9 July 1981; Rodricks interview, 12 Oct. 1982.

33. HSPH joint interview, 9 July 1981; Rodricks interview, 12 Oct. 1982.

34. HSPH joint interview, 9 July 1981; Rodricks interview, 12 Oct. 1982.

35. Costle at the HSPH joint interview, 9 July 1981.

36. R. Albert, *et al.*, "Rationale Developed by the EPA,"pp. 1538–39 (Interim Procedures and Guidelines).

37. Bruce N. Ames, "Identifying Environmental Chemicals Causing Mutations and Cancer," *Science*, Vol. 204 (1979), 587–593.

38. The current state of the art as well as references to the earlier literature may be found in: R. Tennant *et al.*, "Prediction of Chemical Carcinogenicity in Rodents from *in vitro* Genetic Toxicity Assays, *Science*, Vol. 236 (22 May 1987), 933; Herman Brockmar and David DeMarini, "Utility of Short Term Tests for Genetic Toxicity in the Aftermath of NTP's Analysis of Seventy-Three Chemicals," *Environmental and Mollecular Mutagenesis*, Vol. 11, No. 4 (1988), 421–435; M.D. Shelby, E. Zeigler, and R.W. Tennant, "Commentary on the Status of Short Term Tests for Chemical Carcinogens," *Ibid.*, 437–441.

39. Nicholas interview, 13 Oct. 1982.

40. Rodricks interview, 12 Oct. 1982.

41. Rodricks interview, 12 Oct. 1982.

42. Rodricks interview, 12 Oct. 1982.

43. Keller interview, 13 Oct. 1982; Bingham interview, 1 Sept. 1982; letter from Donald Kennedy, former Commissioner, Food and Drug Administration, to Stephen R. Thomas, 26 Oct. 1983.

44. Rodricks interview, 12 Oct. 1982.

45. Nicholas interview, 13 Oct. 1982,

46. Saffiotti interview, 29 Oct. 1982.
47. Rodricks interview, 12 Oct. 1982.
48. Saffiotti interview, 29 Oct. 1982.
49. Albert interview, 29 Oct. 1982.
50. Keller interview, 13 Oct. 1982.
51. Albert interview, 29 Oct. 1982.
52. Albert interview, 29 Oct. 1982.
53. Keller interview, 13 Oct. 1982.
54. Rodricks interview, 12 Oct. 1982.
55. Office of Science and Technology Policy, "Identification, Characterization, and Control of Potential Human Carcinogens," staff paper, 1 February 1979. A version of the paper appeared in the *Journal of the National Cancer Institute*, Vol. 64, no. 1 (January 1980).
56. Douglas Costle, telephone interview with Stephen R. Thomas, 28 Oct. 1983.
57. Keller interview, 13 Oct. 1982.
58. *Federal Register*, Vol. 44 (1979), 39858.
59. "Regulatory Council Statement on Regulation of Chemical Carcinogens," *Federal Register*, Vol. 44 (28 Sept. 1979), 60038.
60. Keller interview, 13 Oct. 1982.
61. Rodricks interview, 12 Oct. 1982.
62. Susan King at the HSPH joint interview, 9 July 1981.
63. Letter from Donald Kennedy to Stephen R. Thomas, 26 Oct. 1983; Bingham interview, 1 Sept. 1982.
64. Bingham interview, 1 Sept. 1982.
65. Brief for the Federal Parties, Wade H. McCree, Solicitor General, *et al.*, *Industrial Union Department, AFL-CIO v. American Petroleum Institute et al.*, and *Ray Marshall v. American Petroleum Institute*, on writ of cert. to the U.S. Court of Appeals for the Fifth Circuit, Supreme Court of the United States, October term, 1978 (18 May 1979). As part of argument that benefits could not be quantified, the brief noted: "The Secretary's view that there is much to be learned . . . before risk assessments could be devised to meet the court of appeals' demands [for measuring cancer incidence at the 10ppm level, or any other level] is consistent with the Interagency Regulatory Liaison Group's Report . . . [which] concluded that the state of the art permits only a 'rough estimate . . . which may be useful in setting priorities among carcinogens and in obtaining a very rough idea of the magnitude of the public health problem posed by a given carcinogen.' "

7

The Steel Industry and Enforcing the Clean Air Act

In this chapter we look at several features of the Environmental Protection Agency's efforts to bring the steel industry into compliance with the Clean Air Act.[1] When Costle arrived at EPA, no industry had done less to clean up than steel, and none was more critical of the nation's pollution control programs. Given the enormous political and economic importance of the steel industry, the size of its companies, and the location of its plants in polluted areas, getting the industry to clean up was a herculean task. It was also vital if EPA was to improve air quality in many critical areas.[2]

Coordinating EPA's enforcement efforts across offices and regions, working with the states, and responding to the concerns of other government agencies severely tested EPA and its management. Over the course of four years, progress was made; some compromises were agreed upon and some important regulatory innovations were introduced. Progress, however, was slow, limited, and depended on pressure from outside EPA. Thus, this case reveals both the strengths and the limits of bureaucratic pluralism as a way of organizing the regulatory process. It also shows how both Congress and the agency modified federal and state relationships under the Clean Air Act to get better compliance from a multistate industry.

As the case demonstrates, regulations only reveal their true meaning through the enforcement process. The inevitably ambiguous language of the rules is defined only as decisions are made about what constitutes a violation in specific cases. In that process, arcane technical specifications that spell out what devices are to be installed and how their operation is to be measured often have significant policy implications. Given the variety of processes and settings that must be dealt with, even rules written in meticulous detail will not eliminate the need for case-by-case technical judgments.[3]

Because it involves such substantial discretion at the operational level, enforcement is notoriously hard to manage. This is true quite apart from the special complexities of the Clean Air Act and the special problems of dealing with the steel industry. The great detail involved makes it difficult for bureaucratic superiors, even those with technical training, to know whether the requirements their staff proposes are adequate (or necessary) to meet cleanup objectives. It is equally difficult for them, as supervisors of an ongoing relationship, to know whether initiating coercive measures is defensible or wise. In addition, if enforcement decisions are to be technically appropriate and survive judicial review, engineers and lawyers must work together in their formulation. This creates immediate problems in managing the necessary integration.[4]

The technical complexity of enforcement, together with its adversarial character, often results in lengthy delays. Even apparently simple determinations, like how to measure current emissions, may be very difficult when a firm is recalcitrant. The available monitoring devices are often imperfect and plants vary greatly in design and operating characteristics. It is even more difficult to determine—and agree to—the likely impact that an as yet untried cleanup device could have on emissions, never mind on local air quality. Add in the need to respect procedural niceties and it is no wonder that identifying violations, negotiating agreements, and instituting judicial actions all tends to be very time consuming.

The Clean Air Act divides enforcement responsibilities between the states and the EPA. The EPA administrator (as we saw in Chapter 3) sets nationally uniform ambient air quality standards. The states have to set emissions limits for existing sources that, when implemented, will insure that those ambient standards are met. The regulations that specify these emission limits are embodied in state implementation plans (SIPs) that, when approved by the EPA, have the force of law.

Writing a SIP is a formidable task and the states had very little time to do it when the Clean Air Act was first passed in 1970. Neither the states nor the newly created EPA knew very much about how to translate ambient standards into limits on each and every source. Therefore, relatively crude rules—like requiring all sources to diminish emissions by some specified percentage—were often employed.[5] After only a few years' experience, no one had much confidence in the technical soundness of the initial SIPs (or for that matter in the compliance data the states and regional offices gathered). In addition, many large sources, including steel operations, were not controlled by the SIPs but, as we will see in a moment, by separate administrative orders. This was one reason Congress was determined, in the mid-1970s, to toughen the enforcement mechanisms of the statute.

Concentrated in the Middle Atlantic states and the upper Midwest, steel was a formidable adversary for regulators. Both management and labor had good access to state capitals. Steel companies, moreover, were skilled at tying up environmental agencies in the cumbersome administrative procedures that state and federal enforcement actions generally involved. They also were often successful in winning judicial relief because of the arguably serious economic consequences of harshly interpreted regulatory requirements.[6] Case-by-case battles aside, the industry also criticized the general ambient standards, particularly that for total suspended particulates (TSP), which the companies claimed was unduly crude and stringent in light of the available evidence on health effects.[7]

Despite all this resistance, in the early Ruckelshaus and Train years at EPA some companies did enter into agreements and did spend substantial sums of money on pollution control. The deadlines, however, in those agreements were regularly extended to accommodate the companies' own proposed investment schedules. And the penalty for delay was often only the inconvenience of negotiating yet another, more lenient, schedule.[8]

Steel making was and is a dirty business. Often located in crowded areas, it also brought great economic benefits. Studies show that environmental cleanup costs were not, and are not, a principal factor in causing plant closings and layoffs.[9] Still, as the economy in general, and steel in particular, fell on ever harder times, the political vulnerability of pollution control programs increased substantially.

When Costle's tenure began, effective enforcement against an industry as powerful as steel required new legal means and renewed political resolve. The 1977 Amendments to the Clean Air Act, as we will see, provided EPA with just some of those badly needed resources. In a sense, the story that follows is about how the agency used those tools and participated in developing a new political consensus. The Costle EPA began with prosecutorial enthusiasm, and by the time he left office, the agency had navigated a series of complex negotiations with those inside and outside government to balance cleanup against national economic objectives. Moreover, in these years we find a significant centralization of enforcement initiatives within EPA, as well as a shift in power to EPA from the states. Making federal and state regulators a match for a set of multistate firms within a multistate industry was no easy task.

We take up first the interplay between SIP revision and consent decree negotiation after the 1977 Amendments. This part of the story examines how difficult it was to reconcile the decentralized, detailed decision-making characteristic of enforcement with the need to integrate and coordinate policy across various units within EPA, and between EPA and the states. Then we explore the difficulties that arose when the agency tried to incorporate a regulatory reform measure within the structure of the Clean Air Act, and how the administrator managed his rulemaking authority—and internal conflict within EPA—to enhance his leverage on the steel industry. Our analysis concludes by considering how the Tripartite Committee, which included representatives from the government, industry, and the United Steelworkers of America, affected EPA's ability to bring the industry into compliance. This part of the story brings us back to the problems of integration, but in a political context quite different from that in which the Costle EPA began.

THE CLEAN AIR ACT AMENDMENTS OF 1977

As it began to consider what ultimately became the 1977 Amendments to the Clean Air Act, the Senate Public Works Committee was well aware that there was widespread noncompliance with the SIPs and that many regions had not met ambient air quality deadlines. Members and staff were particularly concerned with the way the EPA had used Section 113 of the 1970 statute to issue administrative orders specifying compliance requirements for particular plants.[10] Those administrative orders tended to be significantly less burdensome than the SIPs they supposedly enforced. As Shep Melnick, a careful student of the process, has put it:

> In effect the EPA offered polluters a deal: The Agency would lower its demands if they would comply without a long struggle . . . Between 1972 and 1977 the EPA wrote hundreds of administrative orders that informally rewrote state plans by specifying compliance schedules.[11]

Furthermore, while the law provided for criminal penalties, EPA mostly brought civil actions when its orders were violated. Thus, to quote Melnick again, "the worst that could happen to a source that refused to cooperate with the EPA would be a court injunction prohibiting it from exceeding its emission limitations in the future."[12]

The states were not in fact in a position to be any tougher. Both their technical capacity and their overall bargaining position were quite weak. Even more than the EPA, they were reluctant to risk the political repercussions of significant plant closings, especially since the hastily adopted SIPs from the early 1970s were widely regarded as technically flawed and overly ambitious.

Such practices left considerable discretion in the hands of enforcement personnel at EPA's regional offices, who passed on these orders, and in the hands of state officials who did much of the actual bargaining. Furthermore, often the SIP was not revised to reflect these agreements. Enforcement lawyers argued that leaving in place the tougher, although often less specific, SIP requirements preserved the Agency's future bargaining power.[13]

The agency did succeed in putting many sources on compliance schedules. Nonetheless, the strategy sacrificed the unrealistic 1975 deadline for achieving national ambient standards. Some schedules stretched into the 1980s.

By 1977, the Congress, and especially Senator Muskie's subcommittee, were unhappy with the delay produced by EPA's approach. There was also much criticism of the dual system of rules, one in the SIPs and the other in plant-by-plant agreements. The 1977 Amendments specifically forbade the agency to agree to compliance dates beyond a new deadline of 1982. Also, any administrative orders and consent decrees inconsistent with the SIP had to be treated as SIP revisions and go through the extensive state and federal approval process.[14]

The new amendments also made some important changes in the law that applied to so-called nonattainment areas. Many parts of the country, including the most heavily industrialized and populated areas, violated one or more of the National Ambient Air Quality Standards. A strict interpretation of the 1970 statute would have prohibited any new industrial sources in such areas. In 1976, however, the Agency realized that it could not possibly enforce such a limit. In consultation with Congress and with the help of some environmentalists (including David Hawkins), EPA developed an alternative approach. It promulgated regulations allowing new sources to locate in nonattainment areas if they arranged for more-than-offsetting reductions in other emissions. These might occur either at a firm's own facilities or at other plants in the area. The new sources also had to be especially clean, meeting a newly formulated standard, the "lowest achievable emissions rate" (LAER).[15]

This approach was now added to the statute. States could provide the offsets through central planning, or leave that up to the sources themselves, allowing would-be entrants to "buy" offsets from firms in the area. Either way, the states were required to revise their SIPs to ensure that all ambient standards would be met by the end of 1982. Furthermore, offsets would only be allowed if all other major sources of the same owner in the same state were either in compliance or on compliance schedules. This was an important new enforcement tool, for it offered a way to leverage companies, not just individual plants.

But the new amendments went even further. The SIPs in nonattainment areas would only be approved if they required all existing sources to install "reasonably available control technology" (RACT). Although the agency had already used this standard in judging SIPs, the amendments themselves did not define what the requirement meant.[16] And given variations in the technology and the control costs of existing steel operations, RACT was going to be difficult to define operationally.

After the 1977 Amendments, EPA became convinced that the time was ripe to put renewed pressure on the industry. Steel's sorry record had been rehearsed in congressional hearings for several years. The industry's pleas for special treatment had not succeeded. The agency now proposed to conduct centralized consent decree negotiations in Washington in the hopes of producing a highly visible public victory.

Centralizing the negotiations also helped to insure a consistent resolution of the many difficult technical issues involved in defining RACT and LAER. "We wanted to avoid mistakes," one official said, "and really get a handle on the possibilities by accumulating experience."[17] Furthermore, because the companies were "well set up to beat everybody by tooling for the weakest position," centralization was a way for EPA to protect the regions and the states (who had to embody RACT in new SIPs) from industry pressure.

Steel companies, too, liked the idea. The new statute created uncertainty, and if nothing else companies needed to know what to expect of the new regime at EPA. No one knew what RACT or LAER meant or what the as-yet-unrevised SIPs would require. Negotiations in Washington might bring speedy and consistent clarification. Moreover, several financially pressed companies were seeking loan guarantees from the federal government. To satisfy their bankers and to strengthen their case before Congress, they had significant incentives to put their regulatory affairs in better order.[18]

REWRITING STATE IMPLEMENTATION PLANS: ENVIRONMENTAL PROTECTION AGENCY GUIDANCE TO THE STATES

Even before final passage of the 1977 Amendments, EPA foresaw that they would require the states to submit revised implementation plans no later than January 1, 1979. By specifying new deadlines and by imposing new requirements on various sources, the bill nearing final passage made existing SIPs obsolete. In addition, it imposed a very tight time schedule.[19]

To help insure that the new SIPs satisfied the RACT requirement, EPA could have issued guidance documents to the states, telling them exactly what would be acceptable for various categories of existing sources. Like the SIPs, such documents are subject to notice-and-comment rule-making procedures. Guidance documents therefore must be defensible in court on their technical merits. Insuring this requires a substantial commitment of the agency's engineering resources. Alternatively, the states could just proceed. Although this might produce more initial variation, the actual SIPs still had to be approved by EPA.

The states were ambivalent about how they wished to be treated. They resented any implication that they were not to be trusted; but they also did not like having to guess what would be required of them. Some state regulators liked to have their hands tied, at least on occasion, but others did not.

EPA's Office of Air Quality Planning and Standards (OAQPS) was responsible for helping the states formulate SIPs. In June 1977, Hawkins and Walter Barber (the director of OAQPS) met with the state air program directors to discuss what guidance the agency would provide for the new round of SIP revisions. The state directors were especially concerned about the RACT requirement for hydrocarbons, where attention and controversy was focusing on how to allocate the cleanup burden between mobile and stationary sources. In contrast, they seemed to feel they could deal with particulate matter on their own. Hawkins accordingly directed the preparation of a guidance document on hydrocarbons, but not for particulates.

Hawkins later regretted this decision. He believed at the time that EPA's experienced regional engineers, who would first review state plans, had enough detailed knowledge of particulate control technology to do the job. Hawkins assumed that the signals the regional offices would give to the states regarding RACT for particulates would lead the latter to propose SIPs that headquarters would then be able to approve. This assumption, however, risked leaving state regulators more or less on their own when they confronted steel companies over SIP revisions.

Not writing a detailed guidance document did avoid the delay that rule-making could have otherwise produced. As it happened, steel companies complained about the agency's failure to issue a guidance document precisely because they were primed to challenge it in the courts.[20]

Meanwhile, in the states and regional offices, and in OAQPS, many people saw the SIP revision process as an opportunity to reconsider the technical adequacy and enforceability of rules that had been hastily pulled together in the early 1970s. Engineers in the Division of Stationary Source Enforcement, however, had very different ideas. This group included Bernard Bloom, a dedicated antagonist of the steel industry. First as an official in Allegheny County, and later as a regulator for the state of Pennsylvania, Bloom fought company engineers. He was, in addition, a veteran of the protracted struggle with U.S. Steel over the Clairton Works near Pittsburgh.[21] With a prodigious command of technological detail, Bloom was more than a match for both the head of the Division of Stationary Source Enforcement, Richard Wilson, who was neither a lawyer nor an engineer, and for the staff at OAQPS. For such exotica as baghouses, cast house hoods, and opacity standards, the man was Bloom.

Both Bloom and his colleagues in Enforcement worried that SIP revisions could compromise the consent decree negotiations that they were conducting in

Washington. They were happy to avoid the threat of yet a third arena—litigation over some guidance document. But even two processes were one too many. As Edward Reich, deputy for Stationary Source Enforcement, put it in a memo to Walter Barber:

> First, we wish to avoid renegotiation of hard won in-place compliance programs by petitions to modify existing consent decrees as a result of SIP relaxations. Secondly, we wish to avoid signals that say to companies now at the bargaining table, in effect, that delay will be beneficial in that proposed SIPs will be less.[22]

The EPA could not control this problem simply by "hanging tough" on the enforcement side. A consent decree, after all, enforces a SIP and its terms can always be reopened if one of the parties can convince a relevant judge that such a step is appropriate. Indeed, decrees often specified that requirements would automatically be relaxed to accommodate any later SIP revision.

While the conceptual relationship between SIP revision and company-by-company negotiation was easy enough for EPA officials to discern, it proved much harder to manage. SIP revision was in the hands of the states, with the plans to be reviewed, initially, by regional representatives of the Office of Air Programs. The lead in compliance negotiations was being taken by the Office of Enforcement in Washington. Coordination had to occur quite far down in the organization, at the working level where people understood the technical details upon which both sets of decisions were based. Because the two foci of regulatory attention were the responsibilities of different assistant administrators, however, this was not easy to accomplish.

The agency's organizational problems were exacerbated by the fact that the Clean Air Act recognizes that the states are not just administrative units of the national government, but political entities in their own right. It was the states, as the developers of the SIPs, that allocated the burden of cleanup among various sources. By allowing the states to determine RACT for particulates, Hawkins facilitated just the sort of decentralization that the act and the regionalization of EPA operations were meant to accommodate.

What Hawkins did not anticipate was that the absence of a guidance document signaled to the regional offices and the states that RACT for particulate matter was a state's call. The result was a "dynamic for relaxation."[23] Several layers down in the agency, the view was, "whatever the state said was reasonably available control technology for particulates, EPA would accept without further analysis." In several states, steel companies were able to nudge regulators into the resulting policy vacuum. As Hawkins viewed it:

> The steel industry . . . had gotten three or four states to relax their current control regulations for particulate control and to come up with a new definition of 'reasonably available control technology' which was more lenient than they had on the books, and these were areas that weren't meeting particulate standards . . . So, we were seeing a retreat on the level of control in these areas, and I looked into that and found that it was traced back to this reinterpretation of what our policy meant.[24]

The industry viewed the SIP Amendment process quite differently. It was an opportunity to purge SIPs of onerous requirements that "required real technical leaps of faith." Even though some of the tougher provisions were not being enforced, the industry knew they were there and feared they could cause future trouble. The companies began to put pressure on state air program directors, and through them on EPA regional offices. To quote Hawkins again:

> Some of the state directors had early on told regional office people, 'We're going to need more back-up from EPA, if we're going to withstand these assaults on the regulations' . . . When the steel industry started to put the heat on, the mid-level policy people in the [EPA] regional office, not feeling that they would be backed up by headquarters, did not back up their engineers.

In recounting this incident from a headquarter's perspective, we have so far assumed that "relaxation" of source requirements, mistakenly agreed to by several regional offices, is a well-defined notion, and that it was bad for EPA to have let this happen. That was certainly the view of Costle and Hawkins. But for the engineers in the regional offices reviewing the SIPs, the issues were not so clear-cut. Their training and experience led them to be concerned with fairness, practicality, and technical progressiveness. They wanted the rules to represent good engineering and they often saw their opposite numbers from the companies as professional colleagues as well as opponents. A senior engineer in one of the regional offices put it this way:

> Congress never really defined . . . what it intended by 'reasonably available control technology.' They never really made it clear to the agency to what extent economies should be considered. Do you put a freeze on the picture . . . or do you go around and talk to vendors and try to find out what they have on the drafting boards. The law doesn't recognize the technical difficulties and at times, technical people don't recognize the legal difficulties.[25]

As this quotation illustrates, the environment in which regional discretion is exercised makes it difficult to control from headquarters. Regional engineers are likely to be more impressed than their Washington superiors with particular facts about particular sources. For them, regulatory toughness does not translate unambiguously into the choice of one engineering approach over another.

From a contrasting point of view of Hawkins, and the Washington Enforcement Office felt that the states had a standing need for federal help. Backbone and technical consistency came from Washington. States varied in their command of the necessary technical expertise, just as they varied in their ability to make unpopular decisions stick. Given the pervasive technical uncertainty, Hawkins was afraid that states would too often see their way clear to accept industry arguments.

In addition, Hawkins had a national program to protect. An approved state plan was a precedent other states and companies could use; another case, so to speak, in the common law of RACT. In contrast, state regulators, even more than regional EPA engineers, worried mainly about how a given decision would affect their own state.

These contrasting perspectives are well illustrated by the controversy that arose over the Illinois SIP. In 1978, the Illinois state environmental agency began to rewrite the SIP for particulate matter in nonattainment areas, with steel plants the major targets. In January 1979, after the hearing required under state law, state officials decided to meet with steel industry people "to see where we were," as Dan Goodwin, director of the Illinois Air Program, recalled.[26] Four companies were involved: Interlake, Republic, U.S. Steel, and National. They proceeded together, working closely with a lawyer from a Chicago firm well versed in the technical details of SIP development and enforcement.

Both sides stood to gain if an acceptable deal could be worked out. Industry spokesmen suggested that they might be prepared to commit themselves publicly not to litigate an agreed-upon rule. In return, the state agency would—in Goodwin's words—make all its concessions up front. In May, EPA regional administrator John McGuire met with the state regulators and agreed that the proposed regulations, soon to be submitted to the Illinois Pollution Control Board, would in due course be approved by EPA as a SIP revision. McGuire's commitment was part of the record at the board's final hearing in June, 1979.

A controversy soon arose, however, provoked by Bloom and his colleagues in Enforcement in Washington. The issues at stake are well illustrated by the question of what RACT would mean for cast houses. Such emissions could be controlled either by sealing the openings of the building, or by installing hooding inside the building and using a gas cleaning device. Illinois proposed to let the company decide which approach to take. The state would simply measure the emissions coming from the building to test compliance. "While there was no [EPA-approved] reference method," for making such measurements, Goodwin observed, "we thought our approach was adequate."[27]

Although, as noted, the regional administrator had been kept apprised of—and approved—the emerging agreement, there had been little or no contact with Enforcement in either the region or in Washington. In the end, this was a decisive error. Bloom's problem was that there was no enforceable specification of the equipment that was to be installed. Instead, compliance was to be defined by what he viewed as an unreliable measurement that could always be lowered for a test—and allowed to rise when no one was looking.[28] Bloom wrote a memo to the lawyers in his division making what Goodwin took to be inflammatory charges. Reich, who received the memo, convinced Richard Wilson, and Wilson in turn convinced Hawkins, that the EPA faced the prospect of a significant relaxation of control requirements on steel.[29]

In Goodwin's opinion, Bloom and Hawkins were afraid that another firm, elsewhere, might be able to satisfy the proposed test method without installing any controls.[30] Although this was not the case in Illinois, the Illinois SIP might thus serve as a precedent to allow some other company to get away with less abatement than was technologically feasible. This possibility looked to Hawkins like a concession of an important principle. Goodwin's insistence that the principle was not at stake in Illinois was not enough for Hawkins, especially since Bloom also argued that the Illinois emission limits were themselves inadequate to produce attainment.

By late summer, it was clear that the Office of Air Programs was prepared to overrule the region, in part on the basis of technical information far more extensive than Illinois had at its disposal. For his part, Goodwin, as the Illinois program head, concluded that the rules had changed regarding whose information and whose judgments would prevail:

> Bloom had a mass of information on what was going on in steel mills around the country. We had no way of being sure we knew all the possibilities. Had we had better information, our agreement with industry might have been more technically precise and easier to enforce. What Hawkins perceived as political weakness is really a lack of the mass of information the feds have. Not just expertise, but data . . . Political muscle was not a problem.[31]

In October 1979, Costle, at Hawkins' urging, overruled the region's approval of the Illinois SIP. To Goodwin, whose attention had been focused on the air quality advantages of getting the industry's commitment not to litigate the SIP, the EPA's reversal was "an act of incomprehensible bad faith. Costle emasculated McGuire, who then had no credibility with the steel people."[32] Goodwin believed that this damaged the state's new and promising working relationship with the industry. He, like others in Illinois, also resented the implication that he had been taken in by wily polluters, and that ambient standards would not be met.

Illinois was not alone. Each of the other states whose actions triggered the opposition of Enforcement people in Washington had its own particular story. The 1970 Clean Air Act's reliance on the states to allocate the cleanup burden among existing stationary sources was meant to mobilize the expertise and political will of the states, thereby accepting and even encouraging some variation. The EPA's regional structure was designed to provide attentive and understanding review of those state plans. But Washington's interest in national programs and precedents put the EPA in conflict with state regulators who had different priorities and inevitably led EPA to try to reduce the states' discretion.

From the perspective of Washington enforcement lawyers, an agreement that looked good from the vantage point of Springfield might wind up weakening the federal agency's position in its negotiations with the steel companies. If the states and the regions could not define RACT with enough toughness and consistency, then they would have to yield control of the process.

In the months after the Illinois debacle, Bloom and his staff produced a huge quantity of data on the cost and performance of different cleanup devices when retrofitted on various steel-making sources. (RACT was supposed to take account of costs.) These data, which the agency did not have in 1977, was used to establish performance norms for each major steel-making process—that is, emission rates based on what sources of similar age, size, and configuration had achieved— which the states could use in writing their SIPs.

In September 1980, EPA announced in the *Federal Register* the availability of the fruits of this work: a compendious summary of the performance of the latest control technology. The notice did not establish any regulatory requirements, and it was not a proposed rule-making. The agency would, the announcement said, use the tabular summary as the "starting point" in reviewing a state's proposal. The data base also proved useful in consent decree discussions since EPA now had cost and performance information about other plants in the industry as they confronted each individual company.

Still, all this information did not eliminate the need or the possibility of case-by-case discretion. Consider the following example from an internal document describing EPA's use of the data:

> One source had installed a new scrubber on a coke plant quench car, that emitted a rate that was twice that of all other systems in that state. An emissions limit based upon the general performance of systems in the data base could have required total replacement of the existing equipment at this specific plant. However, the record indicated that the higher, submitted emissions limit should be approved at the site.[33]

There was still room for maneuver, to say the least.

Perhaps ironically, Bloom's work ultimately facilitated the kind of regulatory innovation he opposed, namely the "bubble" alternative to source-by-source RACT controls. Initially, as we will see in the next section, the bubble could only be used where a revised SIP was in place. When that proved infeasible, Hawkins agreed to using RACT as the baseline that bubble plans would have to meet. And Bloom's data base helped make the determination of such baselines practical.

THE BUBBLE POLICY AND STEEL

Bringing the steel companies into compliance with the Clean Air Act was made both easier and harder by the EPA's attempts to develop new ways to express cleanup requirements. Regulatory innovation was in part a response to criticisms from economists, who had long argued for finding ways to lower cleanup costs. The idea was plain enough. Companies had better information than regulators about how cleanup costs could be minimized. The problem was to create incentives and opportunities for them to use that knowledge to propose cheaper ways to meet abatement objectives.[34]

The bubble policy was an attempt to do just that for a plant with more than one source for a particular pollutant. The idea could even be extended to several plants in a given area. A facility's emissions would be treated as if they came from a single stack, emerging from an imaginary bubble that enveloped the entire plant. Total emissions would have to comply with an enforceable limit determined by asking what emissions would have been if the company had complied, source-by-source, with the operable state implementation plan. Whatever was done within the bubble would be the company's business. A company could then figure out the least costly way of meeting its total cleanup requirements.

Implementing the bubble policy meant integrating it with the existing planning and enforcement mechanisms. This, as we will see, was a particularly tall order for noncomplying sources in nonattainment areas. To those within EPA who were immersed in the technical and legal details of implementing existing policy, the merits of the bubble would have to be demonstrated in practical terms. New ideas were a source of uncertainty and the burden of proof was on the policy innovator.

Implementation of the bubble was complicated by the fact that steel was the first industry to push for its adoption. From the start, regulatory reform for the sake of economic efficiency became entangled with regulatory relief for the sake of steel.

The costs of environmental cleanup—and the merits of using the bubble to minimize that burden—became part of a wide-ranging discussion about the national interest. EPA was understandably suspicious of the kinds of economic arguments that had been used over the years to justify the industry's footdragging. Now, just when the agency found itself with stronger statutory tools and an apparently favorable climate in Congress, the industry came out in favor of a supposedly more rational, cost-effective approach that threatened additional delay and implied extensive new negotiations.

In June 1978, the EPA committed itself to prepare a bubble policy—that is, a *Proposed Rule* for publication in the *Federal Register*. This was done with the knowledge that Armco Steel was already preparing a specific proposal for its Middletown, Ohio plant.[35] Bill Drayton was a strong internal advocate for such economically oriented reforms. Three offices, Air Programs, Enforcement, and Planning and Management joined in the drafting, with Drayton's office taking the lead.

The fact that a steel plant was first in line troubled Drayton:

> The Armco proposal from one point of view was almost the worst possible case . . . Support from the steel industry was positively harmful to my efforts to sell the bubble inside EPA, to environmentalists, to liberals more generally, and to the liberal/environmentalists on our committees in Congress.[36]

Senator Muskie, in particular, made no secret of his belief that the idea was contrary to the Clean Air Act. Moreover, a central feature of the Armco proposal—so-called open-dust trades—was particularly objectionable to Washington environmental lobbyists.

If the bubble concept was to survive its first applicant, the agency's leaders had to disassociate the past record and present motives of steel companies from the merits of the new approach. But at the outset the agency was sharply divided. David Hawkins was skeptical at best. Yet, given his background, he was an important source of the agency's credibility, particularly with the environmental community.

Strategic considerations complicated Costle's management problem in another way. The EPA administrator obviously had some obligation to follow the president's program, and as will be shown, there was pressure from elsewhere in the administration to "do something" for steel. Still, different agency positions were possible, particularly when the president's own position was unclear and those who claimed to speak on his behalf did so with noticeably different voices. At one extreme EPA might have argued that the language of the Clean Air Act left no room for striking a balance between cleanup and other worthy objectives, and that the agency's proper role was to be an environmental advocate and to vigorously enforce the law. The opposite extreme would have been simply to accede to industrial demands. Costle instead chose a middle course and tried to use the bubble as a way to retain room for maneuver within an administration that needed the support of both steel workers and environmentalists.

Armco's Middletown plant was located in a relatively rural part of Ohio that was a nonattainment area for particulates. To bring the plant into compliance using approved technology would have required capital expenditures of between \$14 and \$16 million.[37] The company thought that was needlessly high and its engineers and managers spent months developing a bubble proposal. By summer 1978, Armco had held informal discussions with EPA technical people, and in December the company's board committed itself to the plan.

Pollution from smokestacks had, for the most part, been taken care of at the plant. The issue was "fugitive emissions" from door seals, roof vents, and so on. Armco argued that the particular technological fixes specified in the Ohio SIP were based on obsolete engineering practices hastily adopted years earlier.[38]

Armco developed a new plantwide inventory of particulate emissions. Using emission factors developed by the Midwest Research Institute under contract to the EPA, Armco was able to show, to its satisfaction, that open dust sources (unpaved roads, open areas, and storage piles of raw materials) contributed sixty percent of total TSP emissions, process fugitive emissions comprised twenty six percent, existing stack emissions fifteen percent.[39]

Armco then did a cost analysis that showed dramatic results. For example, storage piles were estimated to produce 473 tons of particulates annually. Spraying these piles would produce a sixty percent reduction (284 tons) at an annual cost of \$704 per ton. In contrast, a blast furnace at the plant contributed an estimated 309 tons of particulate matter. Installing approved controls would get 95% reduction (294 tons) at an annual cost of \$11,905 per ton.[40]

Using this analysis, Armco proposed concentrating on reducing "open dust" emissions. This would be done by reducing traffic inside the plant, by cleaning, treating, and paving roads, by planting groundcover, and by spraying storage piles. This plan, the company estimated, would cost about half what it would cost to comply with existing process-by-process requirements.

In January 1979, the EPA published its "Proposed Policy Statement Recommending Alternative Emissions Reduction Options" and specifically rejected open-dust controls. "[E]missions from open dust sources like roads and storage piles," the statement said, "may not be traded against emissions from stacks or against fugitive emissions from industrial processes." Moreover, the EPA rejected the kind of analysis that Armco had used as the basis of its bubble:

> The problems of determining emission rates from these open-dust sources, modeling their impact on ambient air quality in concert with process emission sources, and assuring that all actual and potential ambient air quality standard violations are protected against, are sufficiently complex to effectively preclude acceptable demonstration of equivalency of controls.[41]

Open-dust trading posed a special challenge to the agency because it called into question the very definition of the TSP standard, which the steel industry itself had criticized for years. TSP is a crude index, based on the mass of solid material in a given volume of air. The standard is expressed as micrograms per cubic meter, and includes specific averaging times and the number of times per year the concentration can exceed the standard. The standard does not discriminate among particles of different chemical compositions or different sizes, even though they may have quite different effects. Yet smaller particles—specially the "respirable" ones that penetrate far into the gas exchange areas of the lung—are much more significant from a public health point of view.[42]

None of these considerations was new. For a long time the Agency had focused research attention on small particles, especially those carrying sulfates, and new methods had been developed for sampling the low end of the particle size range. Because open-dust controls would diminish large particles, rather than smaller particles from process emissions, Armco's proposal focused attention on a controversial feature of the existing standard, which regulators could hardly ignore.

The counter-argument was that Armco's proposal clearly fell within the existing rules of the game. If the existing standard was scientifically defective—as the American Iron and Steel Institute had argued for years—then it was defective, bubble or no bubble. In fact, Armco argued, its inventory showed that even for particles less than five microns, open-dust emissions contributed three times more than process fugitive emissions.

The EPA was also troubled by how to evaluate the validity of Armco's emission factors and modeling efforts. The company was asking for familiar and monitorable process controls to be traded for untested approaches whose effects would be difficult to measure. The burden of proof was, as EPA was quick to point out, on the company, and it was easy to conclude it had not been met. Said Hawkins:

> Computer modeling for these sources is in its infancy, and we simply don't know if it's any good . . . Even the Midwest Research report, to which Armco appealed, admitted the possibility of substantial error . . . [43]

For Hawkins there were also substantial political and credibility risks:

> I have this nagging image of a political cartoon which shows a steel plant in the distance with soot billowing out big, black clouds while in the foreground a citizen's group is screaming for action and an EPA administrator is trying to quiet them by saying, "But they're watering the roads!"[44]

Other features of EPA's proposed policy likewise left Armco's plan out in the cold. Acceptable bubble proposals would have to come from plants that were already in compliance or on an approved compliance schedule. Bubbles in non-attainment areas were out of the question in any case. Apparently Hawkins' concerns and Costle's desire to avoid the suspicion that the bubble was a pretext for accommodating a recalcitrant industry had overcome Drayton's efforts to create a genuinely workable program.

The compliance requirement at first seems irrational. If the company were in compliance, it would have to have already made the cost-ineffective expenditures the policy was designed to avoid. The same was true with regard to keeping to a compliance schedule if that involved (as most did) the commitment of capital long in advance for control equipment. The problem was Hawkins' fear of delay. Without such a requirement, the agency might find itself on the defensive when it tried to enforce against a source that was negotiating for a bubble without enough good faith to satisfy the agency. As Costle noted:

> [E]ven if we write down that bubble negotiations have to be completed within six months, what happens if they are not . . . We will try to go into court and get penalties and an injunction. And the steel lawyers will respond, 'Well, wait a minute, both sides have already acknowledged that there is a better solution to this problem. Yes, there are some details to be worked out, and yes, we are a little behind schedule, but, my God, we are moving toward a way of meeting our obligations at lower cost! Now let us finish this process, your honor.'[45]

Furthermore, Hawkins' realized, a solution like the one Costle described might set a dangerous precedent:

> If you create a precedent that the willingness of a company to come in and discuss a bubble is a basis for relaxing an already agreed upon deadline . . . that precedent is cosmic . . . I just felt that we needed a system that was more rigorous than that.[46]

Enforcement, too, had its particular concerns, specifically that the bubble not be allowed to "float." A bubble would float if a company were left free to alter over time the controls on different operations, provided it still satisfied the agreed-upon aggregate limits. At first glance floating seemed reasonable enough. A company would then be able to minimize abatement costs as its process mix changed. But compliance with such a bubble could only be verified by simultaneously monitoring all the sources that it covered. Unfortunately, much of the necessary testing was simply not feasible, especially at the kinds of complex plants where the bubble was most attractive. Instead, Enforcement wanted an agreement in advance, specifying what would be done stack by stack, process by process. Variations caused by manufacturing decisions would have to fall within agreed upon limits—which would be written in the usual way, with averaging times and test methods specified.[47]

In all these negotiations, Hawkins repeatedly put the burden of proof on the Office of Planning and Management (OPM) to show that a bubble policy could be made compatible with the requirements of the law. A tenacious and resourceful participant in the rule-making process, his primary goal was to achieve as much cleanup as possible, and he tried to make the bubble policy serve that end. Stated that way, of course, Hawkins' objective seemed to be at odds with what the bubble was intended to accomplish, namely, reducing the cost of achieving some specified level of cleanup. And so it was.

For example, the bubble clearly contemplated less control from some sources within a plant than would otherwise be achieved through process-by-process RACT, provided compensatory controls were put on elsewhere. But if one thought, as Hawkins did, that pollution should be abated by all technologically available controls, then foregoing some controls (which had already been found to be "reasonably available") was an indefensible retreat from a hard-fought statutory mandate. From his point of view, the very existence of bubble proposals meant simply that RACT was not yet tight enough on those sources that the company proposed to clean up beyond what RACT required.

Drayton and Hawkins also differed, although less explicitly, about the amount and kind of discretion that ought to be given to state agencies. The bubble approach significantly increased that discretion. The evolving requirements did put the burden of proof on the companies—as everyone involved in rule-making agreed was indispensable. But that did not change the fact that allowing company officials and engineers to take some initiative increased the information processing and analytical requirements on the government's side.[48]

Could the states "hang tough" in bubble negotiations in the face of company efforts to chip away at the requirements? Hawkins doubted that state regulators had either the will or the technical capacity to ensure that a company's proposal would in fact, achieve equivalent cleanup. Using RACT on a process-by-process basis would be much easier, as Enforcement also believed.

Following the EPA's publication of its draft policy in January 1979, the American Iron and Steel Institute (AISI) pointed out that it was "hard to imagine a situation wherein a source already in compliance would suggest an alternative plan, and difficult to visualize one where a program already in progress under a compliance schedule or a court decree would at that time seek an alternative." The AISI recommended that bubbles be allowed regardless of compliance status if they were part of an agreed upon plan to come into compliance. Such a provision, the industry argued, might actually encourage companies to agree more readily by holding out the carrot of bubble-based savings.[49]

Armco's reaction to EPA's January 1979 policy statement was to try to keep the pressure on the agency. Armco's management mounted an extensive national lobbying campaign with support from elsewhere in the industry. The company behaved as though the burden was in fact on the agency to show that Armco's bubble was unacceptable.

In fact, by the following fall and winter, EPA was prepared to be more flexible. A number of circumstances had changed, and the agency had time to work through some of its own internal differences. As Hawkins put it:

> By then we were aware that a complete prohibition of open-dust trades was no longer politically possible, given the plight of the steel industry and the effects of the lobbying effort. We now had to make a new presumption. What kinds of compromises could we make?[50]

To justify the new flexibility, EPA sought to emphasize that the industry was not uniformly defiant of legal requirements. The agency now said it was prepared to work with those companies that made good faith efforts. Having visited the Middletown plant, Hawkins acknowledged that "I became more favorably disposed toward permitting the Armco bubble . . . since it was clear that Armco's basic controls were good and they wanted the bubble to control the last margin of pollution."

The upshot was that in December of 1979, EPA issued a revised Policy Statement that relaxed several key requirements. First, it allowed states to consider open-dust trades in some circumstances. Applicants would have to demonstrate equivalence from an air quality point of view—as Armco had sought to do—and approach the agency with evidence in hand. Such a trade could be approved once a source showed (as Hawkins later put it, "real-world, empirical data that said, 'here's what we did and here's what happened.' " No one made a secret of the fact that this feature of the new statement was designed to accommodate the Armco proposal.[51]

In addition, the agency decided to back off from its skepticism of the kind of modeling that Armco used to support its proposed open dust trade. As long as there were "basic" controls on the "traditional" sources (as Hawkins described the intent), then the EPA was prepared to accept the uncertainties of untested modeling. There would have to be some controls everywhere, but not RACT everywhere. The company took this to be a concession to the cooperative attitude it had adopted toward the agency. Problems remained, but both Armco and EPA could begin to foresee steel plants actually qualifying to use the bubble.

The revised policy statement also accepted the AISI suggestion and permitted bubble proposals from companies not in compliance, as part of their plan to come into compliance. This meant that in some circumstances the agency was willing to extend existing compliance dates as part its approval of a plan.

Left in place was a constraint that to Hawkins was the most important one, namely, a source could bubble only in areas that could demonstrate that they would achieve attainment by the 1982 statutory deadline. Successful SIP revision thus remained a condition for bubble implementation. Furthermore, since each bubble proposal would be a modification of a SIP, there would be an opportunity for EPA to ensure that a bubble did not make the prospects for attaining ambient standards any worse.

In due course, this linkage with SIP revision came under fire because of the procedural delays it produced. A bubble, like any SIP modification, had to satisfy whatever notice-and-comment requirements state law imposed, then start all over again to meet the requirements of federal administrative procedure. Eventually, in the last year of Costle's tenure, the agency addressed this problem. Under prompting from New Jersey, the agency approved that state's generic bubble rule for hydrocarbons as part of the state's SIP. This rule allowed the state to accept specific plant proposals, while avoiding the SIP revision process for each individual case. In January 1981, the agency proposed to extend this idea in two more ways. Federal review would not be required for two kinds of bubbles: those involving sulphur dioxide and particulates, the pollutants relevant to steel. Plants would have to show that trades were equivalent and that the sources were either similar or close together. The change also applied to minor sources of any pollutant (under 100 tons per year).

But there was a deeper problem with allowing a bubble only if the SIP could be shown to produce attainment by the new statutory deadline. Unfortunately few, if any, revised SIPs had yet been approved for non-attainment areas. Until the new SIPs were in place, however, it was impossible to say how much each source would have to clean up. Drayton's OPM pushed hard to find a way through this dilemma, and EPA was pushed hard to solve this problem in the context of the so-called Tripartite negotiations that will be discussed shortly.

Just before leaving office, in January 1981, Hawkins and Costle agreed to new conditions that allowed bubbles in nonattainment areas. The change was dubbed the "RACT bubble." A source could bubble as long as it produced total emission reductions that were equivalent to what the source would have achieved if it had implemented RACT on all emission points within the bubble. Thus, even where the new SIPs were delayed (which was almost everywhere) a RACT analysis gave sources a target bubble proposals had to reach. With the work that Bloom had done on RACT, this was something the EPA could determine without waiting for the SIPs.

Why did Costle initially propose conditions for the bubble that made the reform virtually worthless in practical terms? The answer appears to lie in EPA's internal pluralism. Costle felt he needed Hawkins' acceptance, and Hawkins' convictions made him a stubborn adversary. The first *Proposed Rule* in January 1979, was clearly Hawkins' doing. Others might have regarded the Clean Air Act's deadlines as overly optimistic, but to him they were an important part of the pressure to make cleanup happen. He feared that the existing SIPs, even when fully implemented, would still not meet ambient air quality goals. He therefore wanted to avoid any step that would make it more difficult to tighten controls later on. He believed that the bubble increased the risk of this eventuality. Such agreements, he argued, could be construed as giving a plant a presumptive right to continue indefinitely at a level of cleanup that was less than what was both possible and necessary.

One might counter that a firm could always resist further controls, bubble or no bubble. In fact, it might prove easier to tighten limits on a bubbled plant—since a source would still be free to meet the added reductions at the minimum cost. But Hawkins was more worried about the reverse contingency and he held his ground as long as he could.

The RACT bubble was, in the circumstances, not a bad outcome from Hawkins' point of view. The RACT data that the agency had accumulated provided "hard numbers" by which alternative plans could be judged. In the end, Hawkins estimated that the RACT bubble could achieve something like 90% of what could be had by insisting that SIPs actually be in place first.

But more was needed to quell Hawkins' fears. The companies' acknowledgment of their continuing obligation to meet applicable SIP limitations would help reduce the chance that a bubble would be used to protect sources against additional abatement obligations. The companies explicitly accepted this responsibility, and it was included in both the report and the recommendations of Tripartite in September 1980. By then, Washington environmentalists had reluctantly joined in a policy consensus, as we will discuss in a moment. The RACT bubble, a few months later, was for all intents and purposes the final part of the package.

In sum, the evolution of the bubble policy rested upon a great deal of hard interoffice work at EPA. It seems unlikely that the advocates of cost-effectiveness could have overcome Hawkins' opposition without outside pressures. The story thus illustrates both the weakness and the strengths of the pluralist policy game. The first round of internal EPA debate was less of a good-faith effort to formulate national policy than a contest over whose views should prevail.

Drayton relied on economic theory to show why reform was needed. Early on he did not seem to appreciate that the agency's problem was steel company noncompliance, not economic inefficiency. Hawkins relied on the existing statute and assumed the worst, to show why reform was shortsighted. His arguments, resourcefully put forth, seemed insurmountable. He meant them to be. On the other hand, Hawkins' early position, although extreme and impractical, did keep the bubble from becoming irretrievably identified with the special pleadings of the steel industry.

Once the Tripartite discussions clarified Costle's wider strategic options, his lieutenants were able to adjust their thinking and produce a reasonable proposal. As events forced them both to try to produce a workable plan, they had to come to terms with practical difficulties and potential opportunities. The bubble could not save the industry, or indeed very many jobs, but from EPA's perspective, the case had to be made (as pressure mounted to "do something" about steel) that environmental protection was not the villain.

That pressure came from a series of negotiations through which the administration tried to produce a policy that satisfied some of its competing, but politically important, constituencies. That larger process, within which the story of enforcement against steel companies was played out, was the so-called Tripartite Committee.

THE TRIPARTITE COMMITTEE:

Background and Early Discussions

The Carter Administration formed the Tripartite Advisory Committee in 1978, as a successor to the Treasury group headed by Assistant Secretary Solomon. Chaired by the secretaries of Labor and Commerce, Tripartite included the Secretary of the Treasury, the EPA administrator, the Special Trade representative, several steel CEOs, and Lloyd McBride, the president of the United Steel Workers. In due course, Tripartite organized five working groups to examine specific areas: research and development, capital formulation, trade, worker assistance, and environmental protection. The Working Group on Environmental Protection included Michele Corash, then EPA general counsel; George Stinson, president of National Steel; and John Sheehan, the legislative director of the Steelworkers. Although participation varied, working group meetings were regularly attended by company CEOs or their representatives and by high officials from the various member agencies. The staffs of the working groups, drawn from government, the steel industry, and labor, laid the ground work for the meetings of the working groups themselves.

Although the problems facing the steel industry were unmistakable, the exact role of the Tripartite Committee was unclear. Not even its senior membership could guarantee that the committee would have more than *pro forma* work to do or that it would be able to produce concrete policy suggestions. Nevertheless, several features of the politics of steel gave Tripartite prominence and made it useful to key players inside and outside the government.

Layoffs in the Midwest were politically very damaging to an administration that was already vulnerable on economic issues. Many Democratic congressmen were members of the self-styled "steel caucus." Moreover, political support for environmental protection might erode if the public and the Congress came to believe that the costs of ill-advised regulation were a major source of the industry's plight. Carter needed the support of both workers and environmentalists. In these circumstances Tripartite was, if nothing else, a way for the administration to take steel's problems seriously. It also served to organize the discussion of policy options, which had previously been the business of a variety of different agencies in the government.

As it confronted the problems of steel, the administration also confronted its own internal disagreements over the proper role of government, with particular regard to troubled sectors of the economy. Under the banner of "industrial policy" some wanted to single out specific industries (or companies) for special governmental help. Others advocated making the economy as a whole more internationally competitive by means of centrally directed investment strategies. On the other side, many of the administration's more orthodox economists were very skeptical of such efforts. Like Charles Schultze, chairman of the Council of Economic Advisors, they opposed industry-by-industry tinkering, on the grounds that it would lead to cartelization, protectionism, and anticompetitive subsidies. Government, they believed, would wind up weakening valuable entrepreneurial incentives and threatening the dynamic efficiency of the economy. These critics were not impressed with the counter-argument that, since various governmental policies already had pervasive, if unplanned, microeconomic effects, the nation would do better to choose such effects explicitly and by design.[52]

As Costle recalled the context:

> You had real ambivalence within the Carter Administration; the free traders and the laissez fairists . . . were saying, . . . that the industry will shake down over time, but they shouldn't be protected; remember the Chrysler bailout and the Lockheed bailout? . . . We ought to let some plants shut down. And the politicians in the Administration said, 'Jesus, it's an election year. Come on, you can't shut down Youngstown and . . . say, that's economics folks.' So you had political schizophrenia within the administration, plus honest intellectual uncertainty about what constitutes an industrial policy.[53]

Into this broad and ill-defined debate, Tripartite brought the specific concerns and proposals of the steel industry. Those concerns included the financial burden of complying with environmental laws. Indeed, Tripartite was initially viewed by some as a way to pressure EPA on both the bubble and similar issues.

Costle had to navigate adroitly between his "foreign" and his "domestic" concerns. His senior deputies needed time and incentives to settle their differences over the bubble. At the same time, the agency needed to avoid appearing intransigent to others on Tripartite who did not appreciate the technical and legal complexities that Costle's senior subordinates were struggling with. Striking the right balance was made more difficult by the fact that the steel industry had long argued that the bubble would have greater advantages than EPA thought. Accordingly, the bubble had become the test of EPA willingness to be "reasonable" and to move beyond old suspicions.

As 1980 progressed, the issue of open-dust trades came to a head in the environmental working group. At a July working group meeting, the steel CEOs expressed great impatience with the agency. Despite the approach of the 1982 deadline, and despite good progress on consent orders, the agency's January proposed bubble policy was, as we have seen, quite inflexible. As Corash later recalled, "the Special Trade Representative, Rubin Askew, joined industry and let me have it . . . Ray Marshall (Secretary of Labor) and Secretary Klutznik (Commerce) also put pressure on Doug on the bubble issue."[54]

Another particularly knotty problem of concern to the working group was how to define *source*. The law allowed less stringent controls on existing sources (RACT), than on new facilities or major modifications (LAER). Defining a source as an entire plant would mean that LAER would not be needed for some facilities that would otherwise be classified as "new sources" or "major plant modifications." If source was defined more narrowly, on a process-by-process basis, LAER would be required in such cases.

The steel industry wanted plantwide sources. The EPA was divided, with Corash and Drayton privately supporting the plantwide position to maximize incentives for the modernization of facilities. Commerce Secretary Klutznik, Stuart Eizenstat of the president's domestic policy staff, and Charles Schultze of the Council of Economic Advisors also favored defining source on a plant-wide basis. Costle, however, would have none of it. As Corash put it, "Doug is independent—pressure him and he'll get angry." One day before departing for his summer vacation, Costle signed off on a proposal that defined a source as a particular pollution point within a facility and he left town.[55] The consequences were predictable. According to one participant:

> All hell broke loose: Eizenstat and Schultze were furious. Had it not been three months before the election, they would have gone to the President and Doug would have walked—but at this time, it would have been bad politics.

In July 1980, Tripartite's chair called for reports from the several working groups. All except the one on environmental protection had by then agreed on findings and recommendations. According to Corash, "EPA and industry could not agree to resolving their differences and each submitted statements which reflected their rhetoric. EPA came out of this process badly."

The Steel "Stretch-Out" Proposal

As the Summer of 1980 progressed, economic and electoral imperatives began to push all parties toward compromise. The key idea was to merge progress on cleanup with job-preserving new investment, both under the scrutiny of regulators. Unemployment in the industry was above fifteen percent and there had already been widely publicized plant closings. Michele Corash and Jack Sheehan, who led the effort, understood that if the industry were pushed too hard, the companies might win concessions by threatening additional shut-downs.

Sheehan and the union had tried in the past to make an issue of how capital was being drained away from the domestic steel industry (whose relative profitability was low) into diversification and investment abroad. For Sheehan, the key was to use environmental flexibility as a mechanism for pressuring companies to invest at home.

Corash and Sheehan realized that such a plan would be hard to sell to Costle, who would not support any proposal that seemed to compromise the tough stance the agency had been taking with the industry. Just as he would be vulnerable if EPA ignored economic hardship in steel communities, Costle could also be hurt if he appeared ready to sell out to steel. The Washington environmental community had been apprehensive all along about Tripartite's focus on pollution control issues.

To sell the deal inside EPA required some moderation in the long-standing animosity between the industry and the agency. Corash recalled that Costle warmed to the idea of a stretchout in part because he had developed a working relationship with David Roderick, the new chairman and CEO of U.S. Steel. As Corash put it, Roderick pursued a "personal campaign with Doug, wooing him, using sound reason, taking the blame for the past, in a face-to-face approach."[56]

Costle found Roderick easier to deal with than Roderick's predecessor at U.S. Steel, Edward Speer. As Costle once put it, Speer liked to try to get EPA in the corner and have the referees look the other way. From his personal experience, Costle became convinced that Roderick knew EPA intended to drive a hard bargain and was prepared to work with others in the industry to fashion a workable program. In short, Costle and Roderick encouraged one another to think of themselves as statesmen. That way, the abstractions and the posturing of the past could give way to real negotiations.

By early September, the working group on environmental protection was prepared to specify that steel faced a uniquely difficult situation:[57]

> No other industry is so capital intensive and so in need of capital intensive modernization. Because of the nature and quantity of uncontrolled steel wastes and the physical magnitude of corrective actions needed, no industry has made or continues to face capital expenditures of such a magnitude and representing so great a percentage of its capital needs.

Moreover, these extraordinary capital requirements were occurring at a time when the industry:

> . . . is experiencing one of the most serious capital shortfalls in its history—a shortfall which could threaten the future of substantial portions of the industry.

The goal was still full compliance, since, "with additional time in some cases, the industry can and should meet all Clean Air Act requirements." Agreement on environmental issues, however, should not occur in isolation. Instead such proposals "must be a part of a broader industrial revitalization policy designed to promote modernization of the industry." Interestingly enough, although similar proposals were made with regard to water pollution control, these were of much less concern to industry and it abandoned these ideas before the working group completed its deliberations.

The bubble policy, however, was a major concern. By the time the working group issued its report, the central question was whether to allow bubbles in nonattainment areas, where most steel plants were located. As we have seen, the agency was divided on this matter. Eventually it yielded—by agreeing to the "RACT-bubble"—but in September 1980 that matter was not yet resolved. The report was candid and explicit about the issues at stake:

> EPA is concerned about the propriety of any relaxation of controls in areas where the strategy for timely attainment of federal ambient air quality standards has not yet

> been developed . . . Moreover, once the SIPs were in place, sources using the bubble might well be called upon for further emission reductions.

The working group reported the industry's reply, which was probably of some help in moving the agency toward the RACT-bubble proposal: "In view of the substantial savings which bubble transactions may confer on a plant, the industry is willing to accept this risk and obligation."

By doing little more than reporting on the status of these discussions, the Tripartite working group allowed the EPA to continue rulemaking on the bubble without having to concede to Tripartite an explicit role in this most precious form of agency autonomy. Thus, Costle and Hawkins retained control over the conditions for bubble usage, an issue of great concern to environmentalists who were bound to have trouble enough swallowing any proposal to amend the Clean Air Act.

The heart of the group's stretch-out proposal was a statutory change granting additional discretion to the EPA administrator. "Limited postponements" of the deadline for compliance might be granted (for up to three years, until the end of 1985), if the administrator was satisfied that four conditions were met:

1. The delay is necessary to allow the company to invest in domestic iron and steel production operations.
2. The funds made available by the delay will be so invested—although not necessarily in the same plants where the delay occurs.
3. The company can show it will have the capital to meet air pollution control requirements by the deferred deadline.
4. All of the company's facilities will either be in compliance or subject to consent decrees covering the phasing of new investment and having adequate interim control and monitoring provisions.

These conditions accomplished several objectives. The first responded to the steelworkers' concerns that the companies would simply make investments outside of domestic steel facilities. The second required an estimate, acceptable to the administration, of the expenditures that would be required for compliance, and the commitment of a comparable amount to modernization. The third condition was tantamount to a commitment by the industry to comply with existing standards and to accept any additional emission limitations that might be necessary to comply with revised SIPs. The final condition meant that the trading of compliance delay for modernizing investments would be enforceable by federal district courts, with the required expenditures spelled out. Thus, companies that wanted an extension had to agree to comprehensive consent decrees on all of their facilities.

The working group also agreed that once the proposed amendment was in place, then "pre-existing consent decrees may be subject to renegotiation so long as all of the conditions for deadline extensions contained in the amendment can be met." Such renegotiation would effectively complete the process of the nationalization of steel enforcement.

The second and third conditions sketched in the working group's report seemed to imply that steel companies would be prepared to forego other statutory changes covering the same ground—changes in ambient standards, for example, or making deadlines into indefinite goals. Environmental groups wanted this implication spelled out explicitly in any legislation, along with the implied commitment to comply with the new deadlines.

Just how much would this plan accomplish? Normally companies tended to make high estimates of control costs to argue for relief. Now they had every incentive to claim that control would be cheap, as that lessened the amount they had to commit for modernization to be eligible for the stretch-out. This meant that forecasts made during the earlier Tripartite discussion of the amount of capital that would be freed for modernization were bound to turn out to be much higher than those contained in the ultimate stretch-out plans.[58]

Even those early estimates, however, suggested that the capital made available by the stretch-out plan was not likely to be economically decisive. Tripartite estimated that the industry's capital needs through 1983 would exceed all internal and external sources by about 6.6 billion dollars in constant dollars. Capital requirements for air pollution control were projected at 700 million dollars. Similarly, at Senate hearings early in 1981, Roderick of U.S. Steel testified that, with stretch-out, the industry would spend 500 to 700 million dollars that would not be available otherwise (i.e., only perhaps ten percent of the projected shortfall.)

THE POLITICS OF "STRETCH-OUT"

The stretch-out idea did not suddenly emerge in a creative burst in the summer of 1980. The EPA's negotiations with steel companies had already featured a willingness to let deadlines slip in return for enforceable commitments to easy-to-track capital investments that promoted cleanup. This had occurred in negotiations with U.S. Steel in western Pennsylvania and with Wheeling/Pittsburgh. Legislating this approach by way of a more comprehensive deal with the industry and the steelworkers relieved some uncertainty about whether such decrees could survive litigation.[59]

The Tripartite environmental working group's report was ready on September 11, 1980, and it was quickly passed on to the White House. When he saw it, Stuart Eizenstat became very upset. The leading environmental groups had just endorsed Carter for reelection. The timing of the apparent concessions to industry was an acute embarrassment. As Michele Corash put it, "The environmentalists viewed this as the steel industry's thumbing its nose at the EPA and [the law]."[60]

Having been left out of Tripartite, environmentalists now had to be dealt in. Fran Dubrowski, Richard Ayres, and Bill Butler from the Natural Resources Defense Council initiated a meeting with the CEOs from U.S. Steel, National Republic, and Bethlehem. The Steel executives stressed that under the plan, continuous environmental improvement would occur and that modernization was environmentally superior to inefficient and costly add-on pollution control devices. Moreover, the companies also agreed not to push "steel specific" amendments in the next congressional reauthorization of the Clean Air Act.[61]

The environmental leaders also met with White House staff and Jack Sheehan of the Steelworkers. Sheehan, in particular, emphasized that the industry simply could not meet the statute's 1982 deadline. He echoed industry's claim that modernization was economically necessary and also the best method of abating pollution. Insisting on shutdown, he argued, would be neglecting both an economic and an environmental opportunity, and interfering with a better enforcement strategy than EPA had ever had before.

Dubrowski and Ayres are said to have emerged from these various meetings "silent and sullen." Having pledged support for Carter's reelection, it was hard to criticize the package that had been arrived at in negotiations between the industry and the EPA. Their leverage lay in their ties with congressional allies who would have to act on a legislative proposal. Thus, they were able to exact at least some explicit commitments from steel, as we have already discussed. This aspect of the politics was captured nicely by Jeff Miller, acting assistant administrator for enforcement during these negotiations:

> Doug said, 'In order to convince Congress, we have to convince environmentalists that this is a good thing, and that means, first and foremost, that you're finally saying out loud that you are going to comply and you will bind yourselves to comply and do so in such a way that you cannot back out it.'[62]

As a result of these discussions companies did commit themselves to using deferred compliance funds at existing facilities rather than for "greenfield" plants. While the country would see some plants closed and jobs lost, the job opportunities and tax base of at least some communities would be maintained. Environmentalists argued that this clarification was one result of their belated involvement and showed the value of their participation in any future Tripartite-like deliberations.

These several understandings were reached quite rapidly and on September 30, President Carter announced his program for the steel industry. On the stretch-out proposal for Clean Air Act compliance, his language was carefully chosen to seal the arrangement:

> I am accepting this and the Committee's other proposals because doing so will remove the need for further amendments to the Clean Air Act as it affects the steel industry reauthorization by the next Congress. My decision is also premised on my understanding that the steel industry has committed to comply with the Clean Air Act . . . The amendments would only allow a stretchout—not a postponement—of investments on controls.[63]

Congress passed the stretch-out bill substantially along the lines that Tripartite originally proposed. One feature of the legislation, however, deserves special mention. To be eligible, companies had to be in compliance with existing federal judicial decrees except for *de minimis* violations. The idea was that violations had to be significant if they were to matter. The phrase comes from an old common law adage, *de minimis non curat lex* "the law does not concern itself with trifles."[64]

But how should the agency decide whether a violation was *de minimis*? The legislation itself did not say. The strictest possible interpretation could conceivably disqualify all applicants, which was hardly the intent of the statute. Although EPA wanted to retain discretion in making such decisions to be able to bargain with the companies, both the companies and the environmentalists wanted to limit that discretion, but in opposite ways. (In fact, nine of ten eventual applicants under stretch-out showed consent decree violations of some kind.)

The playing out of this story shows a familiar feature of contemporary administrative law, the cleverly wrought legislative history. During the hearings on the bill, steel companies had suggested that "substantial compliance" with decrees should be enough. The *de minimis* wording came from the National Resources Defense Council (NRDC). The Senate Environment and Public Works Committee Report said: "*de minimis* violation of an emission limitation is a violation resulting from circumstances beyond the control of the source owner or employee which causes no measurable increase in emission from a source."[65]

In 1981, the Gorsuch EPA wanted to grant stretch-out to as many companies as possible, and proposed using a company's "overall compliance record" as the measure of the significance of violations. But since consent decrees were involved, the Department of Justice had to concur. The Department of Justice read the *de minimis* condition more stringently. When the EPA objected on the ground that most companies would have difficulty satisfying Justice's interpretation, the assistant attorney general replied that "the Act was the result of a carefully drafted compromise . . . the choice of words was not inadvertent." Thus, while the statute conferred some discretion, it was not limitless. The words meant something, and environmental strategists understood the value of legislative history in filling out that meaning.

As it turned out, four of the ten companies that applied for deferred compliance under stretch-out were disqualified on this ground. Their proposed modernization projects were valued at 587 million dollars, out of a total of 828 million dollars from all ten applicants. But according to the senior EPA lawyer in charge of stretch-out, three of the companies—Jones & Laughlin, Kaiser, and National, would have been disapproved even under Gorsuch's proposed permissive view of what constituted a *de minimis* violation. Only the fourth company, Inland, with projects valued at $167 million, might have passed the more lenient test.[66]

The *de minimis* issue was not, then, the only reason for the modest impact of stretch-out on actual new investment in steel. Still, this incident is one more illustration of how significant matters of policy could be settled with an ambiguous technical phrase and an undebated gloss in the legislative history. The NRDC got what it wanted. The industry was united in support of the stretch-out amendment. Companies seriously out of compliance with existing decrees were in no position to call attention to the problems they might eventually have with a *de minimis* test, whatever it meant. U.S. Steel, whose president was the leading spokesman for making accommodations with environmental protection, turned out not to have a problem meeting this condition.

TRIPARTITE EVALUATED

In its own terms, how well did stretch-out work? The 49.4 million dollars actually committed under deferred compliance was a far cry from the early estimates of more than a half billion dollars. We have already mentioned two reasons for this:

the *de minimis* requirement and the incentive companies had to revise downward their estimates of the costs of cleanup to minimize their enforceable capital commitments, a process in which the Gorsuch EPA was prepared to cooperate.

Consider, for example, the case of U.S. Steel. The company's initial application called for expenditures of 195 million dollars, yet commitments of only 13.68 million dollars were actually approved. Of the almost 181.5 million dollar difference, 35.5 million dollars came from five cases where EPA allowed less costly hypothetical controls to be used as the baseline. Another 49.4 million dollars was eliminated when EPA determined that various pieces of equipment would not have been needed for compliance. On six other projects, according to the General Accounting Office (GAO), the EPA recomputed the costs, based on "(1) experience with similar projects and (2) discussions with U.S. Steel and suppliers of the equipment that was proposed."[67] Some 24.3 million dollars were mooted when the company, with its application pending, completed some abatement projects. The recession and continued shrinkage of steel markets also played a role. The 17.2 million dollars was intended for five facilities that the company in fact shut down in the latter half of 1982.

According to industry testimony, economic conditions also helped to explain why some companies did not apply at all. Bethlehem and Republic told the GAO that "negotiating consent decrees with EPA would be too costly and time-consuming and that committing funds deferred from air pollution control equipment to modernization was too financially difficult because of the poor economy." However statesmanlike Roderick of U.S. Steel had become, some companies still found it more attractive to remain out of compliance than to modernize under the terms of stretch-out.

Stretch-out did, however, give EPA added negotiating leverage in dealing with companies wanting to qualify. The biggest and most widely publicized example of this was the completion of the agreement with U.S. Steel, which had hitherto proven elusive. Wheeling-Pittsburgh's decree was also apparently induced by stretch-out.

All things considered, was stretch-out for steel a sensible approach to the industry's difficulties? This question raises issues of continuing controversy about "industrial policy"—what it means and whether it is wise.

The Tripartite Committee's recommendations included a diverse set of proposals about taxes and trade, job training, and research and development. But stretch-out was considered on its own merits by Congress. In part this was because the idea was so superficially attractive. Unlike taxes, import quotas, or subsidies—to say nothing of loan guarantees—stretch-out seemed to cost nothing. Hence, it was easy to consider it in isolation from the more provocative kinds of industrial policy.

There were critics of the idea. Both the Office of Technology Assessment (OTA) and the Congressional Research Service (CRS) argued that the key objective of the plan, namely plant modernization, was either unwise or vaguely defined. Joel Hirschhorn, project director of OTA's study "Technology and Steel Industry Competitiveness," emphasized that "by itself regulatory spending cannot explain the industry's declining profitability nor its capital problems."[68] The OTA view was that there was a better use of the funds freed up by delayed compliance than investment in plant and equipment. Chronic low profitability and an unfavorable competitive position could only be reversed by directing the funds toward R&D on "high risk innovative steel making technologies, including pilot and demonstration activities." The industry had failed in the past to engage in long-range strategic planning, the OTA report argued, and needed specific encouragement to do so in the national interest.[69]

According to OTA, the Tripartite plan also failed to respond to the diverse needs of a diverse industry. In particular, some companies in the nonintegrated scrap-based segment and the alloy/specialty steel segment showed high profitability and high growth rates. Such "mini-mills" also operated more cleanly. Ironically, however, such facilities and companies received the least help from the plan since their pollution control costs were lower and they were more likely to be at or near compliance.

Among the major integrated companies, OTA believed only the efficient would survive. But from this point of view, Tripartite's stretch-out proposal was deficient. Some of the least competitive companies would be assisted most. And unless funds were directed cost-effectively, and kept away from supporting superficial changes in production modes, plant closings would only be delayed.

The OTA argued that to undertake ongoing planning, Tripartite's membership should be expanded to represent the diversity of the industry. A revamped and presumably standing Tripartite committee could conceivably determine what "modernization" would actually look like for steel—just the topic of which the working group chose reticence.

The CRS, although less argumentative than the OTA, raised substantially the same points: "Capital investments must be targeted to facilities that have potential for market growth."[70] But on just how this would happen, using what technology, and with what impact on air quality in various possible locations, Tripartite was silent, the CRS found.

The silence to which critics objected, however, was not accidental. Tripartite had consciously avoided getting into the business of making industrial policy. The administration had within it staunch opponents of any such undertaking. Nor was Tripartite's membership—especially the membership of the environmental working group—eager to anticipate the decisions about plant locations and technology that steel companies might make. Downsizing the industry would be painful, and neither EPA nor the steel workers wanted responsibility for the resulting suffering.

Environmentalists, for their part, did not want to replace the regulatory constraints of the Clean Air Act, however modified, with an as yet undefined cooperative planning process. As long as steel met the requirements of the Act—and they joined Sheehan in applauding the new enforcement leverage that stretchout provided—from their perspective steel investment choices could be left to the market.

Thus, by the time he gave it, Hirschhorn's OTA testimony was beside the point. Before the Senate committee, the people who had shaped the Tripartite compromise dealt easily with OTA skepticism. Stinson of National Steel simply asserted that the industry's problem was not having the money to exploit known technology. There was, he said, nothing special about the technology of the foreign competition. And anyway, other parts of Tripartite recommendations had to do with capital formation, not this one. McBride disputed OTA's analysis of American productivity. Reductions in capacity had already eliminated much inefficiency. For the rest, he joined Stinson in support of more modernizing in the proposed way.

In sum, the criticisms raised by the OTA and the CRS were pushed aside because everyone participating in the Tripartite had already gotten something. The resulting agreement was quite stable in the face of somewhat speculative suggestions about the need to begin again with broader terms of reference.

REFLECTIONS

Over these years we see a steady movement toward Washington in the enforcement of the Clean Air Act against the steel industry. The federalist decentralization of the statute was modified to cope with the persistent noncompliance of the industry and with its local power and influence.

The 1970 statute had allowed the states great flexibility, while giving them complex tasks for which they had little data and even less time. The EPA and the states had initially implemented the law crudely, based on seriously incomplete information.

In response, the 1977 Amendments reduced the discretion available to the states in drawing up their State Implementation Plans. The amendments thus offer some evidence that congressional oversight is capable of adapting the machinery of regulation to experience. But new requirements like RACT and LAER put new burdens on the EPA, and created new tensions between the agency and the states. It took time for EPA to figure out how to use its new statutory leverage, and growing technical competence to wring additional concessions from steel.

As the EPA pushed harder, steel's countermoves emphasized the bubble policy—which sharply divided the EPA, and increased its political vulnerability. Conflicting principles—economic efficiency and maximum environmental cleanup—both had their internal partisans, and reconciling the resulting conflict was a difficult task.

How should Costle's leadership be evaluated at this juncture? One possible assessment is that he should have been more forceful in trying to overcome EPA's internal pluralism. From this point of view, Tripartite could be seen as having rescued EPA from itself, leading it to produce an adequate bubble policy it could not have produced on its own.

On the other hand, the Carter Administration as a whole was afflicted with wildly contending views regarding steel. In these circumstances, Hawkins' stubbornness bought EPA some time while the stretch-out coalition took shape. And perhaps it was easier for Hawkins to agree to workable bubble regulations once steel was prepared to make other concessions within the Tripartite framework.

Tripartite represented what might be called managed pluralism: the attention of several advocacy agencies was focused on broader consequences through their structured confrontation with other interests. As such, it did succeed in expanding the terms of the debate about the steel industry, but not very far. A more ambitious program of industrial planning was forestalled by the strictures of environmentalists and economists alike, who wanted nothing to do with an ongoing process of microeconomic intervention. At the same time, EPA's leaders were able to keep the agency's mission and authority largely intact, while making only modest concessions to competing interpretations of the national interest.

In part this modest impact occurred because Tripartite was stuck with the problem definitions and solutions brought forward by the participating parties. Thus, the unions insisted that modernization exclude greenfield plants. Remedies tended to favor the larger, older, integrated firms, and did little to facilitate reorganization of the industry or to aid the growing specialty steel segment, which many observers believed was the best hope for a competitive industry. The pragmatic focus on the status quo underrepresented a future in which there was a different steel industry.

Tripartite's underlying assumption was that more adequate policy could and should be made by asking those most affected to hammer out a deal they could live with. But such an arrangement ratified existing power relations (as both environmentalists and skeptical economists feared) and guaranteed that only some questions would be relevant (as OTA and CRS learned).

Tripartite was both a useful and a flawed device. It managed to move a few steps beyond pluralism, while at the same time revealing its limitations. One of pluralism's deepest— and perhaps most debatable—assumptions was underlined by its structure, namely, that the American government (or "the state") is and ought to be just like any other interest group: constrained by the constitutional dispersal of power, lacking in distinct authority, merely another player in the search for public purposes. Instead, we believe that in a democratic society, government has a deeper role to play. That role, as explained in Chapter 1, involves expanding the capacity of a free people, both individually and jointly, to successfully govern themselves. Whether the Tripartite experience, and the enforcement activities of EPA directed at the steel industry accomplished very much in that regard, is open to question. Did regulators and regulatees learn to trust each other any more? Was the technical capacity and political credibility of public institutions strengthened? Do citizens have a better sense of the tradeoff between economic considerations and pollution control? We see only modest gains, at best, in the case we have just completed.

NOTES

1. For the discussion of RACT and the regions, and of certain aspects of the bubble policy and Tripartite, we have used material from Christopher E. Dunne, "The Environmental Protection Agency, the Steel Industry, and the Clean Air Act: An Ongoing Minuet," Harvard School of Public Health, April 10, 1982. Mr. Dunne is not responsible for the interpretations and conclusions we draw from his summary of the interview with Costle and Hawkins or from his own interviews with other EPA officials.

2. Robert W. Crandall, *The U.S. Steel Industry in Recurrent Crisis: Policy Options in a Competitive World* (Washington, D.C.: The Brookings Institution, 1981); Council on

Environmental Quality, *Environmental Quality—1978* (Washington, D.C.: Government Printing Office, 1978), pp. 71–76; Arthur D. Little, *Steel and the Environment*, 1975.

3. Marc J. Roberts and Susan O. Farrell, "The Political Economy of Implementation: The Clean Air Act and Stationary Sources," in Anne F. Friedlaender, ed., *Approaches to Controlling Air Pollution* (Cambridge, MA: MIT Press, 1978), 152–181.

4. How the legal process itself directs and disciplines that integration is one of Edward H. Levi's topics in *An Introduction to Legal Reasoning* (Chicago: The University of Chicago Press, 1949). For examples of the exercise of discretion in a non-scientific but no less technical context, see Suzanne Weaver, *Decision to Prosecute: Organization and Public Policy in the Antitrust Division* (Cambridge: MIT Press, 1977).

5. Roberts and Farrell, "The Political Economy of Implementation," 152–181.

6. R. Shep Melnick, *Regulation and the Courts: The Case of the Clean Air Act* (Washington, D.C.: The Brookings Institution, 1983), especially 217–238.

7. The most persistent critic was Earle F. Young, American Iron and Steel Institute vice president for environmental affairs. According to Douglas Cooper, an environmental physicist formerly at the Harvard School of Public Health, AISI relied primarily on two studies to make the industry's case: G.C. Cheng, et al, "Steel and the Environment - A Cost Impact Analysis," (Cambridge, MA: Arthur D. Little,Inc., 1978), and A.E. Bennett, et al, "Health Effects of Particulate Pollution: Reappraising the Evidence," for the American Iron and Steel Institute, 1979.

8. Melnick, *Regulation and the Courts*, 155–192.

9. Office of Technology Assessment, *Technology and Steel Industry Competitiveness*, US Government Printing Office, June 1980.

10. United States Senate, Committee on Environment and Public Works, Senate Report 95–127.

11. Melnick, *Regulation and the Courts*, 170.

12. Melnick, *Regulation and the Courts*, 169.

13. Melnick, *Regulation and the Courts*, 173.

14. Section 113(d) of the Clean Air Act Amendments of 1977; for Congressional background, see Senate Committee on Public Works, Subcommittee on Environmental Pollution, *Clean Air Act Oversight, Pt 2*, 93rd Congress, 2nd Session, Committee Serial No. 93–H42, 1033–1044; Senate Committee on Public Works, Subcommittee on Environmental Pollution, *Implementation of the Clean Air Act - 1975, Pt 2*, 94th Congress, 1st Session, Committee Serial No. 94–H10, 1495–1535. See also Melnick, *Regulation and the Courts*, 198ff.

15. 41 FR 5525ff (21 Dec. 1976); Note also "The Trade-off Policy: Solution to the Clean Air Act?" *Harvard Environmental Law Review*, 1 (1976); Bruce Yandle, "The Emerging Market in Air Pollution Rights," *Regulation*, July/Aug. 1978. For an account of the early implementation of the offset policy, see Stephen R. Thomas, "Publics, Markets, and Expertise in Environmental Planning" (Ph.D dissertation, Harvard University, 1980), chapters 5–7; also Richard A. Liroff, *Air Pollution Offsets: Trading, Selling, and Banking* (Washington, D.C.: The Conservation Foundation, 1980); Jorge A. Calvo y Gonzales, "Markets in Air: Problems and Prospects in Controlled Trading," *Harvard Environmental Law Review*, 5 (1981).

16. Memorandum from Roger Strelow, assistant administrator for air and radiation, to regional administrators, "Guidance for Determining Acceptability of SIP Regulations in Non-attainment Areas," December 9, 1976, *Environment Reporter: Current Developments*, December 17, 1976, p. 1210. This memorandum said that RACT "should represent the toughest controls considering the technological and economic feasibility that can be applied to a specific situation," cited in Richard A. Liroff, *Reforming Air Pollution*

Regulation:The Toil and Trouble of EPA's Bubble (Washington, D.C.: The Conservation Foundation, 1986), p. 53n.

17. Richard Wilson, deputy assistant administrator for mobile source enforcement, USEPA, interview with Stephen R. Thomas, Washington D.C., 7 Dec. 1984. In the Costle EPA, Wilson served as deputy assistant administrator for general enforcement.

18. A good description of this mood may be found in Robert J. Samuelson, "The Woes of Wheeling-Pittsburgh—The Crunch Before the Slump," *National Journal*, 14 Jan. 1978, 53–57.

19. The outline of this story about RACT guidance comes from a joint interview with Costle and Hawkins on the EPA's air program, Harvard School of Public Health, Boston MA, 3 Mar. 1981.

20. Wilson interview, 7 Dec. 1984.

21. Bernard Bloom, former environmental engineer, Technical Support Branch, EPA Division of Stationary Source Enforcement, interview with Christopher E. Dunne, Washington D.C., 24 July 1981.

22. Memo from Edward Reich, director of the division of Stationary Source Enforcement, to Walter C. Barber, deputy assistant administrator for Air Quality Planning and Standards, "Steel Mill Related SIPs," n.d.

23. The phrase is Costle's.

24. Hawkins, joint interview, 3 Mar. 1981.

25. Joseph Kunz and Thomas Maslany, EPA Region III, interview with Christopher Dunne, Philadelphia PA, 18 Nov. 1981.

26. Dan Goodwin, director, Air Program, Illinois EPA, interview with Stephen R. Thomas, Springfield, Illinois, 2 Sept. 1982; Memorandum from Dan Goodwin (writing as vice president of the State and Territorial Air Pollution Program Administrators) to Nancy Maloney (EPA) and others, 26 May 1981, 1–4.

27. Goodwin interview, 2 Sept. 1982.

28. Bernard Bloom, interview with Stephen R. Thomas, Chevy Chase, Maryland, 8 Dec. 1984.

29. Memo from Reich to Barber, n.d.; memo from Richard Wilson to David Hawkins, assistant administrator, "The 1979 SIP Revisions and the Steel Industry,"n.d.; Bloom interview, 8 Dec. 1984; for Bloom's objections, see "Review of Environmental Protection Agency Part D State Implementation Plan Decisions Concerning Reasonably Available Control Measures for Steel Plants: Proposed Evaluation Methodology," typescript, n.d. (approximately Sept. 1981).

30. Goodwin interview, 2 Sept. 1982.

31. Goodwin interview, 2 Sept. 1982.

32. Goodwin interview, 2 Sept. 1982.

33. "U.S. EPA's Methodology for Evaluating RACT Limitations Contained in Part D Submittals," typescript, n.d. (approximately Sept. 1981), 2.

34. Classic statements may be found in J. H. Dales, *Pollution, Property and Prices* (Toronto: University of Toronto Press, 1968) and Allen V. Kneese and Charles L. Schultze, *Pollution, Prices and Public Policy* (Washington, D.C.: The Brookings Institution, 1975).

35. The main points of the story of the Middletown bubble, though not the slant, come from Harvard Business School case, "Armco and the Bubble Policy," Parts A, B, and C, numbers 0–381–226, –227, and –228; see also Richard Liroff, *Reforming Air Pollution Regulation: The Toil and Trouble of EPA's Bubble* (Washington, D.C.: The Conservation Foundation, 1986), pp. 75–78 and passim.

36. Drayton, cited in "Armco and the Bubble Policy" (C), Harvard Business School Case No. 0–381–228, 9.

37. U.S. Environmental Protection Agency, *Environmental News*, 20 Oct. 1980, cited in "Armco and the Bubble Policy (A)," 1.

38. EPA, cited in "Armco and the Bubble Policy (A)," 6–7.

39. Armco, "Fugitive Emissions Estimating Methods" (1977), cited in "Armco and the Bubble Policy (A"), 8.

40. Armco, cited in "Armco and the Bubble Policy (A"), 8.

41. U.S. Environmental Protection Agency, "Proposed Policy Statement Recommending Alternative Emission Reduction Options," 18 Jan. 1979, cited in "Armco and the Bubble Policy (C"), 2.

42. Douglas W. Cooper, "Case Study: A New NAAQS for Particulates?" Harvard School of Public Health, unpublished, 1980; the technical work summarized in this teaching material may be found in D.W. Cooper et al., "Setting Priorities for Control of Fugitive Particulate Emissions from Open Sources," Report EPA-600/7–79–186, USEPA, Washington D.C., 1979; John S. Evans and D.W. Cooper, "Contributions of Open Sources to TSP," EPA Second Symposium on the Transfer and Utilization of Particulate Technology, Denver, 1979; F.P. Perera, "The Adequacy of the National Ambient Air Quality Standard for Total Suspended Particulate Matter," A NRDC White Paper, NRDC, Washington D.C., 1979.

43. Hawkins, as quoted in "Armco and the Bubble Policy (C)," 8.

44. Hawkins, as quoted in "Armco and the Bubble Policy (C)," 8.

45. Costle-Hawkins joint interview, 3 Mar. 1981.

46. Costle-Hawkins joint interview, 3 Mar. 1981.

47. Jeffrey Miller, former acting assistant administrator for Enforcement, interview with Christopher E. Dunne, Washington D.C., 16 Nov. 1981.

48. Hawkins, remarks at an EPA-sponsored symposium on controlled trading and the states, September 1980, Research Triangle Park, North Carolina.

49. Letter of Earle F. Young, Jr., assistant vice president for Environmental Affairs, AISI, to Barbara Ingle, Office of Planning and Evaluation, USEPA, 22 Feb. 1979, as cited in "Armco and the Bubble Policy (C)," 10.

50. Letter Young to Ingle, cited in "Armco and the Bubble Policy (C)," 10.

51. Costle-Hawkins joint interview, 3 Mar. 1981.

52. Charles L. Schultze, "Industrial Policy: A Dissent," *The Brookings Review* (Fall 1983). For general discussion, see Robert B. Reich and John D. Donahue, *New Deals: The Chrysler Revival and the American System* (New York: Random House, 1985), chapter II; Michael L. Wachter and Susan M. Wachter, eds., *Toward a New U.S. Industrial Policy?* (Philadelphia, PA: University of Pennsylvania Press, 1981).

53. Douglas Costle, telephone interview with Stephen R. Thomas, 28 October 1982.

54. Michele Corash, former EPA general counsel and chair, Environmental Working Group of Tripartite, interview with Christopher E. Dunne, Washington, D.C., 17 July 1981.

55. Corash interview, 17 July 1981.

56. Corash interview, 17 July 1981.

57. "Report of the Steel Tripartite Advisory Committee Working Group on Environmental Protection," 15pp typescript, 11 Sept. 1980. The discussion in the rest of this section comes from the Working Group's report, except where noted.

58. Noted by the General Accounting Office Report, "The Steel Industry Compliance Extension Act Brought About Some Modernization and Unexpected Benefits," GAO/RCED-84–103, 5 Sept. 1984, 7. (Hereafter referred to as *GAO Report*.)

59. Wilson interview, 7 Dec. 1984.

60. Corash interview, 17 July 1981.

61. Senate Committee on Environment and Public Works. *Hearing: Report of the Steel Tripartite Committee*, 96th Congress, 2nd Session, 4 Dec. 1980, 8 (Fran Dubrowski), 14f (Michele Corash), 55 (Lloyd McBride). George Stinson did not challenge this testimony, nor did he confirm that the industry had committed itself.

62. Miller interview, 16 Nov. 1981.

63. Carter statement, "A Program for the American Steel Industry, Its Workers and Communities," 30 Sept. 1980.

64. The effect of the *de minimis* provision is discussed in the *GAO Report*, 11–13.

65. *GAO Report*, 7f, 15–17.

66. *GAO Report*, 9.

67. *GAO Report*, 16.

68. *Hearing: Report of the Steel Tripartite Committee*, 8f.

69. OTA's report was *Technology and Steel Industry Competitiveness* (June 1980). Hirschhorn summarized the report in his testimony, *Hearing: Report of the Steel Tripartite Committee*, 70–91.

70. Congressional Research Service, "Some Environmental Aspects of the Steel Revitalization Proposal," prepared by Joseph P. Biniek, 5 Nov. 1980. This document is printed in *Hearing: Report of the Steel Tripartite Committee*, 116.

Part Two Conclusion: The Lessons of the Cases

In Part Three we will offer a broad explanation of both the cases and the events of the Reagan years. Here we simply summarize the major findings from each case and draw attention to the role of leadership in accounting for them.

OZONE

EPA's efforts to reset the ambient ozone standard led to an important educational failure. The Clean Air Act made some incorrect presumptions about the nature of the dose-response function. Yet the agency did not communicate its own more sophisticated understanding to either the Congress or the people. The unresolvable ambiguities in terms like, "safety," "health," and "most sensitive group," went unremarked and unexplained.

There was also an important integrative failure. EPA made little effort to reconcile health protection with practicality. As a result, many major urban areas will not be able to devise a feasible control program enabling them to meet the current ozone standard. Nor is it obvious that the health benefits of having only one, instead of four exceedences a year, are worth their cost, even where that could be achieved. The Clean Air Act excludes economic considerations. But, in the absence of any threshold for risk, some balancing between costs and benefits had to be implicit in the standard setting decision—a reality EPA neither acknowledged nor forced the Congress to confront.

RCRA

The central failure of RCRA was strategic. Because EPA never developed a strategy for dealing with hazardous waste, it was never able to impose any coherence on an admittedly enormous regulation writing task. The agency focused first on land disposal more or less by default. Yet no effective federal program was established to provide the new, carefully engineered facilities that this approach required. Nor was EPA able to organize itself to manage the RCRA regulating-writing task. In four years of effort, the agency failed to produce many of the key provisions. The reasons for the delay include the agency's poor internal integration, as well as flawed congressional draftsmanship and an impotent federal court.

Underlying all this was a fundamental deliberative failure. EPA did not learn what it should have learned from trying to write the regulations. It was unable to understand the source of its own difficulties or to develop new ideas. The conceptual flaws in the act itself—the ambiguity of the idea of "hazardousness" and the lack of a clear objective—were never clearly articulated. Nor was such an understanding used to provoke further Congressional discussion. Only today is the essential strategic issue being confronted—as localities resistent to land fills or incinerators search for other ways to deal with hazardous waste.[1] This is a discussion EPA should have initiated more than a decade ago.

SUPERFUND

The Superfund legislation too was conceptually deficient. Many important policy issues were never addressed. The Congress did debate distributive and fiscal questions such as liability, victim compensation and funding. But it largely ignored the problem of setting priorities or cleanup levels, and establishing a meaningful role for the states.

Without EPA's help, Congress was too uninformed to consider the latter questions. But the agency was too busy skillfully threading its draft through the intricate maze of the legislative process to provide such help. The drafting team did not want to risk real debate because it knew how it wanted the process to come out. Deliberation, and the clearer public understanding that it might have produced, were sacrificed to legislative victory and enhancing the agency's role. Furthermore, the design of the program did not seek to promote decentralization or to enhance the role of the states. This allowed citizens to continue to treat hazardous waste as a problem someone else was going to solve for them.

IRLG

The difficulties with the "cancer policy" produced by IRLG were both conceptual and educational. The task seemed plausible. Uniform rules that specified how agencies should classify compounds as carcinogens appeared to serve reasonable goals like predictability and consistency.

Unfortunately not much is known about the low-dose carcinogenic effects of many compounds. Scientists must guess at these on the basis of quite limited data. Contrary to EPA's approach, the appropriate question is not, "Is substance X a carcinogen?," but rather, "Given what little we know, should we treat X as a carcinogen for certain specified purposes?" Those purposes, and the values they reflect, are thus inextricably part of a decision that involve a good deal more than just science.

Furthermore, because chemical carcinogenesis is not a uniform phenomena, any set of uniform decision rules will be more conservative in some cases than in others. No matter what society's attitude toward the trade-off between risks and benefits, the varied and unmanaged risk aversion produced by uniform rules will necessarily often fail to reflect those preferences.[2]

Instead of clarifying these points, the IRLG document on risk assessment papered over significant policy differences among the participating agencies. It served neither to educate nor to stimulate informed discussion about how cautious the society should be about cancer risks, and what economic and social costs to incur in light of various degrees of risk and scientific uncertainty.

STEEL

The steel case demonstrates the adverse affects which pluralism has upon integration and deliberation. Factionalism within the agency led it to conduct the "RACT" negotiations along two completely different tracks. Headquarters ran its own effort to determine what should constitute "reasonably available control technology" at the same time that the regions were negotiating this same matter with individual states. As a result, agreements which the regions had made with the states were repudiated. The inability of the agency to speak with one voice gave industry added reasons to be recalcitrant and to engage in forum shopping. Adversarial relations within the agency encouraged greater conflict between the agency, the states, and industry.

The internal disputes over the "bubble" illustrate the danger of substituting bargaining for deliberation. One cannot assume that the relative power of groups in a negotiation provides a good surrogate for the public interest. In this instance, Hawkins and his allies initially forced the agency to produce a bubble policy that was unimplementable. Although EPA did eventually "muddle through" to a better result, largely as the result of outside pressure, the ensuing delay was costly. Inefficient investment decisions were made in the interim and opportunities for cost saving were missed.

The Tripartite process illustrates a different problem which results from substituting bargaining for deliberation. Because each party to the negotiations defined its interests quite narrowly, there was no stimulus to devise an innovative program which would have better addressed the concerns of both the industry and the environmentalists.

THE ROLE OF LEADERSHIP

A central thread running through these evaluations is the failure of leadership. Key conceptual ambiguities were unresolved. Crucial strategic decisions were avoided. Internal factionalism was abetted. The importance of civic education was underestimated.

We are not asking agency leaders to be martyrs. They must face up to political reality. They cannot abandon the competition for resources and recognition which is characteristic of all large organizations. But the quest for security and self-aggrandizement does not constitute the only, or indeed the most important, objective of public servants, whose very title implies a good deal more.

Political science shares the blame for EPA's shortcomings. By depicting the activity of governing as mere bureaucratic infighting, it has offered a self-fulfilling prophecy. Officials become less likely to act responsibly as they come to believe that they are not expected to do so.

Such a misperception led Costle to miss an historic opportunity to change the course of environmental policy. As the first Democrat to head the agency, he possessed political capital that his predecessors lacked. The Congress was appropriately skeptical of the Nixon Administration's environmental intentions. Therefore, Ruckleshaus would have found it difficult to oppose the excessively rigid and unrealistic statutory framework Congress had devised, without appearing to lend credibility to those suspicions. Carter did not inspire similar doubts. As his agent, Costle was in a better position than his predecesors, or his Republican successors, to urge Congress to make fresh and probing re-examination of its environmental handiwork. Just as it took a Republican to open the door to China, so it would have taken a Democrat to make a credible case for environmental realism, to ask the right questions.

NOTE

1. See Rochelle Stanfield, "Drowning in Waste," *National Journal*, 5/10/86, 1106–1110
2. See Chapter 8.

Part III

8

The Reagan Administration

During the Reagan administration, the Environmental Protection Agency underwent three changes of leadership, each of which provides illuminating points of comparison with the Costle regime. This chapter provides an overview of EPA under Gorsuch, Ruckelshaus, and Thomas, and describes the subsequent history of each of the five specific policies that we examined in previous chapters.

Reagan's environmental agenda was clearly revealed during the 1980 campaign. The general problem of American life was too much government. The solution was less government. With regard to EPA, the Republican platform stressed the need to ensure that the costs of regulation were justified by their benefits and that environmental protection not "become a cover for a 'no-growth' policy and a shrinking economy."[1]

Reagan's own asides on the campaign trail were far more scathing. He was quoted as saying that if EPA had its way, "you and I would live like rabbits," and he claimed that trees and plants were the chief cause of air pollution.[2] These comments were widely quoted and criticized. For the most part, however, candidate Reagan ignored environment issues, focusing instead on inflation and affirmative action where his stands appeared to be far more popular than those of the incumbent administration.

Reagan chose to interpret his landslide victory as a mandate. Certainly his call for tax and budget cuts, and a more aggressively anticommunist foreign policy struck a responsive chord with the electorate. His environmental views, however, were clearly less popular. During the 1970s, in the face of severe energy shortages and tax revolts, public support for environmental programs, and the spending they required, had remained consistently high. Polls taken during and after the 1980 election showed no diminution in the public's desire for strong environmental protection efforts. This dissonance between the poll data and Reagan's position, however, obviously did not hurt him in the election.[3] The public's mind was on other things.

Despite this ambiguity, the administration chose to respond to its more conservative supporters and business interests and pursue a program of regulatory relief and the transfer of greater decision making authority to the states. The greatest obstacle in Reagan's path was his lack of congressional control. The Democrats had retained a sizeable majority in the House. Although the Republicans had control of the Senate for the first time in more than two decades, the vagaries of the seniority system mitigated against environmental retrenchment. The chairmanships of the Senate Committee on Environment and Public Works and of its Subcommittee on Environmental Pollution, were assumed by staunch environmentalists, Stafford of Vermont and Chafee of Rhode Island.

The new Majority Leader, Baker of Tennessee, had served on the Environment and Public Works Committee and had played an active and supportive role in the drafting of several landmark environmental statutes. He, too, was most unlikely to support administration efforts to erase his handiwork.

Public support for environmental protection implied that low visibility measures were preferable to highly visible symbolic struggles so that the president would not be closely identified with specific retrenchment efforts. The administration's approach toward reauthorization of the Clean Air Act showed such caution. After extensive high level discussion, the administration chose not to put forward its own legislation. Rather it submitted a statement of general principles and left the initiative with Congress. This allowed it to avoid creating a focal point for congressional opposition and to sidestep charges of a "sell out" to industry.[4] The eventual demise of those members of the administration—Gorsuch, Watt, and Lavelle—who became identified by the public as being antienvironment, illustrates why Reagan so keenly desired to avoid acquiring such a label for himself.

Three means were available for inconspicuously curtailing environmental programs: the appointment of loyalists to key posts, reorganization both inside and outside EPA to allow those loyalists to control the agency, and the budget. All these were used to insure that EPA moved in the direction the White House intended.

These measures also had a fiscal benefit. The administration was committed to large tax cuts, increased defense spending, and to not cutting entitlement programs. Hence, only a few small programs (like EPA) remained to absorb the spending cuts that even a token effort at budget balancing required.

APPOINTMENTS

Reagan's initial environmental appointments reflected an insistence on ideological commitment and personal loyalty. Secretary of the Interior, administrator of EPA, and director of the Bureau of Land Management all went to close political allies. To an administration fearful that key agency officials would "go native," the fact that these three appointees were known to be enemies of environmentalism and were personally on hostile terms with many environmentalists, was a distinct advantage.

Anne Gorsuch, the new EPA administrator, had made her reputation as a state legislator in Colorado. She was one of the leaders of a small band of extremely conservative legislators who called themselves "the crazies." Among her fellow crazies was Stephen Durham, whom she would appoint as administrator of EPA's regional office in Denver. James Watt, Secretary of the Interior, was counsel for the Mountain States Legal Foundation, a conservative public interest law firm that spearheaded efforts to block environmental regulation. Robert

Burford, director of the Bureau of Land Management, had been helped by Gorsuch to become Speaker of the Colorado House of Representatives.[5]

The same pattern was repeated inside EPA. Gorsuch's closest confidant and most influential advisor was James Sanderson, a political ally from Colorado, who served as a private consultant. Sanderson had served as regional counsel for EPA in Denver and was currently a lawyer for a Denver firm which represented energy and development interests. It also represented brewing magnate Adolph Coors, a powerful force in conservative Republican politics.[6] Later, Sanderson was nominated by Gorsuch to replace Nolan Clark as associate administrator for Planning and Management but he was never confirmed.[7] John Daniel, who served as Gorsuch's chief of staff, had worked for state environmental agencies in Ohio and Alabama and had most recently served as a lobbyist for Johns Manville, the Denver based firm that was most well known as a manufacturer of asbestos products.[8]

With the exception of Sanderson and Daniel, all other high level appointments were made directly by the White House.[9] Robert Perry, general counsel, had been a trial lawyer for Exxon.[10] Rita Lavelle was a public relations specialist who had worked in California state government during the Reagan governorship and was a political ally of Edwin Meese; Kathleen Bennett, assistant administrator for air, had been a paper industry lobbyist; Nolan Clark, associate administrator for planning and resource management, was a Washington lawyer; Frank Shephard, associate administrator in charge of enforcement, was a Florida lawyer and conservative.[11] Of the entire group, no one had ever worked at EPA's Washington headquarters or on a congressional committee staff with EPA oversight responsibility. Sanderson was the only one who had even been employed by EPA.

For the key post of deputy administrator, the White House chose a man similarly devoid of government experience. John Hernandez had been dean of engineering at New Mexico State University and was a protege of the state's powerful senior senator, Peter Domenici. Domenici had touted Hernandez for the administrator's post and he emerged as an early front runner. He ran afoul of aides to David Stockman however, when, during an interview with them, he advocated upgrading EPA's research and development capabilities and refused to support their assertion that the EPA budget should be cut by fifty percent.[12] Instead, he was offered the deputy post and given wide latitude to act as EPA's "house scientist."

White House control caused considerable delay in filling these positions. Two of the six assistant administrators were not appointed for seven months. One position remained vacant for ten months, while two others stayed open for fifteen months. And one, the head of the Office of Research and Development did not have a permanent incumbent during Gorsuch's entire tenure.[13] These vacancies hampered the agency's ability to perform many of its normal functions, which to some was an advantage. They also provided Gorsuch with a strong rationale for centralizing decision-making in her office and for relying on confidants like Sanderson.

Gorsuch also made the deputy assistant administrators (DAAs) —one level below the assistant administrators—political appointees. Previously the rank of DAA had represented the highest career level within the agency. Incumbent career DAAs were demoted to the rank of manager and reported to the new DAAs rather than to the assistant administrators.

REORGANIZING ENVIRONMENTAL DECISION MAKING

Even in the Carter era, as we have seen, EPA's independence was continually threatened. The Quality of Life review process and its successor, the Regulatory Analysis Review Group (RARG), were both attempts to subject EPA to the scrutiny of the Office of Management and Budget (OMB) and others in the White House. Skillful maneuvering by Costle and his predecessors had served to blunt such efforts. The appointment of Gorsuch and her team ensured that no such resistance would be mounted to Reagan's efforts to subordinate the agency.

First, EPA's access to the president was limited by the interposition of a new administrative structure, the cabinet councils. As a device to facilitate executive branch integration, these councils were to work on those key policy questions that transcended any single department. Initially five councils were established: economic affairs; human resources; food and agriculture; commerce and trade; and natural resources. Each council was composed of the Cabinet officers with responsibilities in that area, with a member of the White House staff as the executive secretary. In addition, a staff structure was established involving staff from each of the member departments, as well as staff from OMB.[14] High level White House staffers were ex officio members of all the councils.[15]

Gorsuch, lacking cabinet status, was not appointed to the Cabinet Council on Natural Resources (CCNR).[16] Its membership included: the chairman of the Council on Environmental Quality (CEQ); the chairman of the Council of Economic Advisors(CEA); and the secretaries of Interior, Agriculture, Transportation, Housing and Urban Development, Energy, and the Attorney General. The CCNR met twenty six times in 1981. On seven of those occasions, revision of the Clean Air Act was discussed. Thus, the EPA administrator could not contribute to the highest level discussions of one of the most important laws administered by her agency.[17]

The EPA's decision-making autonomy was also severely restricted by Executive Order 12291. This required that, except where forbidden by law, all proposed major rulemakings were to be submitted to OMB for review and possible revision before they were promulgated. Although the order did not expressly give the OMB director power to overturn agency rules, the clear intent was to insure that all proposed regulations were in line with administration policy. Furthermore, OMB was also to examine existing regulations to ferret out those that were duplicative, overlapping, and conflicting. The order also mandated the agencies to ensure that all proposed rules had undergone rigorous cost benefit analysis, that rules only be issued when the benefits clearly exceeded costs, and that the least costly option had been chosen.[18]

The OMB director exercised these powers subject to the direction of the White House Task Force on Regulatory Relief chaired by Vice President Bush.

The executive director of that task force was James Miller III, director of OMB's Office of Information and Regulatory Affairs.[19] In practice, his office performed the practical tasks involved in implementing the order. Miller's deputy, James Tozzi, had been in charge of OMB's Environmental Branch since 1972 and was thus extremely knowledgeable about how the agency functioned.[20]

The crucial difference between Carter's regulatory reform effort and the Reagan regulatory relief program was Reagan's routinized and hierarchical structure. All new regulations were scrutinized by OMB and its supervisory task force. And these groups did not simply comment, but were entitled to engage in revision. In contrast, RARG's advisory interventions had been episodic, with many important EPA initiatives receiving no RARG scrutiny at all.

The internal changes that Gorsuch instituted were in keeping with the administration's preoccupation with regulatory relief. She altered EPA's approach to regulation writing, enforcement, and relations with the states. The flow of new regulations was curtailed simply by allowing inertial forces within the agency to operate. Under Gorsuch no one was ordered to "put the pedal to the metal" or to short-circuit complex and time consuming agency procedures in order to get new regulations "out the door." The atmosphere of frenetic activity and organizational ambition that had characterized EPA during the Costle years simply dissipated.

Enforcement presented a more complex problem. On the one hand, Gorsuch felt that enforcement had often been overly zealous and had not been conducted in a spirit of compromise and cooperation with industry. Yet, given public support for environmental programs, the administration did not want to appear to be kowtowing to "big business."

This programmatic ambivalence lead to administrative incoherence. Initially, the Office of Enforcement was abolished and its staff distributed among the various program offices. An Office of Legal Counsel and Enforcement was also created with two deputies and no clear demarcation of authority. This led to an intense struggle for authority and staff resources over a two-year period, to successive reorganizations and to several high level resignations.[21]

This administrative turbulence served to sharply curtail enforcement. Between 1980 and the end of 1982, the number of cases that EPA referred to the Justice Department declined by fifty percent, from 200 to 100, and the number of enforcement orders issued by the agency dropped by one third.[22]

Gorsuch pursued a threefold program to achieve Reagan's goal of shifting federal regulatory activity to the states. First, EPA accelerated delegation of program authority to the states where that was permitted by statute. Second, it promulgated generic regulations that gave the states increased discretion. Third, it reduced federal oversight of state activities.[23]

The delegation effort was quite successful. In the first two years of Gorsuch's tenure, the number of states authorized to administer the Clean Air Act's program to prevent significant deterioration rose from sixteen to thirty-six. Three more states had assumed responsibility for the Clean Water Act's permitting program and four states had assumed responsibility for the Safe Drinking Water Act's underground injection control program with another twenty expected to do likewise in a matter of months.[24]

In a similar vein, in April of 1982, EPA promulgated "generic" regulations for its emissions trading program. As a result, the states were allowed to approve many trades on their own without revising their state implementation plans (SIPs) or submitting to a case by case federal review. The SIPs review process itself was also streamlined to diminish the federal role. Minor changes in SIPs were exempted from the initial proposal stage and allowed to proceed directly to final federal review. State and federal reviews were to be carried out concurrently rather than consecutively. Finally, if no comments were received during the designated period after the revision appeared in the *Federal Register*, full responsibility for the review was given to the appropriate region. As a result of these changes, the backlog of proposed SIP changes awaiting EPA review declined from 643 to 20 between January 1981 and October 1982.[25]

THE BUDGET

The third prong of the Reagan attack on EPA was the administration's efforts to cut the agency's budget. This became the central environmental battleground between the White House and Congress. Between 1980 and 1982, exempting Superfund, which was supported by dedicated revenues, the EPA budget declined, in constant 1972 dollars, from $701 million to $515 million.[26] This contraction coincided with a continuing expansion of EPA responsibilities in the realms of toxic substances, drinking water, and hazardous waste, as programs legislated in the middle and late 1970s finally reached the implementation phase. To meet these budgetary limits, drastic personnel cuts were made. Between 1981 and 1983, the number of positions at the agency (excluding Superfund) declined 22.6 percent. These cuts, combined with Gorsuch's open contempt for the EPA bureaucracy, led many seasoned employees to quit. By 1982 more than 4000 employees had resigned.[27] Thus, close to forty percent of the agency's workforce left in the space of little more than a year.

Ironically, the budget cutting process actually served to undermine Gorsuch's political standing with the White House. The OMB found that ever greater cuts were required to meet its deficit reduction targets and thus continually revised the goals that agencies like EPA had to meet. Despite her own draconian efforts, Gorsuch was continually reprimanded by OMB for seeking sums that it claimed were too high. Placed on the defensive, Gorsuch found herself in the anomalous position of internally defending programs which she had attacked publicly. Her defense of EPA was insufficient to gain her the respect of the agency's traditional constituency but did serve to impair her reputation as a loyal and effective servant of the administration.

GORSUCH'S FALL

Ironically, Gorsuch was forced from office because she wanted to take a more moderate line than the White House with regard to the burgeoning wrangle over Superfund and EPA's assistant administrator for hazardous waste, Rita Lavelle.

In the fall of 1982, John Dingell, chairman of the House Committee on Energy and Commerce and of its Oversight and Investigations Subcommittee, initiated an investigation of alleged abuses in Superfund enforcement. In September 1982, he entered into negotiations with EPA to obtain certain relevant documents. Under pressure from the Justice Department, EPA stalled. On October 21, 1982, Dingell subpoenaed Gorsuch to appear and provide the documents. Gorsuch wanted to comply but the Justice Department refused to allow her to do so, citing executive privilege. On December 19, 1982, the full House of Representatives voted to hold her in contempt for her continued refusal to furnish the documents.[28]

At the same time, the discovery of dioxin in the roadways of Times Beach, Missouri, produced the greatest torrent of "time-bomb" journalism since Love Canal. Public interest in environmental protection greatly increased and all the allegations of EPA mismanagement and lack of enforcement diligence during the previous two years were reexamined and publicized anew.[29]

Displeased with Lavelle's handling of the Times Beach incident, Gorsuch overruled her and approved the buy-out of all the private property in the town. Perhaps in retaliation, Lavelle's staff had sent a memo to the White House accusing Gorsuch's protege, Robert Perry, of "systematically alienating the primary constituents of this administration, the business community."[30] This provoked Gorsuch to demand Lavelle's resignation. When that was not forthcoming, she convinced the president to fire her.

Meanwhile, in response to the outpouring of criticism surrounding the Lavelle affair, the Justice Department announced it was beginning its own investigation of EPA. It informed Gorsuch that since she was now the subject of its scrutiny, it could no longer defend her against the congressional charge of contempt. Privately, Gorsuch continued to urge the White House to release the documents but to no avail. Now she would have to defend herself for loyally following a policy she had opposed. On March 9, 1983, in the face of this final evidence of White House disinterest in her well-being, she resigned.[31]

RUCKELSHAUS RETURNS

Gorsuch was replaced by William Ruckelshaus, the first EPA administrator, who most recently had been working for the Weyerhauser Corporation in Washington State. His impressive reputation, and the desperate need of the administration to restore its environmental credibility, provided him with enormous leverage.[32] As a result, he could advocate positions and take initiatives with greater impunity than perhaps any other member of the administration. Nonetheless, his erstwhile superiors continued their posture of budgetary stringency toward the agency, as well as a generally antienvironment stance on policy issues. Ruckelshaus' approach to agency leadership reflected an appreciation of the opportunities and the constraints these circumstances created.

Perhaps Ruckelshaus' most important contribution to EPA was his willingness to accept the job. This gesture did much to restore the morale of middle level workers who had felt they needed to apologize for working at the agency.

In addition, the new administrator quickly reversed the adversarial posture of the previous leadership toward Congress and the media. He met frequently with congressional staff and members and with reporters and he testified ungrudgingly at hearings.

Ruckelshaus also brought in a new leadership team of proven administrators. Al Alm, the new deputy administrator, had been a member of the task force that created EPA and served there for several years before becoming a faculty member at Harvard's Kennedy School of Government. Unlike his two predecessors, Blum and Hernandez, Alm served as the chief operating officer, enabling his superior to perform the strategic functions of a chief executive.[33] According to the *Environmental Forum*, "Alm's day to day influence over EPA matters is unquestioned, making him the most influential D.A. the agency has ever had."[34]

Ruckelshaus appointed another former EPA staffer, A. James Barnes, to be general counsel. At the time Barnes was serving as general counsel of the Department of Agriculture.[35] As assistant administrator for Hazardous Waste he chose Lee Thomas.[36] As deputy director of the Federal Emergency Management Administration (FEMA), Thomas had served as the head of the successful task force on Times Beach. He would later succeed Ruckelshaus as EPA Administrator. Bernard Goldstein, a well known scientist and chairman of the Department of Environmental and Community Medicine at Rutgers University, became assistant administrator of research and development.[37] The caliber of these and other new appointees led at least one knowledgeable observer to comment, "I'm a Democrat, and I hate to say it, but the agency now is in the best hands overall that it ever has been in."[38]

Ruckelshaus also sought to improve the agency's enforcement record. He instituted a new management system for tracking federal and state enforcement actions and berated enforcement lawyers for the insufficient speed with which actions were taken. As a result, the number of enforcement actions and the size of civil penalties imposed rose substantially.[39]

Risk Education

Programmatically, Ruckelshaus' biggest effort was an educational campaign designed to improve public understanding of the nature —and unavoidability—of environmental risks. Here was one initiative of which the White House could thoroughly approve. In his very first policy address after his reappointment, delivered at the National Academy of Sciences, Ruckelshaus went to some length to distinguish between risk assessment and risk management. The former task, he argued, was a scientific task, the latter a political one. Once an objective assessment of risk has been arrived at, it was the right and responsibility of the public to decide how much risk it wanted to live with. Ruckelshaus' ambition was twofold: to improve the technical capacity to perform accurate risk assessment and to increase public involvement in "managing" the risk, that is,in making the conscious choice between danger and cost.[40]

Nine months later, in a speech at Princeton University, Ruckelshaus backed away somewhat from this simple distinction. On this occasion he stressed the need to "expose to public scrutiny the assumptions that underlie our analysis." He pointed out that it had always been EPA policy to choose the "upper bound" of risk estimates and criticized "the stacking of conservative assumptions one on top of another."[41]

Ruckelshaus maintained that the best way to educate the public was to pose the question in more concrete terms.

> My experience with involving the public in these issues is that, if you lay before them all the information you have—not only in terms of what the nature of the risk is, but also what you can do about it—they tend to act sensibly . . . Getting that same kind of consideration made in the abstract . . . is a much harder question or issue to drive to public acceptance.[42]

As a place to test his ideas, Ruckelshaus chose the emissions standard for inorganic arsenic that EPA was then in the process of establishing under Section 112 of the Clean Air Act (which dealt with hazardous air pollutants). One of the various processes that created this pollutant, high arsenic throughput copper smelting, was used at only a single plant in the entire nation, the ASARCO plant just outside Tacoma, Washington.[43] Ruckelshaus was familiar with the problem, having lived for the previous several years in that region. Tacoma, thus, would be the only place in the nation affected by the standard for this particular process.

An EPA risk assessment concluded that airborne arsenic emitted from the ASARCO plant accounted for between 1.1 and 17.6 cancer cases a year in Tacoma.[44] The EPA's proposed regulations recommended the use of a "best available technology"(BAT), consisting of the installation of various forms of hooding systems to control those secondary emissions that were not trapped by filtering equipment on the smokestacks.[45] . These controls were expected to reduce the number of cancer cases to between 0.2 and 3.4 per year.[46]

In defending the proposal, the agency noted that there was no safe level of exposure to arsenic. Therefore, any emissions posed some risk. To achieve greater reductions, the plant would have to switch to low arsenic content ore. Such a conversion would be so costly as to force the plant to close, resulting in the loss of about 500 jobs at the plant, 300 additional jobs in the Tacoma area, and 20 million dollars in lost revenue to local suppliers.[47] Also, greater reductions than those proposed were unnecessary because the conservative assumptions built into the risk assessment made it extremely unlikely that the true risk remaining after installation of the hooding devices would exceed that which was predicted. In fact, the true risk was likely to be considerably lower.[48]

Ruckelshaus determined to give Tacomans a chance to discuss and criticize both the decision the agency had arrived at and the method of balancing economic and health considerations that had been employed. He made a highly publicized visit to the city to announce that three workshops would be held there during the summer of 1983. To facilitate discussion between citizens and officials, part of the workshop would consist of small group discussion with EPA representatives rotating among the groups.[49]

The two workshops held in the city of Tacoma were attended by environmental groups, civic organizations and smelter workers, The third, held on Vachon Island a rural community downwind of the plant, was attended mainly by island residents disturbed by evidence of high levels of arsenic in soil samples taken from their gardens. The meetings were well attended and some commentators attest that the general quality of the discussion was quite high. It is impossible, however, to evaluate the contribution of these discussions to the outcome because the final regulations were never issued. Before EPA reached a decision, declining copper prices forced the ASARCO plant to close, leaving the agency with nothing left to regulate.[50]

Acid Rain

Ruckelshaus' second major policy initiative, acid rain, is a bit more difficult to interpret. Acid rain had become a highly publicized issue by the spring of 1983. It played a very prominent role in Ruckelshaus' confirmation hearings. These were chaired by Senator Stafford of Vermont, one of the states most adversely affected by the problem. The National Wildlife Federation had termed acid rain "the most serious pollutant threat throughout the nation."[51] It had also become a growing source of controversy between Canada and the United States.[52] Even before taking the job, Ruckelshaus apparently decided to focus on this issue. In response to Ruckelshaus' own suggestion, Reagan instructed him at his swearing in ceremony to meet the acid rain problem "head on."[53]

The issue was complex both politically and environmentally. The main sources of acid deposition were in the industrial midwest, especially power plants. The main environmental consequences, diminished forest productivity and lake acidification (with effects on fish life) occurred in Canada, upstate New York, and New England. The Midwest was key to Reagan's electoral strategy. That region's blue collar workers, who had switched to the Republican party in 1980, would not be pleased by paying higher electric bills to provide difficult-to-observe ecological benefits to Easterners.

This kind of regional rivalry seemed likely to polarize the Congress, where acid rain control advocates were calling for a massive program to curtail emissions by fifty percent. The standard administration response had been that more research was required and that, therefore, new control efforts were premature. Ruckelshaus ultimately offered a modest program of limited emissions reductions in a limited area. He proposed that a 3.5 million ton reduction in sulfur dioxide be achieved at specified power plants, most of which were located in West Virginia, Ohio, Pennsylvania, and New York. Following the administration line, he characterized this as, in part, an experiment to acquire more information. The cost of the program would be borne by a tax on sulfur dioxide emissions nationally, as well as by a small tax on oil and gas powered electric generators, or from general revenues.[54]

In September 1983, Ruckelshaus presented his plan to the Cabinet Council on Natural Resources. The council meeting was attended by council chairman

Secretary of the Interior James Watt, Energy Secretary Donald Hodel, Secretary of Health and Human Services Margaret Heckler, members of the White House staff, and OMB Director David Stockman. Stockman attacked the plan, arguing that the costs were great, that no human health effects had been demonstrated, and that only trivial environmental benefits could be expected. He denounced it as a political gimmick for placating environmentalists and thus unworthy of an administration that was taking the high road toward fiscal integrity and economic liberty. Nor would the gimmick succeed in mollifying the administration's critics. Congress would simply use the relatively modest sum Ruckelshaus was proposing as a starting point for a much more expensive and comprehensive program.[55] Stockman swayed most of those present and, after some further consultations with White House aides, Ruckelshaus decided not to pursue any acid rain program.[56]

The White House must have known that its decision to ignore Ruckelshaus' initiative would result in a torrent of adverse publicity, which it did. But it is far from clear whether those who were upset would have supported the administration in any event. Ruckelshaus, meanwhile, was credited with having fought the good fight by those same interests. Some cynics even suggested that he had never expected to win on acid rain to begin with.

Lee Thomas

Ruckelshaus resigned shortly after Reagan's reelection and was replaced by assistant administrator for solid waste and emergency response Lee Thomas.[57]

Thomas was the first nonlawyer and the first governmental careerist to head EPA (the posts that Ruckelshaus and Costle had held prior to EPA were, for the most part, political appointments). Thomas began his career as an official of the South Carolina state government working in the area of criminal justice. He then joined the Federal Emergency Management Administration (FEMA), eventually becoming its executive deputy director. In that capacity, he headed the government's Times Beach task force. The Times Beach affair precipitated the downfall of Rita Lavelle and Anne Gorsuch. The EPA's credibility was so low at that point that the White House preferred to have the representative of a different agency lead the response effort. Ruckelshaus was sufficiently impressed by Thomas' performance that he gave him Lavelle's old post as overseer of Superfund and RCRA.[58]

Thomas replaced some of Ruckelshaus' senior management team but his appointments were similar in that they were people who had considerable prior governmental experience and were not perceived to be closely allied with particular constituency groups. He chose a lawyer, the agency's general counsel James Barnes, to be his deputy. Barnes had previously been general counsel at the Department of Agriculture. Milton Russell, assistant for policy planning and evaluation, was an economist with Resources for the Future who had previously served on the staff of the Council of Economic Advisors.[59]

Ecology Returns

Under Thomas, ecology won back the priority status that Costle's "public health strategy" had deprived it of nine years earlier. This shift was the result of growing scientific and public awareness of the dangers posed by new ecological threats, and a renewed appreciation of the persistence of familiar ones. But it was also due to increased skepticism, within the agency and the scientific community at large, about the risk to human health posed by the chemical "time bombs" which had given rise to the public health strategy in the first place. Although the public at large remained highly fearful of the carcinogenic impact of hazardous wastes, such effects were in fact quite marginal in comparison with other sorts of public health dangers, such as smoking, dietary fat, and lack of exercise, which were far beyond EPA's purview[60] The agency's own effort to assess comparative risks placed the carcinogenic effects of hazardous waste relatively low on its list of environmental cancer risks.[61] Sooner or later public opinion would catch up with the data. Respect for the merits and political prudence both dictated that the agency begin to redefine its mission before this shift in public perception took place.

Mounting public concern for ecology took two very different forms. One was localized. Specific communities and regions became exorcised about particular ecological outrages they were experiencing. For example, in July 1988, more than fifteen miles of Long Island ocean beach was closed as a result of a wave of sewage and hospital equipment that washed ashore. Among the debris were two-inch thick balls of solid sewage. A similar event occurred on the New Jersey shore a year earlier.[62] These situations harkened back to the sort of horror stories that had given rise to the environmental movement in the first place. They were a reminder to the public that whatever the long-term health threats might be, pollution could ruin beaches, kill fish, and create a foul stench right here and now.

Despite continued fears about hazardous waste, communities throughout the nation were finding that their most pressing disposal problems were caused by ordinary garbage. There was nowhere to put it. Low prices for raw materials had rendered recycling, by and large, economically infeasible. Existing dumpsites were filling up, and increased public opposition had made siting new ones extremely difficult. Concern about health effects made siting waste incinerators equally onerous. As the public and Congress became more aware of the vast scope of the solid waste disposal problem, EPA was once again called upon to take the lead in coping with the same problem that had resulted in passage of RCRA more than a decade earlier.[63]

Global Concerns

At the same time, the agency and the public at large were paying increasing attention to issues of global ecology. Such previously esoteric subjects as the depletion of the stratospheric ozone layer, and the greenhouse effect became the stuff of headlines and congressional hearings.[64]

A layer of ozone in the earth's stratosphere blocks some of the ultraviolet radiation that emanates from the sun. If that layer were depleted, more ultraviolet radiation would reach the earth's surface leading to increased incidence of skin cancers in humans.[65] Concerns about the impact on ozone depletion contributed to the decision by the United States not to build a supersonic transport aircraft (SST) in the early 1970s. Suspicions that chlorofluorocarbons (CFCs) were responsible for depletion led to a nationwide ban on their use in aerosol sprays in 1978.[66] Renewed concern for the issue surfaced as a result of an expedition by a team of U.S. scientists to Antarctica in 1986, which reported a drop of forty percent in the Antarctic ozone layer. Although it could offer no cause for this severe decline, the team considered the evidence to be consistent with the theory implicating CFCs.[67] These dramatic findings gave renewed impetus to the effort to negotiate worldwide limitations on CFC production.

Thomas quickly went on record in support of strong international action.

> I've been talking to my counterparts in other countries about the issue and I wrote a letter to several hundred environmental ministers indicating to them I thought this was a very serious issue . . .[68]

In September 1988, forty nations, including the United States and the other major CFC producers and consumers, agreed to a protocol freezing current levels of production and to future cutbacks in production and use.[69]

Global warming refers to a rise in the earth's mean temperature. This phenomenon is most commonly attributed to the increase in worldwide emissions of carbon dioxide, resulting from the burning of coal and wood, and of other gases such as nitrous oxide, methane, and CFCs. In sufficient concentrations, these gases act like a "greenhouse," trapping the sun's heat within the earth's atmosphere.[70] Concern about the issue increased as a result of the unusually warm temperatures experienced worldwide during the 1980s. Among the possible results of continued warming are a rise in sea level and an increase in extreme meteorological events like hurricanes and droughts. Public concern was intensified in the summer of 1988 by a prolonged heatwave and drought that affected the Mid-West and Great Plains and that was linked by the media to the "greenhouse effect."[71]

In some important ways, the rise of global ecological concern in the mid-1980s resembled that of hazardous waste in the 1970s. In both cases, the issue became prominent despite the lack of a clear scientific consensus about its severity and magnitude.[72] But such weakness was compensated for by freshness and sensationalism. A hole in the ozone layer, massive and persistent droughts, and a rise in world sea level conjure up truly apocalyptic visions that are at least on a par with "toxic time bombs."

Global concern has also benefited from important technological breakthroughs. Just as the advent of mass spectrometry enabled scientists to detect minute quantities of toxic substances, so too did new satellites and improved computer technology revolutionize the study of climate change, providing new data and permitting the construction of far more complicated computer simulations of global climate patterns.[73]

Finally, new political actors appeared on the scene with the capacity and the desire to mobilize support behind the issue. Prominent among them was the World Resources Institute founded in 1982 and led by James Gustave Speth, former chairman of the Council on Environmental Quality. The MacArthur Foundation pledged to match all donations to the institute up to a total of fifteen million dollars.[74] This well-endowed organization made the effort to control CFCs and reduce hydrocarbon emissions a central goal.[75]

A prominent congressman, Albert Gore Jr.,chose to identify himself with the issue of global warming. By age forty, Gore had already served in both the House and Senate and had mounted a major drive to obtain the Democratic presidential nomination. He chaired hearings on global warming, was a prominent witness at hearings chaired by others, and introduced legislation to coordinate and promote domestic and international research efforts and to identify strategies to limit the emissions of greenhouse gases.[76]

Important differences, however, also existed between the political challenge that globalism posed for EPA, and that which hazardous waste had confronted it with a decade earlier. By definition, global problems were not solvable by the United States alone. They inevitably became part of the diplomatic realm thereby severely limiting EPA's ability to take action on its own. Also, global issues emerged at a time when the incumbent administration had made clear its intention to oppose costly new environmental initiatives. Although it was not always steadfast in adhering to that position, Reagan signed the Superfund expansion and the administration did support the international protocol on CFCs, it would not have countenanced a major redefinition of EPA's mandate aimed at expanding its regulatory function.

Under Thomas, the agency restricted itself to making the EPA more visible in international forums and upgrading the relative standing of global issues among its priorities.[77] Thomas listed the problem of global consequences as one of the three most important new challenges facing the agency.[78] In the report that detailed the agency's massive effort to reconsider its priorities, EPA designated global warming and depletion of the ozone layer to be among the very most important.[79]

The Strategic Use of Risk Management

As important as this redefinition of the agency's mission was, it was largely reactive, the result of events and pressures over which the agency had little or no control. The major creative step that Thomas took was to expand on Ruckelshaus' use of "risk management" by making it the strategic linchpin for agency decision making. To accomplish this he commissioned a comparative risk assessment of all agency activities to see how close the match was between agency expenditures on different policies and the degree of risk they posed.[80]

The nine-month effort was presided over by Richard Morgenstern, director of the Office of Policy Analysis, but involved personnel from all parts of the agency.[81] The four separate work groups were charged with considering the thirty-one problem areas that were deemed to be within EPA's statutory purview, and ranking them according to the relative risk they represented. Each type of risk was granted equal status. A problem did not gain added significance if it were a cancer as opposed to an ecological risk.[82] Each task force prepared a separate report which, together with a summary overview, was published as *Unfinished Business—A Comparative Assessment of Environmental Priorities.*

The results were startling in terms of how much the rank order of risks differed from current agency priorities and from public perceptions. For example, hazardous waste and groundwater contamination ranked very high in terms of public concern and agency activity, but comparatively low in terms of relative risk to health or ecology. Criteria air pollutants, stratospheric ozone depletion, and pesticides received the highest ranks in the most number of categories. Of these, only air pollution ranked high among agency priorities and public concern.[83]

The organization of the project was noteworthy because it required the different media offices to involve themselves directly in relative risk calculations rather than relegating such considerations to the Office of Policy Analysis. Also the inclusion of ecology legitimized its equal standing with health effects in the comparative risk calculations.

In addition to providing a rationale for agency-wide priority setting, Thomas hoped to use risk management as a means of integrating decision-making on a geographical basis as well. He refined and expanded a program begun under Ruckelshaus that attempted to use comparative risk assessment methods to evaluate the relative danger posed by different environmental problems in a specific locale. Known as Integrated Environmental Management Projects, such efforts were conducted in Philadelphia, Baltimore, Denver, Santa Clara County, California, and the Kanawaha Valley of West Virginia.[84]

These projects were spearheaded by EPA's Office of Policy Analysis, but as time went on the level and quality of the participation by local officials and politicians increased.[85] These efforts generated considerable controversy within the scientific community regarding the quality of the risk assessment methodology employed.[86] But regardless of the specific merits of the individual experiments, they represent an intriguing experiment in policy integration.

OZONE

The Ozone controversy has continued along two quite separate tracks. On the one hand there is disagreement about the standard itself. Some argue that the whole effort to establish a threshold is ill conceived, while others claim that new scientific findings require an even stricter standard than the 0.12 ppm imposed by Costle. On the other hand, sixty-eight metropolitan areas with a population of seventy million failed to achieve the current standard by the 1987 compliance deadline, and the prospect of their doing so any time soon appears remote.[87]

Congress

Congressional deliberation about the ozone standard eventually revealed a refreshing willingness to listen to arguments, considered taboo a decade earlier, that are along the lines suggested in the conclusion of Chapter 3.

At oversight hearings conducted by the Senate Committee on Environment and Public Works in the spring of 1981, George Eads, an economist at Rand who had served in the Carter Administration,[88] argued that:

> the effects of pollutants do not display well-defined thresholds, their health consequences are subtle and highly varied throughout various populations, and the scientific evidence concerning when and in what populations they are likely to appear is often highly ambiguous.

He favored abandoning the "critical population-critical effect" approach since it gives researchers the incentive "to find ever more subtle health effects in ever more obscure populations," and give the EPA administrator

> the impossible task of either ratcheting the standard closer and closer towards zero—a point that EPA has already admitted would impose an intolerable economic burden and has consequently rejected—or of devising a rationale of sorting out—which of the discussed 'health effects' is subtle enough to consider significant, and which population is large enough to deserve protecting . . . To pretend that economic concerns will not enter into the decision-making process [is] foolish.[89]

We do not claim that Eads' argument persuaded everybody, but it did receive thoughtful and respectful reception. Its full effect will only become clear when the new Clean Air Act Amendments are completed and the EPA embarks upon another round of ambient standard-setting.

The Science

At the same time, new evidence about the harmful effects of ozone on pulmonary function have led some to conclude that the current ambient standard—an hourly average of no more than 0.12 ppm—was set too high, and that its focus on short term peaks per one hour diverted attention from effects at lower doses over longer periods of time.[90]

In an effort to be more precise in its decision making, EPA has come to define an "adverse health effect" as a ten percent reduction in FEV1 (FEV1 is the "forced expiratory volume" of air exhaled in one second). On that basis, the agency is concerned about findings contained in several recent studies, one of which included fifty-three boys and thirty-eight girls at a New Jersey summer camp. A third of the subjects "experienced a temporary reduction in FEV1 of sixteen percent on average during a period when ozone concentrations were close to the federal standard, but never exceeded it."[91]

Other studies have focused on the effects of longer term exposure to ozone. In particular, laboratory studies on animals have raised new concerns.

> Various experiments with rodents and primates showed that longer term exposure to ozone concentrations near the ambient [standard] range retard the ability of the animals' lungs to clear out toxic particles and cause inflammation of the lining of the animals' lungs. One study indicated that the function of cells that fight off bacterial infection in the lungs was impaired.[92]

The former chairman and still a member of EPA's science advisory committee, Morton Lippmann, has called the new evidence about the effects of ozone "clear and compelling." Bernard Goldstein, until recently the head of the EPA Office of Research and Development, has concluded that "there is sufficient toxicological data to be concerned about repetitive exposure over the years." The Natural Resources Defense Council will, according to staff attorney David Doniger, work "to get the standard strengthened because new evidence strongly indicates health effects at lower levels." But the American Petroleum Institute has come to a different conclusion:

> The human data that is available suggests that the people most adversely affected by high ozone are healthy people doing jogging and that the effects are fairly transitory. So you could raise the standard slightly without producing any adverse health effects and solve [the ozone deadline problem] because the cost of compliance gets very high.[93]

Meanwhile, the EPA's Office of Air Quality Planning and Standards has estimated that a 0.08 ppm, eight hour standard would cause nine additional areas around the country to be in noncompliance. Furthermore, attainment in the most troublesome regions, Southern California and the Washington-to-Boston corridor, would be that much more remote.

The Politics of Compliance

The essential compliance problem posed by ozone is that any improvement achieved by stricter controls is all too easily overcome by increases in the number of sources. In the Los Angeles basin, ozone concentrations are sometimes three times the national standard, despite the imposition of tight controls, because of a 320 percent increase in the number of motor vehicles since 1954.[94] In 1982, the Gorsuch EPA devised a strategy for accepting implementation plans for four areas in California (Ventura, Sacramento, and Fresno Counties, plus the Los Angeles metropolitan area) even though they would not show compliance by the 1987 deadline. This was dubbed the "Reasonable Extra Efforts Program." Its objective was to keep technology-based cleanup moving without imposing sanctions. In effect this enforcement strategy evaded the literal requirements of the statute.

In the wake of the passing of the 1987 deadline, which had been twice extended by Congress, EPA altered its plan. On August 29, 1988, the EPA announced a ban on all new construction of major sources of air pollution in nonattainment areas. The ban had little practical effect since firms in California and elsewhere had anticipated it and had filed construction applications in advance. But EPA's action was a reminder to Congress, still in the throes of revising the Clean Air Act, that the issue of enforceable deadlines would have to be faced.

Further progress in Southern California must come from further reductions of both stationary sources and automotive emissions. Schemes for altering driving are being considered, including requiring employers to set up car-pooling programs and staggered work hours, and bans on drive-through food service. Indeed, there has been renewed discussion of deploying electric cars and other quite radical steps that may prove necessary to achieve compliance.[95]

Environmentalists are reluctant to resort to litigation to force EPA to compel metropolitan areas to meet the deadlines for fear that Congress will react by abandoning the deadlines altogether. According to the political director of Friends of the Earth, "at least for now, everybody has the inclination to go slowly and go carefully. I'm hoping that as a result, everybody will be a bit more reasonable."[96]

THE RESOURCE CONSERVATION AND RECOVERY ACT (RCRA)

Until the departure of Ann Gorsuch and Rita Lavelle in early 1983, the Reagan Administration pursued a policy of retrenchment vis-a-vis RCRA. The RCRA regulations were among those targeted by Vice President Bush's regulatory relief task force. In the spring and summer of 1981 several of those regulations were either suspended by EPA or otherwise rendered ineffective. These included regulations covering incinerators, pits, ponds and lagoons, and liability insurance.[97] EPA also requested a two-year delay in the promulgation of landfill regulations, but that was denied by the court.[98] It reinterpreted the definition of an "existing facility" to enable those holding interim permits to expand their capacity up to fifty percent without obtaining EPA permission.[99]

On February 25, 1982, Gorsuch suspended the ban on the dumping of bulk liquids into landfills.[100] That ban had been promulgated in May 1980. Public and congressional criticism was so intense that the ban was reimposed less than a month later. The fire of public mistrust was fueled by the continued presence of James Sanderson as the top advisor to Gorsuch. Sanderson served as a legal representative for Chemical Waste Management Inc.(CWM) of Denver. It was alleged that he used his influence at EPA to obtain favorable rulings for CWM and that he had encouraged Gorsuch to approve some of the regulatory suspensions.[101]

In 1984, Congress amended RCRA to create new means for controlling what it perceived to be an irresponsible and unresponsive agency. The amendments deployed a device, called a "hammer," which temporarily deprives the agency of its rule-making authority when it fails to meet a specified deadline, and imposes a statutorily defined rule instead. Thus, if the agency fails to issue regulations for small quantity generators within the time period specified, the law terminates the small generator exemption and requires those exempted to comply with the normal RCRA regulations until such time as the agency promulgates the regulations.[102]

Congress was disturbed by new estimates that the annual production of hazardous waste was not forty million metric tons, as previously reported, but 150 metric tons and that less than one quarter of the 60,000 firms that identified themselves as hazardous waste generators were actually subject to the RCRA regulations.[103] It was also troubled by new research on landfill technology that revealed that even the use of "state of the art," double plastic liners did not prevent hazardous materials from leaching out of the landfill in a very brief period of time.[104]

The 1984 amendments responded to these concerns by placing all sources of hazardous waste under the regulatory umbrella and discouraging the use of landfills. Existing regulations had exempted those who generated less than 1000 kilograms of hazardous waste per month from the record keeping provisions of the act. The new law specified an interim set of record keeping and reporting requirements for those generating between 100 and 1000 kilograms per month and mandated EPA to promulgate a full set of regulations within two years. If it failed to do so, the hammer described above would swing into action.

The act ended two other exemptions as well. The EPA's own estimates indicated that 10 to 15 million metric tons of hazardous waste were mixed into the fuel of various "waste to energy" plants, yet these were not covered by the incineration regulations. In addition, products classified as hazardous were being blended into home heating oil. The new law contained a "burning and blending" provision that required facilities engaged in either of these two activities to notify EPA and required EPA to set standards for those facilities and for those who transport waste to them.[105] The Act also called for EPA to set standards for the design, construction, and installation of underground chemical and petroleum storage tanks. The states were required to enforce those standards and to monitor existing tanks as well.[106]

Most significantly, the new amendments offered a strategy for the conduct of waste management and a method for establishing priorities for the attainment of that goal. The "Findings and Objectives" section of the law was changed to read

> . . . land disposal, particularly landfill and surface impoundment, should be the least favored method for managing waste[107]

To enforce this proposition, EPA was given a series of deadlines to determine whether or not specific groups of chemicals were to be banned from landfills. The administrator was given two years to decide whether dioxins and solvents should be totally prohibited from landfills and, if not, under what circumstances continued land disposal should be permitted. He was given thirty two months to make a similar finding with regard to a list of other high risk chemicals. All other chemicals that the agency listed as hazardous were to receive a similar scrutiny according to a staggered five-and-a-half year timetable. This provision also contained a hammer. If the agency failed to meet the deadline for promulgating the regulations for any of these different categories of hazardous waste, that category of waste would be banned from landfills.[108]

STEEL AND THE REPUBLICAN BUBBLE

Despite the EPA's heated debates over the bubble policy and the intricate politics of Tripartite, the American steel industry has continued both its overall competitive decline and its restructuring toward specialty products and "minimills." Ironically, the developments we reviewed in the Chapter 7 did not contribute significantly to either of these trends.

In 1983, and again in 1986, the steelworkers union agreed to billions of dollars in wage and benefit concessions in attempt to save jobs.[109] But by 1986, it dues-paying membership had continued a decline to 700,000 (counting 225,000 members on layoff) down from 1.4 million a decade earlier. Considering that some members were now employed in other, healthier sectors of metal production, mining and fabrication, the losses in steel were all the more devastating. In 1986 average employment in the steel industry was just 215,000, less than half what it had been six years before.[110]

Geographically, employment has moved west from western Pennsylvania and eastern Ohio to Detroit and Chicago and into largely Southern, nonunion, minimills, which make steel from scrap by continuous casting (a process the bigger American companies, in contrast to overseas competitors, had been slow to adopt). These relatively clean operations now comprise about 20 percent of the American market, and are continuing to gain market share.[111]

Recall that just before the Costle team left office, the EPA approved a set of changes in the use of the bubble, making trading possible in nonattainment areas even in the absence of approved state implementation plans (SIPs). This was the "RACT bubble": though lacking a baseline of regulatory requirements, states could use Reasonable Available Control Technologies (RACT) to define the target, and then use trades with other points to reach it.

Allowing the bubble in nonattainment areas was, as previously noted, troublesome from the start. To opponents, any trading gave the stamp of approval to less cleanup than was technologically possible. It was objectionable, from this point of view, to allow *existing* sources in dirty areas to trade to meet abatement requirements. Worse would be a bubble on *new* sources, since they would otherwise be subject to even closer regulatory scrutiny than existing sources.

Complicating policy development was the fact that the Clean Air Act had no single unambiguous definition of "source." The statute provided for controls on various types of sources in areas where air quality might meet, fail to meet, or exceed ambient standards (PSD areas). Thus, the statute not only specified multiple degrees of control, its several provisions also referred to a multitude of "sources" subject to those controls. For a time, different EPA offices wrestled with this problem on their own. Eventually, the economics and the logic of emissions trading called for consistency even though statutory language alone could not provide clarity.

The courts had been in the picture, and stayed. With the change of administrations in 1981, the arcane conflicts over how widely to apply the bubble continued in litigation involving what a "source" is.[112] For his part, Costle had considered and rejected a policy that would have brought the nonattainment definition of source in line with that used in the Prevention of Significant Deterioration (PSD) program, applicable in areas cleaner than the ambient standard. Alignment between these definitions was favored by the regulatory reform staff at EPA. By March they had persuaded Gorsuch of their position. Opportunities to trade within enlarged "sources" increased at the expense of regulatory requirements that trading now sidestepped or simplified for the sake of economic efficiency.[113]

The Natural Resource Defense Council thereupon sued. In *NRDC v. Gorsuch,* issued in August 1982, the D.C. Circuit Court overturned Gorsuch's single definition of source as applied to nonattainment areas on the basis of a principle it claimed to find in earlier decisions:

> The bubble concept, *Alabama Power* declares, is mandatory for Clean Air Act programs designed merely to maintain existing air quality; it is inappropriate, both *ASARCO* and *Alabama Power* plainly signal, in programs enacted to improve the quality of the ambient air.[114]

The EPA and industry allies appealed to the U.S. Supreme Court. Eight states filed friend-of-the-court briefs defending the Costle-EPA's dual definition. "They argued that netting [within-source trading] would reduce the scope of review of new sources in some jurisdictions, thereby increasing economic pressure on other jurisdictions to adopt it."[115]

In 1984. in *Chevron v. NRDC*, the Court upheld the EPA's single-definition conception of non-attainment sources. The Court observed that neither the statute nor the legislative history was much help, but that EPA's interpretation was reasonable and supported by the public record.[116]

In the early years of the Reagan administration, the atmosphere for initiatives like the bubble was hardly propitious. On the whole, the Gorsuch administration preferred simple cutbacks in regulation and enforcement to the sort of arcane reform which the bubble represented. Nor was its fate improved by the judicial wranglings that were not settled until near the end of Reagan's first term. Nonetheless it survived and prospered.

By early 1985, more than 200 bubbles had been approved, proposed, or were under development, in twenty-nine states. Total savings were estimated to be more than $800 million. The primary metals industry, especially steel, was the main beneficiary. "In the future, as more bubbles are approved at the state level, particularly for volatile organic compounds (VOCs), this leading position of the steel industry in bubbling almost certainly will decline." Of the forty-seven bubbles directly approved by EPA, eighteen were for VOCs, seventeen for particulates, and twelve for sulfur dioxide. About half of EPA's bubbles were in attainment areas and half in non-attainment areas.[117]

A major cause of the bubble's survival was the sustained attention it received from the regulatory reform staff of the Office of Policy and Management under the direction of Michael Levin. Unlike so many other EPA offices, this one was not decimated by reductions in force and voluntary resignations during the Gorsuch regime. The core staff remained in place, providing continuity despite the changeovers from Gorsuch to Ruckelshaus, to Thomas. It marketed the program internally and provided the institutional memory that was essential for the survival of such a technically complicated program.[118]

SUPERFUND

Implementation of Superfund proved the most controversial of all the policies of the Gorsuch regime and was ultimately the cause of her demise. Congress blamed her for the extremely slow rate of progress, accusing her of malfeasance as well as ineptitude. Whatever the merits of those accusations, and they appear to be substantial, the design of the Superfund law would have presented severe challenges even to the most able, law abiding, and aggressive administrator. As we have seen, the fund was too small to provide uniformly high levels of cleanup at all possible sites, yet the law provided no clear basis either for choosing among sites or determining the cleanup level at a given site.

The law's drafters had hoped that the strain would be eased by the stringent liability provision that would encourage potentially responsible parties (PRPs) to settle voluntarily.[119] Unfortunately, the law's enforcement provisions were not drafted to encourage such settlements, and PRPs had no incentive to do so until ordered by the agency. The complex negotiations that resulted were inevitably time consuming. Rather than face long years of litigation, EPA proved willing to accept settlements that critics claimed were too small, and to agree to cleanup levels that some considered insufficiently stringent.[120]

In the wake of Gorsuch and Lavelle's departure, public and congressional clamor for speedier cleanup emboldened the new agency leadership to ignore the letter of the law that required protection of the fund. William Hedeman, director of EPA's Office of Emergency and Remedial Response announced that the Superfund program would henceforth be managed in "an aggressive way . . . regardless of the impact on fund resources."[121] Presumably, if the fund were depleted the agency would simply go back to Congress and ask for more.

Returning to "shovels first," cleanup would commence before negotiations with the PRPs were begun. States would no longer be required to provide ten percent of the cost of the initial planning and design work, thus enabling EPA to proceed immediately. And, regional administrators would be granted the discretion to authorize emergency actions costing up to $250,000 without receiving headquarters approval.[122]

The EPA also took several steps to encourage voluntary settlements. Under certain circumstances, when only some of the PRP's agreed to a settlement, they would now be asked to put up only a share of the total cost, with the remainder covered by the fund.[123] In addition, EPA now proposed to grant PRP's a release from future liability claims where industry consented to spend enough to minimize the risk that current clean up would prove to be inadequate.[124]

The agency also agreed to provide voluntary settlers with another form of protection from future liability, known as a release from contribution. After the government has reached a partial settlement with some PRPs, it often sues the nonsettling PRPs to obtain the rest. These nonsettlers are then entitled to sue the voluntary settlers, claiming that the settlers had not paid their fair share. The agency proposed eliminating this contingency by agreeing in advance to reduce any judgment to the extent necessary to make sure that settlers would not have to make any additional contribution.[125]

How Clean is Clean

The "how clean is clean" question, ignored during passage of the Superfund, became one of the program's most controversial aspects. Environmentalists and their congressional allies were convinced that, as part of its "save the fund strategy," EPA was skimping on cleanup. They sought to limit agency discretion by insisting that cleanup meet all the ambient standards set forth in the major federal air, water and toxic substances statutes.[126]

In February 1985, EPA proposed a revised version of the National Contingency Plan which committed the Superfund cleanup program to meet those standards. The revision, however, also provided the agency with considerable room to exercise discretion. It could waive the requirements if it found them to be technically impracticable, if the effort to impose them created unacceptable health or environmental effects, if they were excessively costly, or if there were " over-riding public interest concerns" in not applying them.[127]

Superfund Reauthorization

In 1985, the Congress was impelled to reconsider Superfund because the fee that served as the source for the fund was due to expire. The EPA's draft called for an increase in the size of the fund from 1.6 billion to 5.3 billion dollars to be spent over a five-year period. This was far smaller than the 7.5 billion dollars that the Senate was recommending or the 10.37 billion dollar figure in the House version.[128]

The draft also was designed to encourage more state participation. The states had shown considerable reluctance to enter into cooperative agreements with EPA, which would enable them to take the lead in fund-financed cleanups. They were also proving recalcitrant in providing legal dump sites to house the material removed in cleanup, and many of them had been unwilling to adopt new revenue sources to pay for their share of cleanup costs. In response, the EPA draft abolished the preemption clause forbidding states from using the same taxing methods used by the federal fund.[129] It also allowed states to propose a plan covering all Superfund sites and negotiate with the agency on this comprehensive basis.[130] On the other hand, EPA proposed raising the states' share of construction costs from ten to twenty percent. In addition, states would be required to either show that they could adequately provide for their disposal needs or face the loss of all federal cleanup assistance.[131]

While the Senate essentially accepted EPA's approach, the House version provided the agency with far less discretion. It required the agency to expand the list of priority sites to 1600 by January 1, 1988, and to begin cleanup studies at 925 sites within five years. Work was to begin on a total of 600 sites by 1990, with a specific number to be initiated in each successive year.[132] It also narrowed the conditions under which EPA's proposed cleanup standards could be waived. Finally, it required EPA to focus on drinking water contamination when listing priority sites and for establishing emergency response plans. Both the House and the Senate refused to increase the state's share of construction cost.[133]

In July 1986, after a protracted conference, the House and the Senate finally agreed on a compromise. It was much closer to the House than the Senate version. It called for a 9 billion-dollar fund and retained the uniform cleanup standards and the timetables for starting cleanups.[134]

A separate House-Senate conference was called to deal with the highly contentious funding issue. The EPA had proposed retaining the feedstock fee and a modest expansion of the fund by a per-ton waste management tax. The House imposed significantly higher rates on feedstocks and on petroleum.[135] The Senate proposed no increase in the feedstock fees, rejected the waste management tax, and added a value added tax of 0.08 percent on the sale or lease of all manufactured products, with the exception of those produced for export.[136]

The final revenue package designed to raise the enormous sum of 9 billion dollars was a compromise. Instead of a value added tax, there was a 2.5 billion dollar income tax on all corporations whose income exceeded 2 million dollars a year. An additional 2.75 billion dollars was raised through increases in oil excises, with importers paying more than domestic producers. The chemical feedstock fee that had financed the original program was raised by 200 million dollars, to 1.4 billion dollars. The government would contribute 1.25 billion dollars from general revenues and an additional 600 million dollars was to come from fund interest income and from cost recovery from responsible parties. The new program for cleaning up leaking underground storage tanks would be paid for by a separate tax of 0.1 cent on motor fuels.[137]

The final Superfund bill quickly and easily passed both houses of Congress. President Reagan restated his misgivings about using a broad-based tax as a source of revenue for this program. Nevertheless, he signed the bill into law on October 17, 1986.[138]

THE INTERAGENCY LIAISON GROUP (IRLG)

To understand what happened to carcinogenicity assessment during the Reagan Administration, it is necessary to track two sets of developments. One represented a more or less lineal descendant of IRLG—a series of interagency efforts to develop a common approach to carcinogen policy. The other took place within EPA itself, and involved issuance of new risk assessment guidelines.

Central to both of these developments was John Todhunter, EPA's assistant administrator for Pesticides and Toxic Substances. Soon after Reagan took office, EPA was faced with the question of what to do about formaldehyde. In November 1979, a laboratory study of formaldehyde in rats and mice had shown a clear excess in the number of tumors in exposed rats. The Consumer Product Safety Commission established a scientific panel to report on human health implications. It reported a year later that the compound should be "presumed" to pose cancer risks despite the lack of any human epidemiological evidence. This was exactly what the IRLG guidelines required in such cases.[139] Further action was stalled by the 1980 election. But, when Gorsuch took over in May 1980, she faced a recommendation by the Office of Toxic Substances staff that the compound be designated a rule-making priority under section 4(f) of the Toxic Substances Control Act (TSCA). She directed Deputy Administrator John Hernandez to explore the situation. This led to a series of meetings, nominally scientific, organized by Todhunter that summer. These were later to prove highly controversial because industry lawyers and other industry invitees were present.

Todhunter wrote a memorandum to Gorsuch in February of 1982 recommending against the 4(f) designation and calling into question the IRLG guidelines.[140]

He explicitly raised the possibility that not all "carcinogens are alike,"and should not be treated alike. Some were "promoters" not "initiators." Some caused mutation (i.e., were "genotoxic") but others did not (i.e. were "epigenetic"). The thrust of such distinctions was that some carcinogenic compounds posed little risk in isolation or at sufficiently low doses.[141]

The EPA also issued new guidelines for assessing carcinogens in the context of its effort to develop water quality regulations.[142] This document continued the line taken in the Todhunter memorandum arguing against the linear extrapolation of risks from high to low doses for epigenetic carcinogens and emphasizing the importance of human epidemiology over animal experiments as the basis for regulatory decision-making. Also important was its emphasis on the "strength of evidence " approach that had been dropped from the IRLG draft during its development. This approach, which maintained that information varied in quality for various suspect chemicals and that different ones should be treated differently, was now explicitly recommended.[143]

A parallel set of activities was occurring on the interagency front. Reagan had disbanded IRLG and the Regulatory Council soon after taking office. Meanwhile, the Office of Science and Technology Policy (OSTP) assembled an interagency committee, in which John Todhunter played a key part, for the explicit purpose of updating the IRLG guidelines. The document they produced went even further than the EPA water guidelines in emphasizing the promoter/initiator and the genotoxic/epigenetic distinctions, and in suggesting the possibility of "safe levels" for at least some apparent carcinogens based on these distinctions.[144]

These efforts at "rolling back" the Carter Administration's cancer principles were significantly slowed by the scandals that enveloped Gorsuch's EPA (in which Todhunter's memorandum on formaldehyde played a part).[145] Under Ruckelshaus' leadership, yet another set of EPA guidelines were drafted in 1984 and made final in 1986.[146] To some extent these guidelines moved back toward the IRLG approach in that they downplayed the possibility of threshold responses and accepted the use of linearized multi-stage models as the best estimating technique. Still, the return was only partial. Greater emphasis was placed on epidemiological than on bioassay data as a source of information and an elaborate "strength of evidence" classification scheme was established.[147] Similarly, more emphasis was placed on the use of pharmecokenetic and mechanistic information to try to understand actual cancer processes. Much of this change, like others at EPA, had come in response to congressional criticism. Yet, the fact that the "swing" back to the IRLG approach was only partial is as interesting from our perspective as the fact that the oscillation occurred at all.

On the interagency front, there was a parallel development. OSTP produced another, more moderate draft of its 1982 document and it was quite well received.[148] Following IRLG precedent it not only appeared as a government publication but was also published by a professional journal.[149]

Despite these developments, substantial variety still exists in the way different agencies treat the problem of determining carcinogenicity. Variations in mission, in legislative authority, in leadership strategy, and in technical judgment have made the initial IRLG goal elusive. On the other hand, there is no question that far more attention is being paid to achieving interagency consistency on these matters than existed before IRLG.

Finally, public education about these issues remains rudimentary. Although the "strength of evidence" system explicitly distinguishes among cases that are "proven," "possible," and "unknown," little effort has been devoted to explaining to the public the limits on EPA's ability to know for certain what is and what is not safe. True, some EPA policy tries to take these distinctions into account, but their import has not been fully conveyed to those outside the agency.

CONCLUSIONS

This overview of the Reagan years and its account of how the Congress and the executive behaved serves to reinforce the importance of strategy, deliberation, integration and accountability. The government's strengths and weaknesses in these four areas help to explain its performance with regard to the evaluatory criteria that we developed in Chapter One: promoting civic education, and building institutional capacity, responsiveness to the public, fidelity to the technical merits.

Strategy

The importance of a clear strategic analysis for facilitating public discussion is highlighted by contrasting the recent histories of RCRA and Superfund. The RCRA reauthorization debates led to a clear statement of the priorities to be accorded various disposal options. As a result, the subsequent public debate about these matters has been more realistic and coherent. In Superfund, the use of massive funds to avoid priority setting has left the public unaware of the hard choices that await.

Deliberation

EPA's recent treatment of risk assessment has increased the sophistication of public discussion. Todhunter did focus attention on critical assumptions regarding cancer risk in a way that has led all parties to pay much more attention to them and to their implications. Similarly with ozone: questions like, what is a health effect, and, who should be protected, are now part of the ongoing debate. Indeed, both Ruckelshaus and Thomas played a very constructive role in reshaping these issues. And, the steady improvement in the science confirms the wisdom of the Clean Air Act requirement that air quality standards be regularly reconsidered.

Integration

Both Ruckelshaus and Thomas facilitated intra-agency deliberation by choosing subordinates who could function as a general management team rather than as ambassadors of interest groups. This strengthened management also enabled the administrator to spend less time mediating internal squabbles and more time formulating strategy and representing the agency to the wider world.

Integration among agencies was not as successful. The "shootout" between Stockman and Ruckelshaus over acid rain was more a pluralist fantasy-made-real than an effective way to structure a consideration of this important problem. Similarly, the history of the steel industry subsequent to Tripartite shows a continued inability to focus on the right questions. The older steel companies, and the union, have kept declining for reasons that have nothing to do with the air pollution costs those negotiations sought to avoid.

Accountability

Overall, however, the Reagan administration was more integrated than its predecessors. It spoke with one voice and its priorities were clear. But this very unity, which should have aided accountability, actually served to undermine it. By often ignoring the spirit, if not the letter, of environmental laws, the Reagan administration denied its accountability to the Constitution. As chief executive, the president's first responsibility is to execute the law. If he objects to it, he is free to use all appropriate forums to try to get it changed. But, he is not free to vitiate it through inattentiveness or subversion. That is what the system of checks and balances is all about. The refusal of Congress to weaken environmental laws was a message to Reagan that, in this regard at least, his own preferences needed to be checked. Only after the excesses of the Gorsuch regime placed the EPA in political receivership did the administration abandon its zealous stance.

Gorsuch was similarly guilty. Her drive to please the president caused her to ignore the other constituencies to which she was accountable. She was inattentive to Congress and to the public, apart from certain business interests. She even ignored the civil servants in her own agency, shielding herself behind likeminded political appointees. Ironically, much of the good that the agency later accomplished occurred in reaction to these excesses.

Civic Education

The EPA did provide some important public education. The *Unfinished Business* report contained important lessons about the relationship between risk and agency expenditure. The IEMP provided similar insights regarding specific localities.

On the other hand, many unpleasant truths continued to be avoided. As the ozone compliance problem in Los Angeles illustrates, continued population growth and economic expansion will lead to increased pollution which will in turn require ever stricter regulation just to keep the air from getting worse. Greater pollution control expense is the inevitable price of better living. The EPA has done little to prepare the public for this harsh news. It has abetted the notion that the ozone problem can somehow be "solved" once and for all. In fact, barring a major war or an economic collapse, it will require ever greater efforts just to maintain the status quo.

Capacity Building

The success of the bubble policy during the Reagan years shows how important it is for EPA to develop and maintain its own institutional capacity. The atmosphere was not favorable toward regulatory reform in the Gorsuch years, but the Regulatory Reform Staff maintained its organizational coherence and was thus able to provide the institutional memory and the internal momentum needed to keep the bubble alive.

But this example is exceptional. The severe budget cuts and personnel reductions that occurred during that period struck many telling blows to agency capacity. These left the agency so emaciated and with such low morale that it had great difficulty implementing even its most routine responsibilities.

In such a context, even the more appealing of Gorsuch's reforms appear suspect. Delegating more program responsibilities to the states and assigning the enforcement task directly to the program offices would appear to improve the capacity of the agency to perform its mission. But, any structural change can be destructive if done for the wrong reasons. However defensible in the abstract, these changes had both the purpose and the effect of reducing regulatory vigilance.

Unfortunately, the response of Congress to these developments was not always constructive. In its zeal to limit executive discretion, it sought to write ever more specific performance standards and to impose ever more rigid timetables. The hammers in the revised RCRA law were a prime example of such an attempt to impose inappropriately detailed controls on EPA. By doing so, Congress risked seriously distorting the agency's priorities. Congress is in a poor position to consider the opportunity cost, in terms of personnel and budget, of rushing one set of regulations to completion.

Instead, Congress should have relied on its consent powers to force the appointment of responsible and responsive agency officials, and on its investigative powers to ensure that the inevitable agency discretion was being used in an acceptable manner.

Responsiveness to the Public

Under Lee Thomas, EPA moved beyond the focus on public health of the Costle years, to a broader span of ecological concerns, both global and local. This redefinition was more in keeping with the agency's original mandate and, as will be argued in Chapter Ten, enabled it to be more strategically responsive, that is, to ask questions capable of framing meaningful public debate and to broaden the president's political coalition.

Fidelity to The Technical Merits

We have already said a great deal about EPA's increased attention to scientific subtlety during the Ruckelshaus and Thomas years. We also find that its behavior regarding global ecological concerns displayed fidelity to the technical merits.

"The global environment" threatens to become to the 1990s what "toxics" were to the late 1970s and early 1980s. It is an issue in which scientific uncertainty is joined to dire possibilities in a way that could easily lead to hysteria among the public, and weak thinking among policy makers. So far, at least, EPA has resisted the impulse to exploit it in order to acquire massive new resources and responsibilities. We applaud its restraint. For the most part, EPA has focused on improving research. In view of how little science knows about global climatic and atmospheric processes, that is appropriate. Only with regard to one such issue, stratospheric ozone, has EPA taken more aggressive action. This exception seems reasonable given that the danger is relatively clearcut, the effects potentially irreversible, and the means for exerting control well understood.

NOTES

1. Norman J. Kraft and Michael Vig, eds., *Environmental Policy in the 1980s: Reagan's New Agenda* (Washington D.C.: CQ Press, 1984) 37.
2. Kraft and Vig, *Environmental Policy*, 35.
3. Robert Cameron Mitchell, "Public Opinion and Environmental Politics in the 1970s and 1980s," in Kraft and Vig, eds., *Environmental Policy*, 52–53.
4. *National Journal*, 22 Aug. 1981, 1515.
5. Jonathan Lash, Katherine Gillman and David Sheridan, *A Season of Spoils: The Story of the Reagan Administration's Attack on the Environment* (New York: Pantheon, 1984), 7.
6. Lash, et al. *A Season of Spoils*, 16–17.
7. Lash, et al. *A Season of Spoils*, 4.
8. Lash, et al. *A Season of Spoils*, 18, 41.
9. J. Clarence Davies, "Environmental Institutions and the Reagan Administration," in Vig and Kraft, eds., *Environmental Policy*, 145; Lash, et al. *A Season of Spoils*, 41–44.
10. *The National Journal*, 24 Nov. 1981, 1902.
11. Lash, et al. *A Season of Spoils*, 49.
12. Lash, et al. *A Season of Spoils*, 11–12.
13. On the slowness of administration appointments, see Chester A. Newland, "The Reagan Presidency: Limited Government and Public Administration," *Public Administration Review*, Jan./Feb. 1983, 4.
14. Newland, "The Reagan Presidency," 7. See also *The National Journal*, 7 Mar. 1981, 399.
15. *The National Journal*, 7 Mar. 1981, 399.
16. *National Journal*, 11 July 1981, 1245.
17. Newland, "The Reagan Presidency," 6; Davies in Vig and Kraft, eds., *Environmental Policy*, 149.
18. See the symposium on Executive Order 12291 in *Environmental Forum*, Sept. 1983, 31–34.
19. *The National Journal*, 14 Mar. 1981, 425.
20. Lash, et al. *A Season of Spoils*, 20.
21. Lash, et al. *A Season of Spoils*, 45–53.
22. Davies in Vig and Kraft, eds., *Environmental Policy*, 153.
23. Michael Fix, "Transferring Regulatory Authority to the States," in George C.

Eads and Michael Fix, eds, *The Reagan Regulatory Strategy* (Washington D.C.: The Urban Institute Press, 1984), 158.

24. "On Delegation to States: Promises and Perils," *The Environmental Forum*, Jan. 1983, 9.

25. Fix in Eads and Fix, *The Reagan Regulatory Strategy*, 170–171.

26. Robert V. Bartlett, "The Budgetary Process and Environmental Policy," in Vig and Kraft, eds., *Environmental Policy*, 129.

27. Lash, et al. *A Season of Spoils*, 60

28. For a good account of these events, see Lash, et al. *A Season of Spoils*, 73–81.

29. Ibid.

30. Lash, et al. *A Season of Spoils*, 77.

31. Lash, et al. *A Season of Spoils*, 80.

32. See "Environmental Protection Agency: Ruckelshaus Returns," Kennedy School of Government Case Program C16–85–638, Harvard University, Cambridge MA.

33. "Interview with Al Alm," *Environmental Forum*, Feb. 1984, 11.

34. "Ruckelshaus' EPA at One Year - How is it Doing?" *Environmental Forum*, June 1984, 9.

35. *National Journal*, 26 Mar. 1983, 660; *BNA Environmental Reporter*, 12 Aug. 1983, 614.

35. *National Journal*, 30 June 1984, 1258.

36. *BNA Environmental Reporter*, 29 July 1983, 519.

37. *Environmental Forum*, June 1984, 9.

38. *National Journal*, 30 June 1984, 1259.

40. *BNA Environmental Reporter*, 24 June 1983, 286–287.

41. *BNA Environmental Reporter*, 24 Feb. 1984, 1829.

42. "Risk Assessment/Risk Management," *Environmental Forum*, Sept. 1984, 20.

43. See "EPA Proposed Hazardous Air Pollutant Standards for Arsenic Emissions From Copper Smelters, Gas Plants" 48 FR 33112 (20 July 1983).

44. *BNA Environmental Reporter*, 22 July 1983, 462.

45. *BNA Environmental Reporter*, 22 July 1983, 455

46. *BNA Environmental Reporter*, 22 July 1983, 462.

47. *BNA Environmental Reporter*, 22 July 1983, 462–463.

48. *BNA Environmental Reporter*, 22 July 1983, 464.

49. For an account of these events, see Robert Reich, "Public Administration and Deliberation: An Interpretive Essay," *Yale Law Journal* 94 (1985):1617, 1632–1634.

50. Reich, "Public Administration and Deliberation," 1617, 1632–1634.

51. Quoted in "Ruckelshaus and Acid Rain," Kennedy School of Government Case Program C16–86–658, (Cambridge, MA: Harvard University, 1986), 3

52. *National Journal*, 4 Apr. 1981, 579.

53. "Ruckelshaus and Acid Rain," 1. The following account of the controversy is taken from this case study.

54. "Ruckelshaus and Acid Rain," 16.

55. "Ruckelshaus and Acid Rain," 17–18.

56. "Ruckelshaus and Acid Rain," 19.

57. Senate Committee on Environment and Public Works, *On the Nomination of Lee Thomas to Be Administrator of the Environmental Protection Agency*, 99th Congress, 1st Session, S Hrg 99–59, 6 Feb. 1985.

58. *On the Nomination of Lee Thomas to Be Administrator of the Environmental Protection Agency*, 54.

59. *National Journal*, 18 May 1985, 1168–1169.

60. See Richard Doll and Richard Peto "The Causes of Cancer: Quantitative Esti-

mates of Avoidable Risks of Cancer in the United States Today," *Journal of the National Cancer Institute* 66 (Jun 1981):1193–1308; and Bruce Ames et. al "Ranking Possible Cancer Hazard, *Science* 239 (17 Apr 1987).

61. Environmental Protection Agency, Office of Policy Analysis, *Unfinished Business: A Comparative Assessment of Environmental Problems*, Feb 1987, I-28 - I-34. (Hereafter referred to as *Unfinished Business*.)

62. *New York Times*, 7 July 1988, 1.

63. *National Journal,* 10 May 1986, 1106–1110.

64. See for example "World Leaders Call for Drastic Action to Slow Earth's Warming," *Boston Globe*, 28 June 1988, 1; "The Heat is On: Calculating the Problem of a Warmer Planet Earth," *New York Times*, 26 June 1988, IV-1; "Will the Planet Remain Habitable?, *New York Times*, 30 June 1986, A19; Senate Committee on Environment and Public Works, Subcommittee on Environmental Pollution, *Ozone Depletion, The Greenhouse Effect and Climate Change*, 99th Congress, Second Session, 10, 11 June 1986, S. Hrg 99–723; Senate Committee on Environment and Public Works, Subcommittee on Toxic Substances and Environmental Oversight, *Global Warming*, 99th Congress, First Session, 11 Dec. 1985, S. Hrg 99–503.

65. See Stephen H. Schneider and Stanley L. Thompson, "Future Changes in the Atmosphere," in Robert Repetto, ed., *The Global Possible: Resource Development and the New Century*, (New Haven: Yale University Press, 1985), 400–406, and *The National Journal* 12 July 1986, 1750.

66. "Future Changes in the Atmosphere," 401.

67. *National Journal*, 1 Nov. 1986, 2638.

68. *National Journal*, 1 Nov 1986, 2638.

69. *The New York Times*, 26 July 1988, 4–1.

70. "Future Changes in the Atmosphere," 411–417.

71. *The Boston Globe*, 28 June 1988, 1; *The New York Times*, 26 June 1988, IV - 1.

72. For a cogent critique of current scientific capacity to forecast the level of future CO2 emissions, see Bill Keepin, "Review of Global Energy and Carbon Dioxide Projections," *Annual Review of Energy 1986*, 357–392.

73. See Brooks et. al., "Ice Sheet Topography by Satellite Altimetry," *Nature* 274 (1978):539–543. For an excellent discussion of computer modelling as applied to projections of global CO2 emissions, see "Review of Global Energy and Carbon Dioxide Projections."

74. "The Global Challenge Fund," insert attached to *World Resources Institute Annual Report 1987* (Washington D.C.: World Resources Institute, 1987).

75. "The Global Challenge Fund," 1, 6–7.

76. See *Global Warming*, 3–6; *Ozone Depletion, The Greenhouse Effect and Climate Change*, 8–10; and "To request the President to take appropriate actions toward the establishment of a cooperative international research program with respect to the greenhouse effect," Senate Concurrent Resolution Unnumbered, 100th Congress, First Session, 6 Jan. 1987.

77. *The National Journal*, 12 Dec. 1987, 3142.

78. Lee M. Thomas, "Environmental Regulation: Challenges We Face," *EPA Journal* 14 (Mar. 1988):2–3.

79. *Unfinished Business*, xv.

80. *Unfinished Business*, xv.

81. *Unfinished Business*, I-iv - I-xii.

82. *Unfinished Business*, I-xiii - I-xiv.

83. *Unfinished Business*, I-xv - I-xvi.

84. For a general review of the IEMP, see "Review of the Office of Policy, Planning

and Evaluation's Integrated Environmental Management Program," Integrated Environmental Management Subcommittee, Science Advisory Board, EPA, July 1987. See also, Marc Landy, *Environmental Federalism: Can It Work?*, Report Commissioned by the Office of Policy Analysis, USEPA, Jan. 1988.

85. *Environmental Federalism: Can It Work?*, 12–13.

86. "Review of the Office of Policy, Planning and Evaluation's Integrated Environmental Management Program," 2–3, 14–22.

87. Philip Shabecoff, "Ozone Pollution is Found at Peak in Summer Heat," *The New York Times*, July 30, 1988, p. 1; Rochelle L. Stanfield, "The Ozone Deadline,"" *National Journal*, September 13, 1986, pp. 2170–2174.

88. Eads served on the Council of Economic Advisors in the last year and a half of the Carter Administration. In that capacity Eads chaired RARG and witnessed first hand the review of the carbon monoxide ambient standard.

89. Clean Air Act Oversight, Hearings before the Committee on Environment and Public Works, US Senate, 97th Congress, first session, 9 June, 1981, p. 198.

90. Marjorie Sun, "Tighter Ozone Standard Urged by Scientists," *Science,* Vol. 240 (24 June 1988), 1724–25.

91. Ibid.

92. Ibid., p. 1725.

93. Rochelle Stanfield, "The Ozone Deadline," *National Journal*, no. 36, 13 Sept. 1986, 2171.

94. Mark Thompson,"Environment: Fighting for Cleaner Air," *Atlantic,* Sept. 1988, p. 22.

95. NY Times 19 Dec. 1988, A-1

96. Rochelle Stanfield, "The Ozone Deadline," p. 2171.

97. House Committee on Energy and Commerce, Subcommittee on Oversight and Investigations, *EPA Enforcement and Administration of Superfund,* 97th Congress, 1st Session, 16 Nov. 1981, Committee Serial No. 97–123, 2, 21.

98. *EPA Enforcement and Administration of Superfund*, 21.

99. *EPA Enforcement and Administration of Superfund*, 33.

100. Steven Shimberg, "The Hazardous and Solid Waste Amendments of 1984: a,k,a "Rita" (as in Lavelle) The Recycling Incineration Act of 1984," *Environmental Forum*, Mar. 1985, 9.

101. Lash, et al. *A Season of Spoils*, 17, 63–66.

102. Shimberg, "Rita," 10.

103. Senate Committee on Environment and Public Works, *Solid Waste Disposal Amendments of 1983*, S Rpt 98–284, 28 Oct. 1983, 2.

104. "Who's Afraid of Hazardous Waste Dumps? Not Us, Says the Reagan Administration," *National Journal*, 29 May 1982, 954.

105. Shimberg, "Rita," 12.

106. Shimberg, "Rita," 16–17.

107. *Hazardous and Solid Waste Amendment of 1984*, House Conference Report 98–1133, 3 Oct. 1984, 5.

108. Shimberg, "Rita," 11.

109. William Serrin, "The Shattered Steelworkers," *The New York Times*, July 29, 1986, p. D1.

110. A.H. Raskin, "The Steelworkers: Limping at 5)," *The New York Times*, June 15, 1986, p. 1; David Warsh, "Economic Principles: The Relocation and Reshaping of US Steel," *The Boston Sunday Globe*, August 11, 1985, p. 75.

111. David Warsh, "Economic Principals: Short Primer on Steel Industry," *The Boston Sunday Globe*, August 3, 1986, p. 69.

112. In 1971, in its first definition of source—for New Source Performance Standards (NSPS), the EPA stayed close to the statute's relevant phrasing: "any building, structure, facility or installation" emitting an air pollutant. But in 1975, the EPA changed its mind. The nonferrous smelting industry (and the Commerce Department) wanted a source to be an entire plant, so that only an increase in total emissions would be subject to NSPS control. EPA agreed: at an existing site a source would be a combination of facilities. This approach, the first bubble, was turned back by the D.C Circuit Court in 1978 (*ASARCO, Inc. v. EPA*, 578 F.2d 319).

But a decision about NSPS did not settle matters in other air programs. In the Prevention of Significant Deterioration (PSD) program in relatively pristine areas, entire plants were treated as sources. In 1979, the D.C. Circuit Court upheld this policy, finding in (Liroff's words) that the EPA "had discretion to define the component terms of *source* and, when developing definitions, could consider differences among the purposes and structure of the PSD, NSPS, and other programs of the Clean Air Act." (Richard A. Liroff, *Reforming Air Pollution Regulation: The Toil and Trouble of EPA's Bubble* [Washington, D.C.: The Conservation Foundation, 1986], p. 120.) The case was *Alabama Power Co. v. Costle*, 636 F.2d 323.

113. David Hawkins (who had meanwhile returned to the NRDC) pointed out three of them. First, the source could now avoid the Lowest Achievable Emissions Rate (LAER) limitation in new source review. Second, the new rule eliminated the enforcement leverage of the Part D offset policy, namely, that all plants owned by the same company in the state have to be in compliance for the permit to be approved. Third, this expansion of trades took them out of the SIP review process, removing them from the reach of another enforcement tool: the condition that a state have an adequate SIP or face a ban on new permits.

114. *NRDC v. Gorsuch*, 685 F.2d 718,cited in Liroff, *Reforming Air Pollution Regulation*, p. 130.

115. *Ibid*. p. 131.

116. *Chevron vs. NRDC*, 104 S.Ct. 2778 (1984).

117. Liroff, *Reforming Air Pollution Regulation*, pp. 62–63, citing official EPA sources.

118. Liroff stresses this point in *Reforming Air Pollution Regulation*. See also Michael Levin, "Getting There: Implementing the 'Bubble' Policy," in Eugene Bardach and Robert A. Kagan, eds., *Social Regulation: Strategies for Reform* (San Francisco: Institute for Contemporary Studies, 1982).

119. Steven Cohen, "Defusing the Toxics Time Bomb: Federal Hazardous Waste Programs," in Vig and Kraft, eds., *Environmental Policy*, 284.

120. Cohen in Vig and Kraft, eds., *Environmental Policy*, 284–288.

121. "A Conversation With Superfund Chief Bill Hedeman," *Environmental Forum*, Aug. 1983, 8.

122. "A Conversation With Superfund Chief," 8–9.

123. Richard H. Mays, "EPA's Superfund Settlement Policy," *Environmental Forum*, Feb. 1985, 7–17.

124. Lee Thomas, Senate Committee on the Judiciary, *Superfund Improvement Act of 1985*, 99th Congress, 1st Session, Serial No. J-99–30, 7 June 1985, S. Hrg 99–415, 29.

125. Mays, "EPA's Superfund Settlement Policy," 12.

126. *BNA Environmental Reporter*, 1 Feb. 1985, 1595.

127. *BNA Environmental Reporter*, 1 Feb. 1985, 1595.

128. "Comparison of Superfund Reauthorization Bills," *BNA Environmental Reporter*, 14 Feb. 1986, 1950

129. "Comparison of Superfund Reauthorization Bills," 1914.

130. "Comparison of Superfund Reauthorization Bills," 1925.

131. "Comparison of Superfund Reauthorization Bills," 1905–1906.

132. "House Approves $10 Billion Dollar Bill For Hazardous Waste Cleanups," *Congressional Quarterly Weekly Report*, 14 Dec. 1985, 2622.

133. "Comparison of Superfund Reauthorization Bills," 1901, 1905.

134. *BNA Environmental Reporter*, 1 Aug. 1986, 483–484.

135. "Comparison of Superfund Reauthorization Bills," 1947–1948.

136. "Comparison of Superfund Reauthorization Bills," 1949.

137. *BNA Environmental Reporter*, 17 Oct. 1986, 908–909.

138. *BNA Environmental Reporter*, 24 Oct. 1986, 955.

139. Graham, Green, and Roberts 10–34

140. J.A. Todhunter, "A Review of Data Available to the Administrator Concerning Formaldehyde and Di(2–ethlyhexyl)Phtalate (DEHP"), memo to Anne Gorsuch, February 10, 1982. See also F.P. Perara and C. Petito, " Formaldehyde: A Question of Cancer Policy" *Science* 216 1982, 1285–1291.

141. See Mark E. Rushevsky, *Making Cancer Policy*, (Albany N.Y.: State University of New York Press, 1986), 37–44. 120–122. See also U.S. Congress,Office of Technology Assessment, *Identifying and Regulating Carcinogens*, OTA-BP-H-42 (Washington D.C.: Government Printing Office, Nov. 1987) 36ff.

142. USEPA, "Additional USEPA Guidance for Health Assessment of Suspect Carcinogens with Specific Reference to Water Quality Criteria" (draft, June 21, 1982) reprinted in *PCB and Dioxin Cases*, Hearings before the Subcommittee on Oversight and Investigations, Committee on Energy and Commerce, U.S. House of Representatives, 19 Nov. 1982, 105–124.

143. *Making Cancer Policy*, 122–125.

144. U.S. OSTP, "Review of the Mechanisms of Effect and Detection of Chemical Carcinogens,"(Washington D.C.: 1982). See also *Making Cancer Policy*, 125–129.

145. See *Making Cancer Policy*, 133–150, and M.Wines, "Scandals at EPA May Have Done in Reagan's Move to Ease Cancer Controls," *National Journal*, 15(25), 18 June 1983, 1264.

146. USEPA, "Proposed Guidelines for Carcinogen Risk Assessment: Request for Comments," *Federal Register* 49, 23 Nov.1984, 46304 and "Guidelines for Carcinogen Risk Assessment," *Federal Register* 51, 1986, 33992.

147. "Review of the Mechanisms of Effect and Detection of Chemical Carcinogens," 69.

148. OSTP, "Chemical Carcinogens: A Review of the Science and its Associated Principles," *Federal Register* 50 14 May 1985, 10372. See also F. Marshall, "Carcinogenesis Without Controversy" *Science*, 224, 8 June 1984, 1078 and *Making Cancer Policy* 143–145.

149. It was printed in *Environmental Health Perspectives*, 67, 1986, 201.

9

The Wrong Questions and Why

EVALUATION

This chapter ascribes EPA's failings to its penchant for asking the wrong questions, and searches for the sources of that tendency. In the next, and concluding, chapter we provide our own view of what the right questions might be.

In Chapter 1 we outlined four criteria for evaluating EPA's performance: responsiveness to the public, fidelity to the technical merits, civic education and capacity building. Of these, EPA's most serious shortcomings involved public education. Whether it was deciding how much to clean up a dump site, classifying chemicals as carcinogens, or choosing what effects of ozone to avoid, EPA did not provide the public with a clear and complete account of the issues these decisions raised. Citizens were encouraged to believe that questions were much simpler than in fact they are.

Typically, no one exposure level represents the dividing line between safety and harm. More is worse, and individuals differ substantially in their susceptibility to different hazards.[1] Determining how much protection to provide, and to whom, is thus not a purely technical choice. Yet EPA repeatedly treated "safety" as if it were a scientific notion definable by experts, rather than a social construct necessarily based on values as well as science.

In addition, the agency's efforts to muster support contributed to an unwarranted level of public anxiety. The best studies suggest that environmental hazards account for only a small percentage of the total cancer burden.[2] The health consequences of hazardous waste, for example, are quite small compared with the effects of cigarette smoking, and deserve to be treated as such.[3]

In this connection, EPA did not help citizens understand that some "cancer clusters" will inevitably occur just due to bad luck.[4] Not every such cluster is necessarily caused by "something." It might just be an unfortunate coincidence. Rather than clarify this point, EPA allowed toxic waste to become the functional equivalent of witchcraft in colonial Salem—the hidden evil force that accounts for all of our troubles.

The public has also been misled about the costs of cleanup. Costs have routinely been discussed as if "someone else" could be made to pay for them. The Superfund feedstock fee scheme was designed to convey the impression that costs would be borne by industry. In fact, most such costs are passed on to the public in the form of higher prices.[5] Nor are the unidentified costs merely financial. Those who live near Superfund sites have not been made to consider that the contaminated soil removed from their neighborhood will have to be put in someone else's neighborhood.

These educative failures were not entirely EPA's fault. Legislative draftsmen, as well as advocates, use the rhetoric of "rights to safety" because it seems to establish an unconditional claim on public attention and resources.[6] But such formulations fail to remind us that perfect safety is generally impossible (or in a few "lucky" cases, merely outlandishly expensive).

The miseducation was political as well as technical. The EPA failed to teach the virtues of citizenship by the example of its own behavior. Instead, it communicated the message that each interest or agency can appropriately pursue its own goals with no broader view of the national interest. When EPA acts like just another interest group, it contributes to the debased notion that there is no distinction between how people do or should behave when making private or public decisions.[7]

We believe that most citizens do realize that their narrow self interest might conflict with what they themselves believe to be fair or just or in society's general interest. The answers that citizens give to policy questions, thus, depend, at least in part, on whether they are asked "What do you want" or "What do you think the society ought to do." Democratic political institutions must be judged, therefore, at least in part on the extent to which they pose policy questions in civic as opposed to self-interested terms. It is through the deliberative discussion of such questions that citizens develop, discover, and clarify their own understanding of their mutual rights and responsibilities.

The "pluralist" pursuit of narrow advantage is not all bad. The United States encompasses many regional, economic, and ideological interests. Any functioning political system has to accommodate that diversity—mobilizing consent in part by its responsiveness to such interest group claims. As de Tocqueville saw when he visited this country more than 100 years ago, the problem is to devise arrangements that meld such "self-regarding" concerns with the "other-regarding" perspective of the citizen.[8]

Instead, our cases often depict pluralism at its purest. In both Superfund and RCRA various powerful legislators extracted concessions—for example, regarding harbor dredging and Alaskan fisheries—without pretending that they were motivated by anything other than constituency interest. Fortunately, not everyone behaved this way, but there was almost no discussion of how to set limits on such self-seeking.

As RARG did during the ozone debate, economists characterize environmental disputes as choices between efficiency and irrationality.[9] As OSHA did during the IRLG cancer policy deliberations, environmentalists contend that it is morally wrong to choose between life and mere dollars.[10] We found few, if any, efforts by the adherents of either of these points of view to take the other seriously and attempt some sort of deliberative synthesis.

Combatants in these policy wars would no doubt retort that victory requires stubbornness and exaggeration, and that environmentalists need EPA to be their bureaucratic champion to countervail other agencies who champion the cause of economic interests. This view is mistaken on three counts. First, it ignores the necessity of giving citizens a sufficient understanding of environmental realities to enable them to participate effectively in political choice. Second, it ignores the possibility that sound ideas can come from suspect sources. As the bubble policy illustrates, industry, despite its self-serving posture, can be the source of important policy innovation. Third, it undermines institutional capacity by preventing the agency from establishing credibility with a broad range of those affected by its actions.

When Costle departed, many in business viewed EPA as the bureaucratic vanguard of the environmental movement, to be fought, or sued, but never to be trusted.[11] Gorsuch shared this perspective and her efforts to "discipline" the agency through wholesale budget and personnel reductions was even more destructive of institutional capacity. The enhanced credibility that the agency achieved under Ruckelshaus and Thomas suggests that institution building along the lines we advocate is quite possible.

Regarding fidelity to the technical merits, the cases show mixed results. The ozone standard decision is defensible if one is willing to put substantial value on avoiding relatively small health effects. In Superfund, RCRA, and steel, meritoriousness is harder to judge because the design of a whole program is at issue. In all three of these instances, a lack of strategic coherence led to serious flaws. True, some landfills were licensed, some waste sites cleared up, but progress was slow, episodic, and needlessly expensive. By compounding uniform "conservative" and "upperbound" assumptions, IRLG produced an unknown and yet highly variable tendency to overestimate risks.[12] Such an approach makes it difficult, if not impossible, for policy makers to determine, and be held accountable, for the degree of risk aversion society actually adopts in various situations.

We have argued that responsiveness to the public must be viewed in strategic terms. An agency's policy is responsive if it represents a coherent and effective synthesis that serves to sustain and enhance the coalition that brought it to power. At first glance, the Carter/Costle record appears favorable in this regard. Carter came to office with significant environmental support. Popular opinion seemed to back a relatively "tough" environmental position which the EPA adopted during the period we have reviewed.

By downplaying both economic and recreational concerns, however, the Carter EPA missed an opportunity to strengthen the Democratic Party coalition on which it ultimately relied. The Democrats depend heavily on the support of those of moderate and less than moderate means, who are far more dependent on public recreation than their richer counterparts. A strong commitment to improved public recreational opportunities might have given blue collar Democrats a reason for overcoming the growing alienation from the party, which drove so many of them into the Republican column in 1980.[13]

This same problem may continue to haunt the Democrats. Although the "pay at any price" cleanup philosophy still appears popular, it has the potential for causing serious voter backlash. For, if voters, especially those concerned about lost manufacturing jobs, become aware of how little improvement in their health they are actually obtaining from those expensive programs, they may well turn against the political party most closely identified with those programs.

These failures were not entirely Costle's fault. Much of the integration needed to produce a broader administration strategy could and should have occurred at a higher level. Still, EPA made a self-conscious decision to downplay ecological, recreational, and cost concerns. To that extent at least, Costle contributed to the Carter Administration's failure to be strategically responsive.

WRONG QUESTIONS

The problems we have encountered have important conceptual roots. Repeatedly EPA avoided dealing with reality as it actually was. By resorting to a series of oversimplifications, the agency tried to reshape the world to fit various legal, bureaucratic, and political imperatives.

The EPA repeatedly ignored the fact that experience is characterized by innumerable dimensions or attributes. Instead of trying to describe where events, objects, or processes were located along these various dimensions, the agency sorted everything into small numbers of distinct, mutually exclusive boxes.[14] Over and over again, EPA pretended that there were well-defined distinctions in the natural world and that these corresponded to the words used in statutes or regulations. In fact, linguistic categories typically fit that reality most imperfectly.[15] This necessarily produces gray areas, fuzzy boundaries, and ambiguous cases.

For example, while some individuals react more to ambient ozone than others, there is no distinct "most sensitive group." Similarly, the harmful effects of chemicals vary in many ways. There is no clear-cut definition of a carcinogen or a health effect. There is no strict dividing line between what is and what is not a hazardous waste.

By not acknowledging the arbitrariness of categories and their boundaries, EPA was able to avoid responsibility for some of its most potentially controversial decisions. Questions of social priority became transmogrified into matters of mere definition, the responsibility of technical experts. *Time and again, the question "What risks are prudent to avoid?" was reformulated as "What risks are in category X?"*

To be sure, EPA must make distinctions. It must choose what to regulate and whom to sue. But this needs to be done in a way that recognizes the absence of an unambiguous scientific basis for assigning dubious cases to one category or another.[16] Instead, decisions about how to treat such cases are partly policy decisions. Agencies have to consider what society would gain or lose by using one definition or another and by putting the burden of proof on different parties. Rather than acknowledging these choices, EPA too often became lost in a fog of meaningless and unresolvable definitional debate.

Also, EPA often exaggerated its ability to anticipate the consequences of proposed actions. It underestimated both the limits of its knowledge and the inherent unpredictability of natural processes. Consider, for example, the difficulties of predicting the impact of a proposed set of steel mill emissions controls on ambient air quality. The available atmospheric models used to forecast pollution levels on the basis of smokestack emissions are very imperfect, and meteorological conditions also vary. Controls that work most of the time will be insufficient in the face of unusually unfavorable weather conditions. Moreover, the understanding of technology and economics is also limited, and mechanical and social systems have their own uncertainties. Hence, one cannot be sure about what level of emissions will actually result from a given set of control devices. It is hard to know exactly how some proposed new technology will work over time, or how much steel (and hence how much pollution) a mill will actually produce ten years from now.

These limitations come together in particularly confounding ways with regard to dose-response relationships. Only a small fraction of those exposed to most pollutants develop any disease.[17] Thus any given dose produces only the *probability* of a response, and this probability is itself uncertain. In addition, response probabilities vary among individuals in ways that are not well understood.

In all these cases, EPA has to decide how much "insurance" to provide against both its own mistakes and nature's unpredictability by requiring controls that might turn out to be needlessly stringent (and costly). Yet the agency was very reluctant to explore these matters explicitly. It invoked the principal of "conservatism" without raising the question of how to decide just how conservative to be. Moreover, invoking unanalyzable professional judgment left the public very confused about the real nature of the decisions that were being made.

The EPA also asked the wrong questions by leaving out parts of the problem. By ignoring economics or the impact on state and local government, it could avoid addressing the problem of tradeoffs and the need to frame decisions in political, as opposed to technical, terms. The relevant legislative language was often irresponsible, as in the Clean Air Act's prohibition against cost considerations. But instead of calling attention to the dilemmas and difficulties such phrases produced, the agency sought refuge in that restrictiveness.

When EPA did worry about money, as in Superfund financing, it viewed the issue as a potential political complication, not a legitimate social concern. If costs could be hidden from consumers, or imposed on evil polluters, they were not worthy of agency attention. As a result, many citizens continue to believe that greater safety is "worth whatever it costs" in part because they still do not realize that they themselves will pay for it.

Thus, the public was given an unrealistic picture of the risks posed by the environment. From hurricanes and heat waves to volcanoes and sunsets, nature is neither fully controllable nor fully predictable. Americans have come to accept the risks associated with such man-made phenomena as automobiles, contact sports and crime. But in the realm of the environment, EPA portrayed itself as an authoritative expert who could offer perfect protection. It did not help the public to understand that environmental risks are only imperfectly understood, and that many cannot be entirely eliminated at any feasible cost.

EXPLANATIONS

To say that the root of the problem was conceptual is not to remove it from the realm of politics and management. On the contrary, the very misconceptions that led EPA to ask the wrong questions were abetted and encouraged by incentive systems and organizational structures that operated within the agency and within the executive branch as a whole.

Most EPA managers were, and were expected to be, advocates of a particular viewpoint. The allocation of responsibility among assistant administrators, and their own training and beliefs, reinforced this pattern. As David Hawkins said about defending the 0.08 ppm ozone standard, he felt a responsibility to present the environmentalist view.[18]

As a result, issues of policy were enmeshed in issues of power. For example, Costle could not put the agency fully behind the bubble policy without "tilting" toward Drayton—risking Hawkins' dissatisfaction and departure.

The formulation of questions became entangled in this competition. When Hawkins and Drayton fought over the ozone standard, they first battled over whether the relevant question included consideration of costs. They both knew that the latter issue would influence the former.

To demonstrate and reinforce their autonomy, units felt compelled to reject the suggestions of others, hindering strategic thinking. Outsiders who pushed the Office of Solid Waste (OSW) to think systematically about RCRA were met with skepticism and defensiveness. The OSW likewise ignored the municipal treatment program's concern to encourage the recycling of sewage sludge, and failed to consider options like source reduction because they fell outside its traditional bailiwick. As a result, the various "Red/Blue Team" and Crystal City meetings, which were intended to be deliberative and strategic, never produced that kind of thinking.

Units were as likely to question one another's facts (or motives) as to work together to develop a common problem definition, or to clarify the nature of their disagreements. For example, the Office of Planning and Management (OPM) was regularly accused of getting its facts from industry and of disloyalty for failure to defend agency positions. Similarly, Enforcement accused the regions of giving in to industry pressure on state implementation plan (SIP) revisions for steel. Such combat was not conducive to the acceptance of new ideas or to the synthesis of an integrated strategy.

This same pattern also undermined agency accountability. Instead of seeing themselves as responsible for broad consequences, agency professionals sought to carry out narrowly defined tasks. For example, the lawyers in the Office of the General Council saw their job as defending the agency against litigation. The *Federal Register* notices they wrote were briefs in defense of agency actions, not efforts to educate Congress or the public about how EPA was interpreting a conceptually inadequate law. These opportunities were not used to help outsiders subject the agency's decision to searching and sophisticated scrutiny.

Costle saw himself as a balancer, a responder to conflicting interests, albeit with a bias in favor of the environmental viewpoint. The elaborate "Red Border Review Process" he developed reflected the presumption that the partiality of individual units meant that they could not be trusted to make policy on their own. Occasionally Costle felt called upon to treat one of his subordinates as a general manager—to make him responsible for a project or activity that seemed to need special attention. But telling Tom Jorling, in frustration, to "put the pedal to the metal" and get out some RCRA regulations was not the same as developing a permanent integrative structure that fostered deliberation and accountability.

These structural problems were reinforced by Costle's lack of high level colleagues capable of constituting a general management team. In the absence of a strong deputy administrator, there was no one to help him with his integrative tasks. Because so much of his time had to be spent on external relations, it is no wonder that several of his subordinates complained of how long it could take to get a decision from the administrator's office.[19]

The same pattern, and the same results, characterized EPA's relationships with the rest of government. The executive branch, the Congress, and the courts all relied on the agency's expertise and, hence, gave it great latitude in shaping the terms of the policy debate. Yet, too often, the agency used this influence to advance relatively narrow institutional and programmatic interests. The Superfund drafting team did not even consider making it a block grant program controlled by the states.

The president and his men acted as if they wanted, or at least expected, Costle to play the advocate. They treated him much the way he treated David Hawkins. They sought to discipline EPA by using various outsiders—The Regulatory Analysis Review Group(RARG) or the Office of Management and Budget(OMB) for example—just as Costle used Drayton against Hawkins. Except for Eizenstat's intervention in ozone, there were few occasions in which EPA was pushed to take a wider view. Carter's reversal, under pressure from the Occupational Safety and Health Administration (OSHA), of the RARG-inspired cotton dust rule did not send a signal to his regulatory agencies that moderation would be rewarded.

The White House attitude toward EPA personnel matters conveyed the same message. It approved a slate of assistant administrators that ranged from several aggressive environmental activists to those with moderate proenvironment credentials and it promoted a local Georgia environmentalist to the crucial position of deputy administrator.

It was natural enough, then, for Costle to view his responsibility in narrow terms and be quite confrontational within the administration. If he was to "back down" on ozone, or on the definition of "source" in the Tripartite negotiations, or on liberalizing the bubble policy, then someone else was going to have to make him do so. If no one could or would, then he "won."

As a result, EPA pursued its own agenda. It initiated the Regulatory Council as a countermeasure to the economists. It battled OMB over the method of funding Superfund and avoided a government-wide task force in order to write its own bill. Indeed, EPA's leaders were prepared to take their own version to Capitol Hill if the White House failed to support key provisions. The EPA may not have been as "pro-environment" as some activists would have liked. Costle did increase the ozone standard and agree to stretch-out steel compliance. But, on the whole, EPA policy pleased the Environmental Defense Fund much more than it did the Chemical Manufacturers Association.

Given this understanding of its responsibilities, it is not surprising that EPA would so often seek to strengthen its own hand by asserting that issues were technical, and that it was the technical expert. If setting the ozone standard was a matter of discovering the compound's "health effects," then EPA could ignore Charles Schultze's cost arguments. If carcinogenicity assessment was a matter of science, than RARG's economic concerns were irrelevant.

The other parts of the government, including the president himself, were not able to counter-balance this advocacy behavior. Given his enormous span of control, the president's time was very scarce. His exposure to EPA was limited. Like Costle, he had very little in the way of a general management team to help him. Indeed, there was no one with line authority between Costle and Carter to assist the president with his integrative tasks. Stuart Eizenstat played this role in the ozone case, but he was only a staff member who lacked real authority.

Given their limited time together, the meetings between the president and Costle should have focused on strategy and priority setting. Instead, they were often preoccupied with details. This was Carter's failure, even more than Costle's. Concern with specifics was no substitute for an overall perspective. Vacillation, as on cotton dust, did not produce synthesis. Furthermore, the president's men sometimes worked actively to keep decisions (and hence responsibility) away from the White House.[20] This only increased the policy vacuum and expanded EPA's discretion. Costle did not want to be "managed" and Carter apparently did not know how to do so in any case.

The problems of integration were exacerbated because various White House staff groups, which might have fostered a broader perspective, often pursued their own agendas. The OMB sought to minimize budgetary impacts, RARG promoted economic efficiency, and the Office of Science and Technology Policy(OSTP) pursued increased spending on basic science. As advocates to the president, rather than his agents, these groups helped little with overall deliberative and integrative tasks.

This interagency fragmentation and weak central management helps explain the many attempts at *ad hoc* integration. Our story is replete with task forces, committees, programs and the like. In general, these devices had modest results. Such mechanisms require vigorous political leadership to constrain their participants. Tripartite did more than IRLG in this respect, but in both cases there was less deliberation than negotiation. The results were at best a narrow compromise and at worst an agreement to disagree. Nor did such arrangements insure that all relevant points of view were represented. Within IRLG, for example, no one sought to enhance economic growth or ensure cost minimization.

Exactly the same problems arose in EPA's relationship to other branches of government. Congress lacked both the expertise and the will to write legislation that took into account the full complexities of the issues at hand. Few members or staff were technically trained.[21] Thus, there were few congressional insiders to call attention to the vagueness of concepts like health or hazardousness.

Congressional committees used ambiguity to resolve conflict or defer hard decisions.[22] Members wrote "tough" language out of frustration or to please advocacy groups, ignoring how programs would work in practice. As a result, EPA repeatedly faced a difficult choice. Being honest would have meant focusing on the conceptual weakness in the legislation. Cooperative vagueness allowed it to "go along to get along." The price of the latter, however, was publicly asking and answering misleading and oversimplified questions.

RCRA, in particular, reflected all the limitations of the contemporary Congress: fragmented power, an activist staff, subcommittee entrepreneurship, and technical incompetence.[23] A very small number of people, functioning in remarkable isolation, produced an ambitious and ambiguous law. Most major issues were not even identified, still fewer were addressed and resolved.

The EPA's response was to view Congress in manipulative terms. Costle sought to deflect and placate Senator Muskie in the ozone hearings. To avoid counterattack, he did not call attention to faulty judgments or flawed assumptions.

The Superfund story is not quite as dismal, EPA provided data, generated options, and drafted language. In addition, multiple committee referrals—lamented by EPA staffers at the time—brought a wider and more diverse circle of legislators into the process and produced a broader discussion of some issues. Still the agency's legislative team avoided substantive discussion wherever that suited its tactical purposes, constricting the resulting debate. Issues of priority setting were not addressed, and the role of the states was diminished as Capitol Hill policy entrepreneurs bought support for centralization with federal dollars.

As a supervisor of implementation, the Congress was more an agent of special interests than a strategic overseer of agency choices. Members frequently criticized the agency for failure to clean up specific sites. But when it did not like a major decision, the ozone standard, for example, Congress failed to articulate its concerns in a way that either reflected or produced explicit and informative discussion.

When legislators posture for the evening news, and make promises that an agency cannot fulfill, but is reluctant to disavow, public understanding surely suffers. Instead of being a deliberative and integrative institution, the Congress, and especially its committees, became little more than just another set of interest groups.

The courts were no more able than Congress to insure that EPA asked and answered the right questions. Some advocates of "the New Administrative Law" have proposed a kind of "policed pluralism," with the courts doing the policing.[24] To prevent "capture" of federal agencies by powerful economic interests, they advocate various procedural safeguards designed to allow the courts to insure agency integrity.

Given the ambiguity of much legislative language, judges seeking to constrain agencies have made increased use of the nonvoted legislative record (debates, hearings, reports) to lend content to otherwise uninterpretable statutory provisions. Ironically, as Melnick has pointed out, this has only expanded the power of committees and their staffs—and hence the role of advocacy groups—within the legislature.[25] A relatively few individuals can influence these nonvoted parts of the record much more easily than they can shape legislative language.

Similarly, the requirement that an agency defend its reasoning in the rule-making process has had a mixed effect. It led EPA to be more careful, but it also encouraged tactical legal considerations to dominate what was published in the *Federal Register*. The results, in both ozone and RCRA, ranged from obscurity to misrepresentation.

Courts were seldom forced to confront the conceptual oversimplifications in the laws they interpreted, still less to devise remedies for them. The plaintiffs in any litigation are far more likely to attack an agency's application of a statute than the coherence of the statute itself. After all, the courts will not use the Constitution to overturn a law just because it is technically flawed. No one in the multisided ozone litigation, for example, attacked the Clean Air Act as unintelligible.

As RCRA showed, the courts were also not an effective guarantor of implementation. Apart from the draconian step of receivership, a federal judge has few tools to compel agency behavior. Problems of inadequate funding for, or poor management of, a regulatory program are not like outright defiance of civil rights laws, or like public actions that injure specific individuals, mental patients or prisoners, for example. Judges have taken over jails, schools and housing projects, but not regulatory agencies.

In sum, EPA's encounters with the Congress, courts and the rest of the executive branch were characterized by a lack of deliberation. The EPA was allowed to ask narrow questions and give narrow answers. Oversimplifications persisted. Too often, strategic issues were avoided or ignored and public debate was concentrated on incompletely or inaccurately formulated questions. The widespread public misperceptions about the nature of environmental issues that prevail today are due in no small measure to these wrong questions.

NOTES

1. See J.D. Brain, Barbara Beck, A.J. Warren and Rashid Shaikh, eds., *Variations in Susceptibility to Inhaled Pollutants: Identification, Mechanisms and Policy Implications* (Baltimore, MD: Johns Hopkins University Press, 1988).

2. See Richard Doll and Richard Peto, "The Causes of Cancer: Quantitative Estimates of Avoidable Risks of Cancer in the United States Today," *Journal of the National Cancer Institute*, 66 (Jun 1981):1193–1308; and Bruce Ames, et al. "Ranking Possible Cancer Hazard, *Science*, 239 (17 Apr 1987).

3. Ames, et al. "Ranking Possible Cancer Hazard."

4. For analysis of this problem see S.W. Lagakos, B.J. Wessen, and M. Zelen, "An Analysis of Contaminated Well Water in Woburn, Massachusetts." Department of Biostatistics Harvard School of Public Health, Study of Statistics and Environmental Factors in Health: Toxic Substances, Technical Report Number 3, Nov. 1984.

5. In general, the extent to which cost changes are reflected in price changes in an industry depends upon three factors: 1. the sensitivity of buyers to price changes,; 2. the extent to which production costs vary with volume; 3. the ability of firms in an industry to cooperate. While a complete analysis of the various industries involved is beyond the scope of this book, it seems unlikely that in highly concentrated industries, which most chemical feedstock markets are, producers would bear much of the cost.

6. For example see the preamble of the Clean Air Act.

7. For a fuller discussion of this point, see Marc Landy, "Policy Analysis as a Vocation," *World Politics* 33 (Apr. 1981): 468–484.

8. Alexis De Tocqueville, *Democracy in America Volume 1*, Phillips Bradley ed., (New York: Alfred Knopf, 1945), see especially 250–256.

9. See Chapter 3 for several different statements of the economic point of view; see also the essays in Robert Dorfman and Nancy Dorfman, eds., *Economics of the Environment* 1st ed (New York: Norton, 1972).

10. See Chapter 6. For an example of this point of view, see Alan Gewirth, "Human

Rights and the Prevention of Cancer," in Donald Scherer and Thomas Attig, eds., *Ethics and the Environment* (Englewood Cliffs, NJ: Prentice Hall, 1983), 170–176.

11. See for example the following editorials in the *Wall Street Journal*: "Regulate Nature," 3 Aug. 1978, 12, and, "EPA Runs Amok," 9 May 1979, 24.

12. In fact, there are cases in which some of the IRLG procedures may actually underestimate risks, especially when dose response functions are "superlinear," see *The Search for Safety*, 43.

13. See Marc Landy, "Muddy Waters," *Working Papers* 10 (Mar./Apr. 1983):60; and Robert Cameron Mitchell, "Public Opinion and Environmental Politics in the 1970's and 1980's," in Norman J. Vig and Michael Kraft, eds. *Environmental Policy in the 1980's* (Washington D.C: CQ Press, 1984), 54–64.

14. See especially Chapter 4.

15. For an excellent discussion of the problem of the imperfect fit between linguistic categories and reality, see George Lakoff, *Women, Fire and Dangerous Things* (Chicago: University of Chicago Press, 1987), especially chapters 12, 13.

16. Ibid.

17. See Louis J. Cassarett and John Doull, eds., *Toxicology: The Basic Science of Poison* 1st ed. (New York: MacMillan, 1975), Chapter 2.

18. See Chapter 3.

19. These opinions were expressed to the authors with the understanding that no attribution would be made of them.

20. These opinions were expressed to the authors by a high level Carter Administration official with the understanding that no attribution would be made of them.

21. Capitol Hill staffers are predominantly lawyers. For those on the committees and subcommittees which we investigated, see Harrison W. Fox Jr. and Susan Web Hammond, *Congressional Staffs* (New York: Free Press, 1977), 43–47; and Michael J. Malbin, *Unelected Representatives: Congressional Staff and the Future of Representative Government* (New York: Basic Books, 1980), 19–24.

22. For examples of this phenomenon see Eric Redman, *The Dance of Legislation*(New York: Simon and Schuster, 1973).

23. See Dennis Hale, ed. *The United States Congress* (New Brunswick, NJ: Transaction Books, 1984); Arthur Maass, *Congress and the Common Good* (New York: Basic Books, 1983).

24. See Richard B. Stewart, "The Reformation of American Administrative Law," Harvard Law Review 88 (1975):1667 and Kenneth Culp Davis, *Administrative Law of the Seventies* (Rochester NY: Lawyers Co-operative Publishing, 1976).

25. R. Shep Melnick, *Regulation and the Courts* (Washington D.C.: The Brookings Institution, 1983), 374–379.

10
Good Questions

In this chapter we offer our view of what questions the Environmental Protection Agency should be encouraged to ask and answer and how it can be encouraged to do so. The discussion is both prospective and prescriptive. It moves from the prior case studies of what EPA has done to a discussion of what it should do in the future. By refocusing the debate, we seek to help EPA, the Congress, and the general public to confront the difficult choices that lie ahead.

Good questions are educationally and strategically provocative. They focus agency and public attention on the choices that are available and the ethical issues these choices raise. By doing so, they can foster both moral responsibility and technical sophistication. Therefore, such questions must reflect critical features of reality that, we have argued, EPA repeatedly ignored. They must reveal the moral and scientific presuppositions that underlie designations such as carcinogen and hazardous substance. They must acknowledge that nature's inherent unpredictability renders outcomes uncertain. And, they must confront the limitations that the available scientific information places upon the ability to take effective action.

Good questions not only reflect scientific sophistication, they also clarify the link between seemingly technical matters and broad issues of rights and obligations. By forcing a consideration of the import and relevance of principles like justice, community, and liberty, they advance the ongoing debate about the nature and purpose of the American Republic.

In formulating specific questions, EPA also must consider its own strategic circumstances. It is but one of many units of government that share overlapping and intersecting responsibilities for natural resources, pollution control, and public health. It cannot and should not try to take the lead with respect to every issue that pertains to those matters. Instead, it has to clearly formulate the central questions regarding each relevant policy domain and then consider how to integrate those concerns, and its response to them, into an overall agency strategy.

Our cases reveal three such domains: (1) human health, (2) equity and income distribution, and (3) quality of life. We began with ozone, precisely because it illustrated the health concern so clearly, a concern that reappeared in RCRA, IRLG, and Superfund. The second domain involves the economic consequences of EPA actions. Examples include the Love Canal buy-outs, victim's compensation under Superfund, the effects of pollution control expenditures on the steel industry, and the impact of RCRA on the profitability of particular industries.

The third domain relates to the physical environmental circumstances of those urban and suburban areas where the vast bulk of the people live. Examples from the cases include the siting of RCRA landfills, and the question of how much to clean up particular Superfund sites. This domain is not as well represented in the book partly because of the cases we happened to select and partly due to the public health focus of EPA during the Carter years. Yet most of the land use, water quality, and siting issues so important historically to the environmental movement are in this arena.

We now seek to describe what better questions would be like for each of these three policy domains and to define what part each domain should play in EPA's overall mission.

HEALTH

There is no doubt that health concerns will and should continue to play a significant role in EPA's strategic thinking. Public concern over these matters remains high, and the agency has many statutory responsibilities with regard to health protection. The issue is, what question should EPA ask as it goes about setting standards and allocating enforcement resources.

The language produced by Congress, the pressures from the public, and the promises made by EPA have too often converged on the question "how can we make this or that safe." In contrast, we have argued repeatedly that, for a whole variety of reasons, this is the wrong question. First, in many cases, safety is simply impossible to achieve. For example, if ozone is "radiomemetic," that means there is some cancer risk at any exposure level. And, given natural backround levels, not to mention any fossil fuel combustion, zero exposure cannot be attained. Second, the idea of "safety" is itself an oversimplification. Ozone shows us that human beings react physiologically in various ways to changes in the environment. There is no unambiguous dividing line between safe and unsafe. Sunlight on beaches and radiation on airplane flights pose health risks. It is a matter of values as well as facts, of policy as well as science, to decide how safe is safe—to decide what effects we will use the regulatory or fiscal powers of the state to minimize or avoid.

Third, high levels of safety may not be worth the cost, in terms of inconvenience, economic effects, or freedom foregone. Hang gliding, cigarettes, red meat, and contact sports are all unsafe measured by the "one in a million" standard. So are automobiles, power tools, and skateboards.

The nation can clean up more or less and spend more or less —realizing that its limited knowledge of both medicine and engineering implies that it cannot always be sure of what results it will achieve.

The question EPA should be asking is, *how much should be spent to achieve how much health protection for different segments of the public in light of the inability to guarantee the success of such efforts?* By focusing on risk reduction instead of elimination, this formulation suggests the impossibility of perfect safety and the reasonableness—indeed the inevitability—of allowing some risks to continue. By explicitly raising the question of costs, it forces the public to consider how much it wants to spend to avoid risks, the magnitude of which are imperfectly understood. By identifying these decisions as public choices, this form of the question also suggests the central role of political values in determining environmental health policy.

This formulation also provides a framework for raising questions about different types of risk and different classes of victim. For example, are identifiable victims more valuable to society than statistical and nameless ones? Academics have often observed that the society will spend more to avoid harm to identifiable victims (e.g., miners in a cave-in) than it will spend to lower the statistical expectation of future lives lost.[1] As the old maxim "women and children to the lifeboats first" asserts, are some victims more deserving of assistance than others? Should voluntary risks be treated differently from involuntary ones? Some argue that levels of occupational exposures to air pollution should be less stringently controlled than those for the population as a whole since workers can quit if they do not like the atmosphere. Does it matter that risk is invisible to its victims, and thus presumably unavoidable? Should frequent accidents, each of which results in a few deaths, be treated as a more or less serious problem than infrequent disasters that produce the same number of expected deaths? Decisions concerning all these issues are embedded in statute and policy, but because they are morally difficult and politically divisive, explicit consideration of them is rare and the public is largely unaware of their ethical and practical import.

Even when properly reformulated, however, health questions should not form the central strategic focus for EPA. Continuing its "public health" orientation would condemn EPA— and environmental concerns in general—to the role of bit players in the grand drama of pursuing improved health status. Pollution control is a much less important lever for improving public health than the control of smoking, drinking, diet, drug use, highway safety, and crime, which are all beyond EPA's control.[2]

Moreover, if EPA is to be primarily a health agency, it should be placed within the Department of Health and Human Services, where it would compete for budget dollars with other health activities ranging from kidney transplants to drug abuse prevention and Medicaid. This could prove very damaging to environmental programs. Lives saved by environmental protection efforts can cost 10 to 100 times more than saving lives through even the most expensive medical interventions—such as organ transplants. Once the public panic over hazardous waste dump sites passes, a health protection rationale would no longer justify the scale and scope of current activities. For those whose instincts suggest that a major environmental retreat would be mistaken, a vision of EPA's mandate that encompasses more than just health protection is necessary.

The agency does have the latitude needed to move beyond the confines of public health. Most of the important statutes it administers stipulate a broader set of concerns. The Clean Water Act, for example, is mostly devoted to nonhealth issues. RCRA includes extensive authority to deal with source reduction and recycling, apart from the health based regulation of treatment, storage, and disposal facilities (TSDFs). Likewise, the "Prevention of Significant Deterioration" program under the Clean Air Act is not limited to health concerns. Indeed much of the concern about "smog" has always involved visibility and transient discomforts that may not have long-term health implications. Thus, statutes do not pose an insuperable barrier to a strategic reorientation of the agency.

REDISTRIBUTION

The EPA's decisions have important effects on the distribution of economic gains and losses to individuals, corporations, municipalities, and even entire regions. To defend the economic interests of communities with existing steel plants, the Tripartite plan prohibited the use of "stretch-out" funds for new facilities. To help electric utilities, the RCRA rules treated scrubber sludges as "special wastes" to be less stringently regulated. The buyout of Love Canal homeowners and the rights of victims to obtain—or sue for— compensation under Superfund involved the question of what sorts of benefits were owed to the individual victims of environmental harms.

In each instance, claims were made that some specific distribution of economic circumstances was "fair," or that specific individuals had certain economic "rights." Yet these questions have often been phrased in ways that presume the legitimacy of the claims they advance. Asking, "How can we compensate person X for his injuries?" is like asking "How can we make X safe?" Such formulations deny the necessity and appropriateness of the political choices that must be made to determine what protection to seek, what rights to recognize, and what compensation to provide.

It is often difficult or impossible to determine who has been affected, and to what extent, by various environmental insults or by efforts to clean them up. Some cancer victims who live near abandoned dump sites do not develop the disease as a result of environmental exposure. Some of those laid off from a factory that was forced to spend large sums on pollution control might well have lost their jobs anyway. For example, many steel mills that were out of compliance with air pollution rules have closed for competitive, not environmental reasons.

But these empirical difficulties are small compared with the ethical ones. In a dynamic society like this one, various individuals, groups, businesses, communities, and regions are continually suffering losses of one sort of another. These are due to many causes including economic change, individual negligence, and just plain bad luck. It is difficult to determine which of these circumstances justifies claims for publicly or privately funded compensation and to what extent. These questions arise in a wide variety of policy arenas apart from the environment, including medical malpractice reform, aid to the handicapped, and compensation for local economic losses caused by the abolition of trade barriers. In each case, the fact that someone suffers a loss does not settle the question of whether he or she should be compensated and how much. Rather, such claims raise the general problem of under which circumstances should unfortunate individuals and organizations be entitled to support from the rest of society.

The debate over the inclusion of victim compensation in Superfund illustrates these difficulties. Among the complex issues requiring consideration were: a determination of what type of suffering entitled a victim to claim an award, whether compliance with all applicable state and federal law constituted a legitimate defense, and how much proof was required to demonstrate that the victim's illness was indeed related to a specific environmental harm caused by a specific waste source.

In recognition of these factual and ethical complexities, we propose the following, admittedly cumbersome, formulation of the distributive question EPA should ask: *What levels of public or private compensation should various parties receive due either to the environmental insults they have suffered or to the adverse economic consequences of environmental protection efforts they have endured?"*

Like our previous proposal on health matters, this formulation calls attention to the policy choices that must be made. It concentrates on the need to decide the amount of compensation, its source, and the circumstances that justify its provision.

When dealing with compensation questions, government has to consider not only the case at hand but the consistency of that determination with those made for other sorts of injuries. The standards for proving "toxic torts," and the recovery allowed, can either be similar or different from that of other tort situations like traffic accidents, product safety, or negligence. Imagine that two coal mines are forced to close. One closes because of competition from cheaper foreign oil. The other closes because the high sulfur coal it produces is no longer attractive to buyers who must comply with air pollution limits. Should the miners who lose their jobs at each mine receive equal compensation from the government? If two poor families each have a child stricken with leukemia, should they receive different levels of assistance if one can, and the other cannot, show that the child's disease might have been the result of an environmental exposure?

These examples raise the question of whether "fault" should continue to be the basis for deciding who is eligible for compensation, or whether a "no fault" system should be instituted that provides relief regardless of who is responsible. In the past, compensation has been linked to liability because it provides an incentive to avoid errors, and it limits public responsibility to fund victims of bad luck. But as "error" becomes harder to define, and more reliance is placed upon prospective regulation to control behavior, the effectiveness and fairness of the entire fault based system is coming under increasing criticism.[3] Indeed in environmental areas, many compensation issues arise not from personal negligence or policy failure, but because the economic and legal system is doing just what the law intended, that is, putting high sulfur coal mines out of business.

There are strong arguments both in favor of and against compensating those who suffer from policy change. The more the government tries to be fair to all, the higher the cost of such compensation. Perhaps, in the long run, these sorts of harms spread themselves out evenly among the populace and, therefore, compensation is unnecessary. And, widespread compensation might tend to retard progress because the effort to avoid the sort of changes that would require compensation would lead to economic stagnation. On the other hand, if citizens know that they will be adequately compensated for the loss incurred as the result of a policy shift, their resistance to that change may well diminish.[4]

We cannot settle these difficult issues here. We ask only that when EPA encounters redistributive problems that they be posed in such a way as to most clearly reveal the crucial factual and ethical matters at stake. Even when properly posed, however, they should not form the central focus of EPA activity. As with health, EPA is destined to remain a minor figure in the distributive realm.

Income taxes, welfare systems, medical care, and economic development programs are all far more important in determining the distribution of wealth than anything that falls within EPA's purview. The need to develop greater consistency in the area of victim compensation insures that EPA will be just one of a number of interested parties involved in tort and liability law reform. Devising a fair and equitable system that encompasses defective products, negligent employers, and erring physicians, as well as midnight dumpers and irresponsible polluters, promises to be a task that will keep battalions of lawyers, legislators and judges busy for a generation or more.

Likewise, EPA possesses too few policy levers to enable it to play a central role in regional development or industrial policy. Although the steel industry wanted a lengthened enforcement timetable and oil wanted special RCRA status for drilling brines, other nonenvironmental policy questions like import fees, trigger pricing, and the strategic petroleum reserve had far more impact on their economic circumstances than did anything that EPA had to offer. The relatively small use that the steel industry made of the stretch-out provisions clearly suggests as much. Furthermore, an environmental agency preoccupied with distributive issues could not be sufficiently attentive to what we believe to be the central questions regarding environmental quality.

QUALITY OF LIFE

The domain of "quality of life" provides the proper strategic focus for EPA. This point of view does not ignore health or distributive justice but it places them in a wider perspective, one that EPA is well situated to offer.

The concerns that comprise quality of life were critical elements of the early political mobilization for environmental improvement. They have reemerged in recent years due to three major factors: a heightened awareness of the damage that pollution does to recreation and aesthetics; a renewed consciousness of the role that land use and locational decisions play in environmental policy making; and an increasing concern for the problem of residuals management (i.e., everything must go somewhere).

To focus on quality of life, EPA should ask: *What should be spent, required, forbidden, or provided to improve the quality of life in this or that place through increased environmental protection efforts?*

This formulation emphasizes that the aim of the agency ought to be to improve the lot of actual people living in real places. It implies that finding places to put garbage and keeping beaches free of debris may well be more important than trying to reduce a statistical cancer risk from slightly more to slightly less than one in a million.

This strategic formulation not only guides the deployment of limited agency resources, it also instructs the choice of whom it is most important to coordinate activities with. It implies that those parts of government that share the agency's focus upon the physical condition of cities and towns—the Department of Housing and Urban Development; the Department of Transportation; and other federal, state, and local housing, transportation, industrial development, and urban recreation agencies—are EPA's most important policy partners.

In stressing the need for EPA to focus on an environment that most people encounter most of the time, we do not denigrate other important environmental objectives like wilderness preservation and wildlife protection. But EPA must do what it is best equipped to do. Those essentially non-human centered matters should be handled by agencies that are better situated to do so, namely the various land management and wildlife agencies of the Department of Interior.

By advocating this focus, we are implying that a cautious approach be taken to global questions like the "greenhouse" effect and stratospheric ozone depletion that have recently become so popular with the media and with environmentalists. These are as yet poorly understood. They require much more research before they can be made the object of more than preliminary control efforts. They may prove to be the most crucial of all environmental issues. It is much to early to tell. If they do, they will be so intertwined with diplomatic, foreign trade, and national security questions that EPA will need to share responsibility for them with a whole raft of foreign policy and defense agencies, as well as the National Oceanographic and Atmospheric Administration. It is unlikely that EPA's role will be sufficiently large in this arena to warrant making global concerns its strategic focus.

As our cases illustrate, EPA deals extensively with quality of life issues. Deciding where to locate new land fills or incinerators, as authorized by RCRA, or how much to clean up old ones, as mandated by Superfund, profoundly influences the character of the affected neighborhoods. When a state agency tries to weigh the enforcement difficulties of a "bubble" plan for a steel mill against the cost savings the bubble offers, it too is shaping the destinies of those who live and work nearby.

We applaud recent efforts by the agency to focus increased attention on such matters. By asking the citizens of Tacoma to weigh the health risks of arsenic emissions from the ASARCO smelting plant against the economic consequences of closing it down, Ruckelshaus was posing the right kind of question. He was encouraging and assisting citizens to determine the future character of their community. The Integrated Environmental Management Project attempts to provide communities with sufficient information about the risks they face and the costs of differing levels of abatement to enable them to establish priorities and make painful choices. It facilitates real deliberation about how a community should integrate a concern for environmental safety with the other objectives it seeks. In both of these instances, EPA's ability to structure the debate and supply pertinent information can help citizens to recognize that the most important questions to be asked and the decisions to be made about the environment are political, not technical. The questions posed by EPA must help citizens to understand and act on this fundamental insight.

This focus will also encourage the agency to be strategically responsive. If the key issues appear to be technical, as in how best to prevent cancer, they provide no real grounds for political debate. But Democrats and Republicans can and ought to provide alternative visions of what quality of life means and how best to achieve it. By articulating their point of view about this vital matter, EPA's leaders can be made meaningfully accountable and can provide a foundation upon which to build stronger and more durable support for the party they represent.

THE ROLE OF STRATEGY

The EPA is responsible for so many activities, under so many pieces of legislation, that internal consistency and external accountability requires a strategy to guide its approach to its various tasks. Otherwise, the different parts of the agency lack a basis for reconciling their narrow bureaucratic concerns with broader objectives.

An explicit strategic focus can also provide a vehicle for Congress, the courts, and the public to review and criticize decisions that would otherwise seem too detailed and esoteric. For example, the question of how best to treat sludge from sewage treatment plants—as fertilizer, by incineration, by ocean dumping, or in land fills—is too opaque for outsiders to deal with unless it is presented in terms of the broader picture of agency objectives. This overall strategy can and should serve as the subject of public deliberation and provides an excellent vehicle for public education.

As the last example suggests, the agency desperately needs such strategic thinking with regard to the residuals management problem. The water, air, and toxic substances programs all seek to shift the cleanup burden from their offices and media to another one. The quality of life focus calls attention to the need for such integrative thinking. By asking, in each specific region, how best to improve citizens' circumstances, it provides a means for evaluating the consequences of alternative residuals management strategies. Conversely, the reality that residuals cannot be wished away, that all options have costs and are imperfect, reinforces the need for this kind of comparative analysis.

In an era of large budget deficits, public dollars will continue to be scarce and highly sought after. The struggle to remain competitive internationally will increase concern about the waste of private corporate resources as well. Asking better questions will force the agency to recognize the importance of cost effectiveness. By focusing on the overall quality of life, EPA will confront the need to reconcile competing claims, and will recognize that unrestrained pursuit of any one environmental objective can all too easily threaten its other goals.

HOW TO GET BETTER QUESTIONS

Structure is both a cause and a consequence of strategy. While an agency's organization chart cannot guarantee that the right questions will be asked, it can help or hurt. In working to develop a better strategy, an agency head's task necessarily includes organizational issues. A quality of life focus implies that the key for EPA is to develop a structure that facilitates the integration of the diverse considerations relevant to that objective. This has to be done on two levels: within each physical system, and across systems in each geographical area. We discuss what these considerations imply in this and the next section,

The results of environmental policy are manifested in the state of various physical systems. What citizens actually encounter are not permits, or reports, or consent decrees, but the condition of the air, land and water. Thus, all of EPA's various activities—from research, to standard setting, to litigation—need to be managed and coordinated with an eye toward getting these ultimate results.

Furthermore, most physical systems are strongly interconnected, so that decisions about one part of them may alter the impact of decisions elsewhere. Ambient ozone levels, for example, depend on both nitrogen oxide and hydrocarbon emissions, so that appropriate controls on one depend on what controls there are on the other. Similarly, one cannot decide how much organic material a stream can safely assimilate unless one knows the pattern of the flow, which may be controlled by dams, irrigation withdrawals, and so on. One way to ensure that such interdependent decisions are consistent with one another is to make the same unit of government responsible for all of them.

Achieving any environmental result efficiently and effectively requires government to combine various kinds of professional and functional expertise. To get attention paid to specific consequences, all those same consequences have to be someone's responsibility. At EPA, the appropriate specialists need to feel first and foremost that they are responsible for the final environmental result, not just for their narrow functions or specific activities. To do this, they must be brought together in the early stages of developing a rule, a standard, or a legislative proposal, thereby insuring that all the options, and their ramifications, can be fully appreciated. Technical experts who understand the relevant natural systems and engineering alternatives must interact with lawyers and management specialists to produce a program that will function effectively in human as well as technical terms. This collaboration should occur in an atmosphere that encourages an open exchange of views on both means and ends, an exchange in which strategic options are clarified. In contrast, the Red Border Review process under Costle too often brought out competitive posturing by bureaucratic adversaries each of whom had their own narrow agenda.

To foster such integration, we propose that EPA adopt the divisional form of organization prevalent in the corporate world, with each kind of physical system serving as the basis for a "division."[5] This would involve expanding the responsibilities and capacities of the current media offices to make them fully responsible for their "line of business." Each of the assistant administrators heading the new units would be be held accountable for the performance of his "line." Such units would include all the types of expert needed to develop and manage a complete program, incorporating them into an organizational context which forced them to confront a broader definition of their responsibilities.

Together, the strengthened assistant administrators in such a scheme would form a general management team. They would be able to help the administrator to think strategically about overall agency policy, and, by supervising internal matters, give him more time to act as ambassador of the organization to the external political world. This would also make the administrator more dependent upon the assistant administrators. They would be expected to develop and carry out agency strategy, rather than pursue ideological or professional objectives of their own. The administrator would want to exercise great care in picking those subordinates, looking for individuals who were well suited for such wide ranging responsibilities.

If the EPA were to move in this direction, staff functions like public information, congressional liaison, personnel, legal counsel, and research would still exist as cross cutting centers of expertise, internal consultation, and technical quality control. This would give the agency a matrix structure. In addition, the administrator and his general management team could form task forces on specific problems by reaching down into the divisions for the expertise and talent required. Such a structure would have the advantage of clarifying who was responsible for what.[6]

The media-based structure existed at EPA's birth and was never fully abandoned. It was altered to make room for new functional units—enforcement, research, and planning. These were viewed as necessary to overcome parochial, program-specific loyalties predating the existence of EPA, and to foster an agency-wide identification and perspective.[7] Perhaps that change was needed at the time, but now it fosters an equally fractionalized and parochial climate.

Decentralization and Regional Integration

The need for regional integration across physical systems and decentralization to the regional level arises, in part, from two characteristics of environmental problems. One is that most physical systems are limited in extent. Because many environmental regulatory decisions have primarily local consequences, not everyone is equally interested in each one. In addition, there are many interrelationships across physical systems because of the problem of residuals management. Thus, decisions about waste disposal with regard to air or water may well affect each other. Not only will the specific mix of intermedia effects vary greatly from one place to the next, but policy must take cognizance of the great economic, ecological and political diversity of a large and varied country.

One implication of these considerations is that EPA needs to strengthen its system of regional offices so that each can serve as the locus for the necessary cross-media integration. Consider the problems of designing a solid waste disposal strategy for a particular community. The regional office should be able to consider how various alternatives would affect actual ecosystems and whether they would conform to current laws and regulations. What would it take for an incinerator to be approved under RCRA and the relevant SIP? What kind of land disposal would be feasible both hydrologically and politically?

But regional integration and decentralization involve more than just EPA's own structure, and more than just considerations of the dynamics of physical systems and the movement of residuals. In the chapters on steel enforcement and Superfund we have already sketched some of the ways in which the division of labor between states and national government has played itself out. As we have indicated, the issues at stake are broader than mere administrative neatness or convenience.

Life is lived in specific communities. That is where locational and land use decisions are made and tradeoffs between competing strategies occur. That is where citizens learn to appreciate, and agonize over, the coexistence of self-interest and public interest. As Tocqueville first recognized, the American states and municipalities are far more than an administrative convenience. They are the schools of citizenship, and their elected officials have a unique perspective from which to consider the political and ethical, as well as the technical content of policy choices.

The EPA must resist the temptation to think of state officials as subordinates. As we argued in the Superfund case, there are important instances where EPA should view its role primarily in terms of providing technical support for, and financial assistance to, state governments, which would retain significant discretion over program design and implementation.

This approach is not without its problems. Almost any form of pollution produces interstate effects, be they physical or economic. National pollution control legislation was, in part, a response to those externalities. National standards and centralized enforcement protect states from the pressures of having to compete for new business investment by lowering their environmental standards.

In addition, state regulators can be vulnerable because of the concentration of extraction and materials processing industries in areas where resource inputs are found. Such industries are politically stronger at the state than at the federal level. Appalachian coal is the classic example.[8] Moreover, some state governments are simply too weakly staffed to be a technical match for polluters.

However, as we discussed in Chapter 2 and in the Superfund chapter, national decision making also has its own distinct biases. Just as certain industries have local power, so certain ideological perspectives, including environmentalism, exert much greater influence in Washington. Thus, no arrangement is neutral, nor is one system "fair" and the other "unfair." All institutional designs are imperfect.

Still, the quality of life focus implies a presumption in favor of decentralization. This means risking underregulation and nonuniformity in return for building greater state governmental capacity. EPA's regional offices should assist states to take the lead in bringing diverse ecosystem concerns together in the places where people actually live. Integration would thus promote the habits and competencies of self-government, which are surely themselves important aspects of the quality of life.[9]

THE EXECUTIVE BRANCH

Our cases also repeatedly and dramatically illustrate the need for improving the means whereby EPA's actions are reconciled with the overall priorities established by the White House. The president lacks the time and the information to perform this duty himself. It is especially difficult because environmental policy has links with so many other issues, including energy development, transportation, plant location, water resource management, recreation planning, and disease prevention.

Regardless of how one rearranges the organization chart, no single agency or department can encompass all of these interdependencies. Some important connections will not be made and coordination with other departments and agencies will still be necessary. As the shift in EPA's focus from ecology to public health demonstrates, the central organizing principle of an agency may change over time. Thus, integrative mechanisms may need to be readjusted periodically and, therefore, should be devised with impermanence in mind.[10]

The Reagan administration's establishment of "cabinet councils" was a promising approach to the problem of executive branch integration. Each council was headed by a single cabinet member and dealt with a specific broad policy area. Its members included all the cabinet level officials whose departments shared responsibility for that domain (economic development, natural resources, manpower, and so on). Since the councils had no formal decision-making authority, but served only as advisory bodies, their existence did not require congressional approval. They were not, therefore, mired in the squabbles over committee jurisdiction that executive reorganization efforts ordinarily provoke. Also, they were sufficiently flexible to be able to change membership, or even to redefine their jurisdictions, in response to changing circumstances.[11]

The councils did not prove to be a great success. That seems to have been due more to a lack of enthusiasm from White House Chief of Staff James Baker and OMB Director David Stockman than to any fundamental conceptual flaw. Since these councils compete for authority and influence with the White House staff, the latter's opposition was perhaps inevitable. Therefore, for such a scheme to work, it would require enthusiastic presidential backing.

Structural reform of the executive branch is only part of the answer. The attitudes and habits of mind of its leadership are equally important, if not more so. For the various departments and agencies of the executive branch to operate properly they must be led by people whose loyalty to the president, the Constitution, and the technical merits counterbalances their inevitable bureaucratic concerns. Therefore, the decisions that a president makes concerning recruitment, retention, and advancement of key members of the executive branch constitute his best opportunity for ensuring that it functions effectively.

THE CONGRESS

Our cases abound with examples of ill-considered congressional action. The original RCRA statute was excessively vague, while the new version has too many cumbersome requirements. Even in its revised form, Superfund fails to give adequate guidance regarding cleanup priorities and ignores the need to foster citizen responsibility. The requirement of the Clean Air Act that forbids the EPA administrator from considering costs when setting ambient standards forces him to either act irrationally or misrepresent his decisions. The root causes of these errors are deeply embedded in the structure of the contemporary Congress and the incentives that its members face. We can offer only modest proposals for improvement.

In our cases, committee and personal staffs were often unduly entrepreneurial, and too often coupled insufficient knowledge with an excessively narrow sense of their responsibilities. The work of such centralized staff agencies as the Office of Technology Assessment and the General Accounting Office displayed, on the whole, more technical sophistication and a greater appreciation of strategic questions. Therefore, we would urge congressmen to establish tighter reins over their personal and committee staffs and place more trust in the products of these and other forms of institutionalized staff support. To that end, resources should be shifted to such organizations and away from personal and committee staffs.

Congress proved much more successful at oversight than it did at lawmaking. Its relentless scrutiny of EPA during the Gorsuch regime was the major cause of her downfall. Fear of its ability to continue to unearth scandal undoubtedly contributed to Reagan's willingness to replace her with a much more effective and acceptable leader. Nonetheless, oversight would be much improved if it was not so prone to wallow in often trivial detail.

Because Congress is so good at investigation, it need not and ought not try to use legislation to accomplish objectives that are properly the role of oversight. The detailed timetables and "hammers" in the revised RCRA and the excessively rigid standards in the new Superfund deprived EPA of the discretion it needs to think and act strategically. The apparent goal was to so constrain the agency that it could "do no wrong." Instead, Congress should give agencies more discretion precisely because it can so effectively monitor performance after the fact.

THE COURTS

We urge the courts also to be more modest, sophisticated, and realistic in scrutinizing the agency. Those who are not well informed should not try to judge the technical merits of administrative rules. The courts have often gotten the issues quite confused.[12] Also, the judicial effort to use a tortured reading of legislative history to establish new environmental rights is an unwarranted intrusion on congressional authority.

Whether one agrees or disagrees with the substance of Costle's ozone standard, we believe the court was correct to defer to the agency in this matter. Whatever deficiencies that standard might have had in the eyes of the Natural

Resources Defense Council or the American Petroleum Institute, the place to settle that dispute was not in the D.C. Circuit Court of Appeals. Likewise, Judge Gesell was wise to recognize that aggressive judicial intervention was unlikely to speed the RCRA regulation writing process or to produce more sensible results. Such judicial reticence is the most effective deterrent to the assiduous forum shopping engaged in by those who are not satisfied by the results they have attained through normal political channels.

FINAL THOUGHTS

This journey through such intricate toxicological, political, meteorologic, and bureaucratic detail was intended to convey a simple idea. A democratic society produces citizens as much as it does goods or services. In a country like the United States, with so much freedom, technological dynamism, and decentralized initiative, there will always be problems adapting to growth and change. Citizens learn about their world and define their political relationships to each other as they struggle with such new problems as how much is owed to the sick and to the especially vulnerable, and what types of risks are to be considered illegitimate. These policy deliberations are the classrooms of the republic.

Leaders who operate at the interface between technical competence and political authority must take responsibility for the quality of these deliberations, which they help provoke and shape. They are also responsible for developing policies that display fidelity to the technical merits, are responsive to the public, and render them accountable to Congress and the courts. By shaping and expounding an explicit strategy, they can give those both inside and outside their agencies a clear sense of what is really at issue and how the agency's efforts should be judged. By asking good questions they can encourage citizens to consider their own responsibilities and obligations, as well as the rights and opportunities that various policy initiatives convey.

The public may not take kindly to being told that there is no such thing as a safe level of exposure to a particular noxious waste, or that cigarettes present greater health risks than all the chemicals dumped by perfidious polluters. Nonetheless, in our lifetime the public has grudgingly come to accept equally unwelcome truths. It no longer believes, as it did in the 1950s, that America can do as it pleases in world affairs. Nor does it tolerate racial segregation in the workplace, the lunch counter, or the schoolroom. To be sure, protest at home and failure abroad precipitated these changes. Still, government played an important role in cultivating these new understandings of international relations and racial equality. It can do the same with the realities of environmental protection. The essential purpose of this study has not been to look backward for the purposes of assessing blame or praise. We admire the energy and public-spiritedness of the agency officials we met in doing this study. Exactly because we have such respect for them we have tried to offer a sophisticated and complex view of the role that public officials can play in getting to the future Americans wish for themselves and their children.

NOTES

1. Thomas C. Schelling, "The Life You Save May Be Your Own," in Stuart B. Chase, ed., *Problems in Public Expenditure Analysis* (Washington, D.C.: The Brookings Institution, 1968).

2. See Richard Doll and Richard Peto, "The Causes of Cancer: Quantitative Estimates of Avoidable Risks of Cancer in the United States Today," *Journal of the National Cancer Institute*, 66,(June 1981):1193–1308; and Bruce Ames et. al., "Ranking Possible Carcinogenic Hazards," *Science*, 236 (17 Apr. 1987): 271–280.

3. See Sean F. Mooney, "The Liability Crisis: A Perspective," in *Tort Reform: Will it Advance Justice in the Civil System?*: the Twenty First Annual Symposium. *Villanova Law Review*, 32, Nov. 1987, 1235–1264. Eugene C. Thomas, "Rights, Remedies, Restitution and 'Reform' ", *ABA Journal*, 73, 1 July 1987, president's page.

4. See for example the discussion of Sweden's policy of protecting workers against the adverse effects of job loss in Robert Kuttner, *The Economic Illusion: False Choices Between Prosperity and Social Justice* (Boston: Houghton Mifflin, 1984) 149–159.

5. The classic discussion of the evolution of this form of organization is in Alfred Chandler, *Strategy and Structure* (Cambridge MA: Harvard University Press, 1962).

6. See Stanley M. Davis and Paul R. Lawrence, *Matrix* (Reading, MA: Addison Wesley, 1977).

7. See Chapter 2.

8. Marc Landy, *The Politics of Environmental Reform: Controlling Kentucky Surface Mining* (Baltimore: Johns Hopkins Press, RFF Working Paper DD-2, 1976).

9. See Marc K. Landy, "Policy Analysis as a Vocation," *World Politics*, 33, Apr. 1981, 478–482.

10. See Marc J. Roberts, "Organizing Water Pollution Control: The Scope and Structure of River Basin Authorities," *Public Policy*, Winter 1981, 89–92; see also, A. Myrick Freeman III, Robert Haveman, and Allen V. Kneese, *The Economics of Environmental Policy* (New York: John Wiley, 1973), 171–175.

11. For a discussion of the cabinet councils see *The National Journal* 7/11/81 1242–1247; 6/28/86 1582–1589 and Chester A. Newland, "A Midterm Appraisal - The Reagan Presidency: Limited Government and Political Administration," *Public Administration Review*, Jan./Feb.1983, 6–10.

12. See the discussion of the U.S Supreme Court's decision on benzene in John Graham, Laura Green, and Marc J. Roberts. *The Search for Safety* (Cambridge MA: Harvard University Press, 1988) 80–114.

Index